Advances in
ORGANOMETALLIC CHEMISTRY

VOLUME 38

Advances in Organometallic Chemistry

EDITED BY

F. G. A. STONE

DEPARTMENT OF CHEMISTRY
BAYLOR UNIVERSITY
WACO, TEXAS

ROBERT WEST

DEPARTMENT OF CHEMISTRY
UNIVERSITY OF WISCONSIN
MADISON, WISCONSIN

VOLUME 38

ACADEMIC PRESS
San Diego   New York   Boston
London   Sydney   Tokyo   Toronto

This book is printed on acid-free paper. ∞

Copyright © 1995 by ACADEMIC PRESS, INC.

All Rights Reserved.
No part of this publication may be reproduced or transmitted in any form or by any means, electronic or mechanical, including photocopy, recording, or any information storage and retrieval system, without permission in writing from the publisher.

Academic Press, Inc.
A Division of Harcourt Brace & Company
525 B Street, Suite 1900, San Diego, California 92101-4495

United Kingdom Edition published by
Academic Press Limited
24-28 Oval Road, London NW1 7DX

International Standard Serial Number: 0065-3055

International Standard Book Number: 0-12-031138-0

PRINTED IN THE UNITED STATES OF AMERICA
95 96 97 98 99 00 QW 9 8 7 6 5 4 3 2 1

Contents

Organotransition-Metal Chemistry and Homogeneous Catalysis in Aqueous Solution

D. M. ROUNDHILL

Structure/Property Relationships of Polystannanes

LAWRENCE R. SITA

Contributors

Numbers in parentheses indicate the pages on which the authors' contributions begin.

MICHAEL D. HOPKINS (79), Department of Chemistry, University of Pittsburgh, Pittsburgh, Pennsylvania 15260

KEVIN D. JOHN (79), Department of Chemistry, University of Pittsburgh, Pittsburgh, Pennsylvania 15260

ATSUSHI KAWACHI (1), Institute for Chemical Research, Kyoto University, Uji, Kyoto 611, Japan

JOSEPH MANNA (79), Department of Chemistry, University of Pittsburgh, Pittsburgh, Pennsylvania 15260

D. M. ROUNDHILL (155), Department of Chemistry, Tulane University, New Orleans, Louisiana 70118

LAWRENCE R. SITA (189), Searle Chemistry Laboratory, Department of Chemistry, The University of Chicago, Chicago, Illinois 60637

KOHEI TAMAO (1), Institute for Chemical Research, Kyoto University, Uji, Kyoto 611, Japan

JOHN S. THAYER (59), Department of Chemistry, University of Cincinnati, Cincinnati, Ohio 45221

Silyl Anions

KOHEI TAMAO and ATSUSHI KAWACHI

Institute for Chemical Research
Kyoto University
Uji, Kyoto 611, Japan

I

INTRODUCTION

The term *silyl anion* denotes the trivalent, octet, negatively charged silicon species. This chapter outlines the preparation, properties, structures, and reactions of silyl anions (*1a–i*). The silyl anions described here are restricted to those that contain the following counter-cations: Group 1: Li, Na, K, Rb, Cs; Group 2: Mg; Group 11: Cu; Group 12: Zn, Cd, Hg. Metal-free silyl anions, whose counter-cations are ammonium (R$_3$NH$^+$ or R$_4$N$^+$), are also described. Other silyl transition metal compounds are excluded; for these species, previous review articles should be referenced (*2*).

This review is intended as an overview of the literature on silyl anions

1

up to early 1994, so there is some overlap with previous review articles ($1a-i$). In this article, silyl anions are formally classified in terms of structure, especially substituents on silicon. All silyl anions so far studied are compiled in Section X.

II

A BRIEF SURVEY OF PREPARATIVE METHODS

In this section, silyl anions are classified in terms of counter-cations, and their preparative methods are summarized.

A. Silyllithium

Silyllithiums are the most popular and most extensively investigated species among the silyl anions, and can be prepared by the following methods:

(1) Reaction of a chlorosilane with lithium metal. This procedure forms lithium chloride as a by-product. Halogen-free silyllithiums can be obtained by procedures (2) and (3).
(2) Cleavage of a disilane with lithium metal or alkyllithium.
(3) Metal–lithium exchange reaction of a silyl metal compound (especially Group 12 metals) with alkyllithium or lithium metal.

B. Silylsodium and Silylpotassium

(1) Reaction of a chlorosilane with sodium or potassium. (More severe reaction conditions are often required for silylsodium and silylpotassium than for silyllithium.)
(2) Cleavage of a disilane with sodium or potassium alkoxide or potassium hydride.
(3) Reaction of a disilylmercury with sodium or potassium metal.
(4) Reaction of a hydrosilane with potassium hydride.

These species are usually more reactive than their lithium counterparts.

C. Silylmagnesium

(1) Reaction of a silyllithium with a Grignard reagent or magnesium halide.
(2) Cobalt–magnesium exchange reaction of a silyl-cobalt compound with a Grignard reagent.

Reaction of a chlorosilane with elemental magnesium gives rise to a disilane as a coupling product, although a silylmagnesium species is postulated as an intermediate.

D. Silylcopper and Silylcuprate

(1) Silylcoppers and silylcuprates are obtained from reaction of silyllithiums with copper salts.
(2) Higher order silylcuprates are obtained by transmetallation between silylstannanes and higher order alkylcuprates.

These species are useful reagents for the Si–C bond formation in organic synthesis.

E. Silylzinc, Silylcadmium, and Silylmercury

(1) Reaction of a hydrosilane with mercury metal.
(2) Reaction of a halosilane with diorganomercury.
(3) Reaction of a silyllithium with the corresponding metal halides.

Bis-silyl derivatives of these Group 12 metals are readily accessible and thermally stable, and can be good precursors to the corresponding silyl–alkali metals via metal–metal exchange reaction.

F. Metal-Free Silyl Anions

In these species, the counter-cation is an ammonium salt instead of a metal. Two procedures are available:

(1) Reaction of trichlorohydrosilane with a tertiary amine.
(2) Cleavage of a disilane with tetrabutylammonium fluoride (TBAF).

III

ARYLSILYL ANIONS

Silyl anions that have at least one aryl group on silicon are described in this section.

A. Preparation

1. Ar_3SiM

The action of Li metal on Ph_3SiCl in tetrahydrofuran (THF) yields Ph_3SiLi, together with LiCl as by-product (3). Ph_3SiRb and Ph_3SiCs are

$$Ph_3SiCl \; + \; 2\,Li \; \longrightarrow \; Ph_3SiLi \; + \; LiCl$$

$$Ph_3SiLi \; + \; Ph_3SiCl \; \longrightarrow \; Ph_3Si\text{-}SiPh_3$$

$$Ph_3Si\text{-}SiPh_3 \; + \; 2\,Li \; \longrightarrow \; 2\,Ph_3SiLi$$

SCHEME 1

also prepared from the chlorosilane, with Rb metal and Cs metal, respectively.

It is believed that this procedure includes three steps (Scheme 1). The first step is the initial formation of Ph_3SiLi. The resulting silyllithium reacts with the excess Ph_3SiCl to give $Ph_3SiSiPh_3$ in the second step. The third step is the symmetrical cleavage of the disilane with lithium metal. This final step is much slower than the initial formation of silyllithium and is effected only by the electropositive alkali metals. Thus, the reaction of Ph_3SiCl with less electropositive Mg gives only $Ph_3SiSiPh_3$, although silylmagnesium $Ph_3SiMgCl$ is postulated as an intermediate (4).

$(o\text{-Tolyl})_3SiM$ (M = Li, Cs) can also be prepared from $(o\text{-tolyl})_3SiCl$ in the same way (3). However, it is doubtful that $(o\text{-tolyl})_3SiM$ is generated via three steps like those just discussed and shown in Scheme 1, since highly crowded hexa(o-tolyl)disilane is hardly accessible by coupling in the second step.

Salt-free Ph_3SiLi, Ph_3SiNa, and Ph_3SiK are obtained by cleavage of the Si–Si bond of hexaphenyldisilane by alkali metals in ethereal solvents such as 1,2-dimethoxyethane (DME) and THF [Eq. (1)] (5,6). Ph_3SiK is

$$Ph_3Si\text{-}SiPh_3 \; \xrightarrow[\text{THF or DME}]{2M} \; 2\,Ph_3Si\text{-}M \; \xrightarrow{2\,Me_3SiCl} \; 2\,Ph_3Si\text{-}SiMe_3 \; + \; 2\,MCl \quad (1)$$
$$M = Li,\, Na,\, K$$

also prepared by cleavage of the Si–C bond in $Ph_3Si\text{-}CMe_2Ph$ with Na–K alloy in ether (Et_2O) [Eq. (2)] (7). Historically, this was essentially the first successful method for the preparation of silyl anions. Ph_3SiK may also be made by reaction of Ph_3SiH with KH at 40°C in DME for 6 hours [Eq. (3)] (8). This method is applicable also to the preparation of alkylsilyl anions (see Section IV-A).

$$Ph_3Si\text{-}CMe_2Ph \; \xrightarrow[\text{ether}]{Na\text{-}K} \; Ph_3SiK \; + \; PhMe_2CK \quad (2)$$

$$Ph_3Si\text{-}H \; + \; KH \; \xrightarrow{DME} \; Ph_3Si\text{-}K \; + \; H_2 \quad (3)$$

A silyl–cobalt complex is a good precursor for silyllithium and silylmagnesium compounds (Scheme 2) (9a,b). Evidence for the generation of

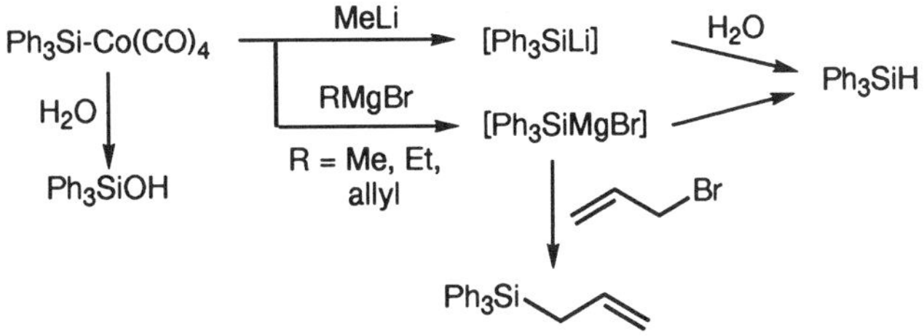

SCHEME 2

these anionic species, especially $Ph_3SiMgBr$, is that hydrolysis of these species with H_2O affords only the hydrosilane Ph_3SiH, whereas hydrolysis of the silyl-cobalt complex gives the silanol Ph_3SiOH quantitatively.

In contrast to the Group 1 derivatives, Group 2 magnesium derivatives have been much less studied so far. As described earlier, the reaction of triphenylchlorosilane with magnesium metal produces the coupling product disilane. It has been found that the action of cyclohexylmagnesium bromide on Ph_3SiCl also yields $Ph_3Si–SiPh_3$, together with cyclohexene and cyclohexane as by-products. This finding suggests that $Ph_3SiMgBr$ is generated in situ (10). Since Me_3SiCl does not undergo similar homo-coupling by cyclohexylmagnesium bromide in refluxing THF, aromatic groups on silicon appear to be necessary for the stabilization of the silyl Grignard reagent. Reaction of Ph_3SiLi with $MgBr_2$ also gives $Ph_3SiMgBr$, although the structure of this species has not been confirmed [Eq. (4)] (11) (see also Section VII).

$$Ph_3SiLi + MgBr_2 \xrightarrow{\quad THF \quad} Ph_3SiMgBr + LiBr \qquad (4)$$

A Group 12 derivative $(Ph_3Si)_2Hg$ is prepared by reaction of Ph_3SiH with $(PhCH_2)_2Hg$ at 165–168°C (12).

2. $Ar_2(alkyl)SiM$ and $Ar(alkyl)_2SiM$

$Ar_2(alkyl)SiLi$ and $Ar(alkyl)_2SiLi$ are prepared by methods similar to those for Ar_3SiLi, that is, the action of Li metal on chlorosilanes or the cleavage of disilane with Li metal in ethereal solvent such as THF (1e). The alkyl substituents can be methyl, ethyl, i-propyl, t-butyl, benzyl, and neo-pentyl groups. Allyl-bearing silyllithium can also be prepared from the corresponding chlorosilane with Li metal in THF at 0°C (see Section VII) (13). Vinyl-bearing silylpotassium is obtained from the corresponding hydrosilane with KH in DME at 40°C (see Section III-A-1) (14). A copper derivative, $PhMe_2SiCu(CN)Li$, and a magnesium derivative, $PhMe_2SiMg$-

i-Pr, can be prepared from $PhMe_2SiLi$ with CuCN and i-PrMgX (X = halide), respectively. Synthetic applications of these reagents are described in Section VII.

When all substituents are different, a chiral center is present on silicon. Enantiomeric silyl anions are generated in three ways, with retention of configuration; for the configurational stability of silyllithium, see Section III-B.

Optically active silyllithium (neo-C_5H_{11})PhMeSi*Li is obtained by cleavage of an optically active disilane with lithium metal, with retention of configuration [Eq. (5)] (15). Formation of optically active silyllithium (1-Np)PhMeSi*Li (Np = naphthyl) and silylmagnesium reagent (1-Np)PhMeSi*MgBr is also observed in reactions of an optically active silyl–cobalt complex ($9a,b$). Treatment of the enantiomer (+)-(1-Np)PhMeSi*Co(CO)$_4$ with MeLi produces, after hydrolysis, the optically active silane (+)-(1-Np)PhMeSi*H, with 70% retention of configuration. Similarly, treatment with a Grignard reagent yields a silyl Grignard reagent of low optical activity ($cf.$ Scheme 2).

$$(-)\text{-(neo-}C_5H_{11})\text{PhMeSi*SiMePh}_2 \xrightarrow[\text{retention}]{\text{Li}} (-)\text{-(neo-}C_5H_{11})\text{PhMeSi*Li}$$
$$[\alpha]_D = -5.03°$$

$$\xrightarrow[\text{retention}]{\text{HCl-}H_2O} (-)\text{-(neo-}C_5H_{11})\text{PhMeSi*H}$$
$$[\alpha]_D = -2.17°$$

(5)

Enantiomeric silylmercury compounds are obtained from reaction of diorganomercurys with an optically active hydrosilane [Eq. (6)] (16). The configuration is retained predominantly at silicon. These compounds, however, racemized readily when sublimed during purification.

$$R_3Si\text{*H} + (PhCH_2)_2Hg \longrightarrow (R_3Si\text{*})_2Hg + 2\,PhCH_3 \qquad (6)$$

Reaction of an enantiomeric hydrosilane with KH yields a racemic silyl anion (17). Thus, treatment of (+)-(1-Np)PhMeSiH with KH at 50°C in DME for 24 hours and subsequent addition of n-BuBr give racemic (±)-(1-Np)PhMeSi-n-Bu, via the formation of the corresponding silyl anion with loss of optical activity. Hydrolysis or deuterolysis also gives a racemic product. These observations clearly rule out a mechanism in which the silyl anion is formed by proton abstraction from the hydrosilane, because retention of configuration would be expected. A possible mechanism involves a pentacoordinate dihydrosilyl anion formed via coordination of H^- in an initial fast, reversible process, and its decomposition to the racemic silyl anion with loss of molecular hydrogen [Eq. (7)]. A gas-phase study supports this mechanism: Pentacoordinate silyl anion H_5Si^- decomposes to yield H_3Si^- and H_2 [Eq. (8)] (18).

$$(+)\text{-(1-Np)PhMeSiH} + KH \rightleftharpoons \left[Me-\underset{Ph}{\overset{Np}{Si}}{\overset{,,H}{\diagdown_H}} \right]^- K^+ \xrightarrow{H_2 \atop \Delta} (\pm)\text{-(1-Np)PhMeSiK}$$

racemization

(7)

$$SiH_5^- + SiH_4 \longrightarrow SiH_4 + H_2 + SiH_3^-$$ (8)

B. Structural Study

^{13}C nuclear magnetic resonance (NMR) studies provide much information on the electronic structure of phenylsilyl anions (19–20). The main interest is the charge delocalization of lone pair electrons from silicon to the phenyl group. The ^{13}C chemical shift changes on phenyl rings on going from chlorosilanes to the corresponding silyl anions give the major trends: The ipso carbon is strongly deshielded, and the para carbon is shielded (Table I, Fig. 1) (19a). As seen in Fig. 1, the charge distribution pattern

TABLE I

^{13}C NMR Chemical Shifts of Silyl Anions and Related Species in THF (ppm)^a

	ipso	ortho	meta	para	Reference
Ph$_3$SiCl	133.8	135.9	128.9	131.5	(19a)
Ph$_2$MeSiCl	134.4	134.0	128.1	130.5	(19c)
PhMe$_2$SiCl	137.0	133.8	128.8	131.0	(19a)
Ph$_3$SiLi	155.9	137.0	126.9	124.6	(19a)
Ph$_2$MeSiLi	160.1	135.4	126.7	123.9	(19a)
PhMe$_2$SiLi	166.0	133.8	126.5	122.7	(19a)
(Et$_2$N)Ph$_2$SiLi	158.5	135.6	126.6	123.9	(19a)
(Et$_2$N)$_2$PhSiLi	160.0	134.9	126.6	123.2	(54a)
(Et$_2$N)PhMeSiLi	164.8	133.7	126.5	122.7	(54a)
Ph$_3$SiK	158.6	136.9	126.7	123.8	(19a)
Ph$_2$MeSiK	163.2	135.1	126.5	123.0	(19a)
PhMe$_2$SiK	170.1	133.6	126.4	121.6	(19a)
(PhMe$_2$Si)$_2$CuLi^b	162.0	134.7	126.3	123.9	(99)
PhMe$_2$SiCu(CN)Li^b	150.0	135.0	126.0	124.6	(99)
(PhMe$_2$Si)$_2$Cu(CN)Li$_2$^{b}	157.4	134.9	126.5	124.6	(99)
Me(PhMe$_2$Si)Cu(CN)Li$_2$^{b}	159.2	135.3	127.2	125.1	(99)

^{a} Cyclohexane used as reference (δ 27.7).
^{b} THF used as reference (C$_\alpha$, δ 26.0; C$_\beta$, δ 68.2).

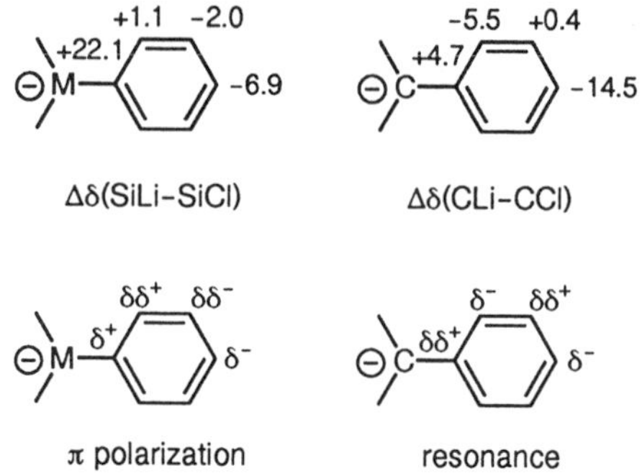

FIG. 1. ^{13}C chemical shift changes (+ sign means downfield shift) on going from Ph$_3$SiCl to Ph$_3$SiLi, and schematic representations of the electron distribution. Data for the corresponding carbanion are also shown, for comparison.

is typical for a π-polarization effect and quite different from the pattern of a resonance effect. Thus, a large amount of negative charge remains on the silicon atom.

The ultraviolet (UV) spectra of (phenylsilyl)lithiums are consistent with these NMR studies (21). The absorption maximum of Ph$_3$SiLi in THF is observed at 335 nm, whereas that of Ph$_3$CLi is observed at 500 nm. The data for Ph$_2$MeSiLi and PhMe$_2$SiLi are similar to that of Ph$_3$SiLi. These findings suggest little conjugation between silicon and the phenyl group.

^{29}Si NMR of (phenylsilyl)lithiums has been investigated (19b). The ^{29}Si resonance of Ph$_3$Si^{6}Li resolves into a triplet due to coupling with one ^{6}Li atom (I = 1, ^{1}J[^{29}Si, ^{6}Li] = 17 Hz) in 2-methyltetrahydrofuran (MTHF) at 173 K (Fig. 2); detailed data are compiled in Table II in Section IX. This result implies a monomeric structure and partial covalent nature of the Si–Li bond. By increasing the temperature or by increasing the cation-solvating power of the solvent (THF > MTHF), the pace of the lithium–lithium exchange reaction is increased, resulting in the disappearance of the ^{29}Si–^{6}Li coupling. The exchange rate is decreased if methyl groups are substituted for phenyl, and a well-resolved triplet can be observed for PhMe$_2$SiLi at 173 K, even in THF (Fig. 2).

Crystal structure analysis of Ph$_3$SiLi·3THF has revealed that both of the lithium and silicon centers are four-coordinate with distorted tetrahedral geometries (22). The Li–Si distance is 2.67 Å and very similar to the averaged Li–Si bond length (2.68 Å) observed in (LiSiMe$_3$)$_6$ (see Section IV-B). The averaged Si–C(phenyl) distances are 1.94 Å, which is much longer than that normally observed in neutral compounds (ca. 1.85 Å). The lengthening may be a consequence of the increased negative charge density on silicon, which reduces the ionic contribution to the Si–C bond

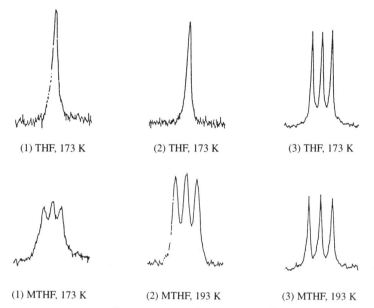

(1) THF, 173 K   (2) THF, 173 K   (3) THF, 173 K

(1) MTHF, 173 K   (2) MTHF, 193 K   (3) MTHF, 193 K

FIG. 2. Low-temperature ^{29}Si NMR of Ph$_3$Si^{6}Li (1), Ph$_2$MeSi^{6}Li (2), and PhMe$_2$Si^{6}Li (3), 0.3 M solution in THF or MTHF (*19b*). Reprinted with permission from *J. Am. Chem. Soc.*, **1985,** *107,* 6408. Copyright (1985) American Chemical Society.

strength. The tetrahedral geometry about the silicon is in clear contrast to the planar coordination at the central carbon of the carbon analog, Ph$_3$CLi·Et$_2$O (*23*). This finding is consistent with the conclusion of the ^{13}C NMR studies mentioned earlier.

Configurational stability of Ph(*i*-Pr)$_2$SiLi has been estimated by means of ^{1}H NMR [Eq. (9)] (*24*). The two methyl groups in each isopropyl group are diastereotopic and anisochronous at room temperature in a wide range of solvents. The nonequivalence is observed up to 185°C. Since rotation around the Si–C bonds appears to be fast on the NMR time scale, the nonequivalence of the isopropyl methyl groups in the anion is evidence for slow inversion about silicon. This result indicates that unimolecular atomic inversion about silicon must be slow on the NMR time scale, and a lower limit to inversion about the trivalent silicon can be set at about 24 kcal/mol, which is consistent with theoretical studies (see Section VIII).

$$\text{(CH}_3\text{)}_2\text{CH}-\underset{\substack{| \\ \text{Ph}}}{\overset{\substack{\text{Li}^+ \\ | }}{\text{Si}}}\cdots\text{CH(CH}_3\text{)}_2 \quad \underset{\text{>24 kcal/mol}}{\rightleftharpoons} \quad \text{(CH}_3\text{)}_2\text{CH}-\underset{\text{Li}^+}{\overset{}{\text{Si}}}\overset{\text{CH(CH}_3\text{)}_2}{\underset{}{\diagdown\text{Ph}}} \tag{9}$$

C. Reactions

1. Stereochemical Aspects

Reaction of an optically active secondary alkyl chloride with Ph_3SiLi proceeds with complete (100%) inversion of configuration. The process is a normal S_N2 reaction [Eq. (10)] (25).

$$\text{Ph} \underset{\text{Cl}}{\overset{\text{H}}{\diagdown\diagup\underset{|}{\text{Me}}}} \xrightarrow[\text{inversion}]{Ph_3SiLi} \text{Ph} \overset{SiPh_3}{\diagdown\diagup\underset{\text{Me}}{\overset{|}{\cdots H}}} \qquad (10)$$

Reaction of an optically active chlorosilane with $Ph_2MeSiLi$ proceeds with inversion of configuration to give an optically active disilane [Eq. (11)] (15).

$$(-)\text{-}(neo\text{-}C_5H_{11})PhMeSi^*Cl \xrightarrow[\text{inversion}]{Ph_2MeSiLi} (-)\text{-}(neo\text{-}C_5H_{11})PhMeSi^*SiMePh_2 \quad (11)$$
$$[\alpha]_D = +7.2° \qquad\qquad\qquad\qquad [\alpha]_D = -5.03°$$

2. Reaction of Silyllithium with Chlorosilane: Halogen–metal exchange and Si–Si Bond Cleavage Reaction

It has been reported that in reactions of silyllithium with chlorosilane, the products depend on the mode of addition (26,27), as shown by the following two examples: Whereas addition of silyllithium to chlorosilane gives the desired cross-coupled disilane [Eq. (12)], addition of chlorosilane to silyllithium forms a mixture of homo-coupled disilanes [Eq. (13)].

$$Ph_3SiCl \xrightarrow{\begin{array}{c}\text{Addition of}\\PhMe_2SiLi\end{array}} \underset{45\%}{PhMe_2Si\text{-}SiPh_3} + \underset{2.3\%}{Ph_3Si\text{-}SiPh_3} \qquad (12)$$

$$PhMe_2SiLi \xrightarrow{\begin{array}{c}\text{Addition of}\\Ph_3SiCl\end{array}} \underset{35\%}{Ph_3Si\text{-}SiPh_3} + \underset{71\%}{PhMe_2Si\text{-}SiMe_2Ph} \qquad (13)$$

Unusual homo-coupling reactions observed in Eq. (13) have been explained in terms of (1) halogen–metal exchange and/or (2) Si–Si bond cleavage reaction. Thus, according to (1), the first step in the reaction may be a rapid halogen–metal exchange in the presence of excess amounts of silyllithium [Eq. (14)] (26). The driving force for this exchange is the generation of the more stable silyllithium Ph_3SiLi. The cross-coupling

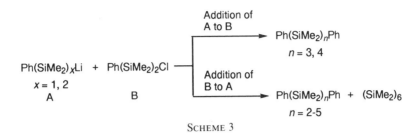

SCHEME 3

reaction has little chance to occur. According to (2), the reaction includes the second reaction after the initial cross-coupling reaction [Eq. (15)]. The cross-coupled disilane is thus cleaved by excess amounts of silyllithium to give homo-coupled disilane (27).

$$PhMe_2SiLi \ + \ Ph_3SiCl \longrightarrow PhMe_2SiCl \ + \ Ph_3SiLi \tag{14}$$

$$PhMe_2Si\text{-}SiPh_3 \ + \ PhMe_2SiLi \longrightarrow PhMe_2Si\text{-}SiMe_2Ph \ + \ Ph_3SiLi \tag{15}$$

Similar results are observed in the reaction of a disilanyllithium with a chlorodisilane (28), as shown in Scheme 3. When a mixture of monosilyl and disilanyl anions is introduced to 1-chloro-2-phenyltetramethyldisilane, the usual coupling products, 1,3-diphenyltrisilane and 1,4-diphenyltetra-silane, are formed. Alternatively, when the chlorosilane is added into a mixture of the anions, the scrambling processes via the Si–Si bond cleav-age occur, to give a mixture of linear and cyclic oligosilanes. These results indicate that the scrambling processes are very fast, but much slower than the coupling of anions with chlorosilanes (see also Section VI-B).

IV

ALKYLSILYL ANIONS

A. Preparation and Reaction

Peralkylsilyl anions having no aromatic substituents cannot be prepared by reaction of the corresponding chlorosilane or disilane with alkali metal (except $t\text{-}Bu_3SiM$). The action of alkali metals on chlorosilane gives not silyllithium but disilane as coupling product, because cleavage of the Si–Si bond of peralkyldisilane does not occur with alkali metals. The energy of the lowest unoccupied molecular orbital (LUMO) of alkyl-substituted disilanes (σ^* orbital of the Si–Si bond) is too high to accept an electron from the metal. At least one aryl group on silicon appears necessary for cleavage of the Si–Si bond.

Cleavage of the Si–Si bond in peralkyldisilanes by organometallic nucleophiles is successful, however. Me_3SiLi can thus be prepared from reaction of $Me_3SiSiMe_3$ with MeLi in $HMPA-Et_2O$ (HMPA = hexamethylphosphoric triamide) [Eq. (16)] (29). This may be the only practical method for the preparation of Me_3SiLi. In a large-scale experiment, however, Me_3SiMe_2SiLi was generated instead of Me_3SiLi (see Section VII).

$$Me_3Si\text{-}SiMe_3 + MeLi \xrightarrow[\text{THF-HMPA}]{} Me_3SiLi + Me_4Si \qquad (16)$$

Me_3SiNa and Me_3SiK are obtained from $Me_3SiSiMe_3$ by treatment with MeONa or MeOK, together with Me_3SiOMe [Eq. (17)] (30–31). A strongly basic (coordinating) aprotic solvent such as HMPA or 1,3-dimethyl-2-imidazolidone (DMI) is essential to this reaction. The driving force for this reaction is the formation of the strong Si–O bond.

$$Me_3Si\text{-}SiMe_3 + MOMe \xrightarrow[\substack{\text{HMPA or DMI} \\ M = Na, K}]{} Me_3SiM + Me_3SiOMe \qquad (17)$$

The alkali metal hydrides NaH and KH are also effective as metalating reagents for hydrosilane and disilane to give trialkylsilyl anions in an aprotic solvent such as THF, DME, or HMPA (8) (Scheme 4) (for the mechanism, see Section III-A-2). The reaction of disilane with 2 equivalents of metal hydride converts both silyl groups into silyl anions in two steps, in contrast to the reaction with MeOM described earlier.

Silylmercury compounds have long been well investigated. Some previous reviews should be referenced (1c,d). $(Me_3Si)_2Hg$ is prepared by reaction of Me_3SiCl in cyclopentane with Na/Hg. These reactions must be carried out in the dark to avoid photochemically induced decomposition (32). Silylmercury compounds serve as precursors for silyl-alkali metal derivatives [Eq. (18)] (33–35). Mercury–alkali metal exchange reaction proceeds in THF or benzene. Benzene gives better yields than THF, although the reaction rate is larger in THF than in benzene. If made in benzene, Me_3SiLi can be sublimed under high vacuum at 60°C. Reaction of a silylmercury with potassium in benzene, however, gives no Et_3SiK but instead Et_3SiPh in quantitative yield [Eq. (19)]. It is hypothesized that

$$Et_3SiH + KH \xrightarrow[\text{DME or HMPA}]{} Et_3SiK + H_2$$

$$Me_3Si\text{-}SiMe_3 + MH \xrightarrow[\substack{\text{DME or HMPA} \\ M = Na, K}]{} Me_3SiM + Me_3SiH \xrightarrow{MH} Me_3SiM + H_2$$

SCHEME 4

the initially formed Et_3SiK abstracts a proton from benzene to yield Et_3SiH and PhK, which couple with each other to give Et_3SiPh.

$$(Me_3Si)_2Hg + 2\,Li \longrightarrow 2Me_3SiLi + Hg \tag{18}$$

$$(Et_3Si)_2Hg + 2\,K \xrightarrow{C_6H_6} 2\,Et_3Si\text{-}C_6H_5 + Hg + 2\,KH \tag{19}$$

Bis(trimethylsilyl)magnesium, $(Me_3Si)_2Mg$, can also be obtained by reaction of $(Me_3Si)_2Hg$ with magnesium metal in DME and recrystallized from cyclopentane to give pale pink crystals of $(Me_3Si)_2Mg \cdot DME$ (36).

Lithium tetrakis(trimethylsilyl)aluminate, $LiAl(SiMe_3)_4$, prepared by reaction of Me_3SiCl with lithium and aluminum in the presence of mercury (37), reacts with $ZnCl_2$ or $Zn(OAc)_2$ in Et_2O to give $(Me_3Si)_2Zn$ [Eq. (20)] (38a–c). The pale yellow crystalline product is isolable by vacuum sublimation and is readily soluble in ethers, benzene, or pentane. The silylzinc compound can be stored for about three weeks at $-20°C$, but decomposes slowly, with separation of the metal, at room temperature. The cadmium analog, $(Me_3Si)_2Cd$, is synthesized in the same way. The silylcadmium is less stable than the silylzinc and decomposes within two days, with precipitation of the metal, even at $-20°C$. This compound is also light-sensitive.

$$LiAl(SiMe_3)_4 \cdot 2DME + MX_2 \xrightarrow{Et_2O} M(SiMe_3)_2 \tag{20}$$
$$M = Zn,\ Cd$$
$$X = Cl,\ OAc$$

A metal-free trimethylsilyl anion is formed from hexamethyldisilane by cleavage with TBAF in HMPA in equilibrium concentration, as revealed by ^1H and ^{19}F NMR analysis of the reaction mixture. Treatment of trisilane with TBAF provides $Me_3SiMe_2Si^-/NBu_4^+$ and Me_3SiF; the fluoride anion attacks at the terminal silicon of trisilane [Eq. (21)] (39). Synthetic application of this species is described in Section VII. Other metal-free silyl anions are described in Section V.

$$Me_3SiSiMe_3 + Bu_4N^+F^- \rightleftharpoons Bu_4N^+[Me_3SiSiMe_3/F^-]$$
$$\rightleftharpoons [Me_3Si^-/NBu_4^+] + Me_3SiF \tag{21}$$

A sterically crowded silyl anion, tri-tert-butylsilylsodium (t-Bu_3SiNa), is produced from t-Bu_3SiBr and Na in refluxing THF (Scheme 5) (40). In this reaction, the coupling product, t-Bu_3Si–$Si\,t$-Bu_3, is not formed, because of the high steric hindrance at silicon. The compound crystallizes as yellow needles from a solution in pentane at $-78°C$. Its ^1H, ^{13}C, ^{29}Si, and ^{23}Na NMR studies indicate the constitution to be t-$Bu_3SiNa \cdot 2THF$.

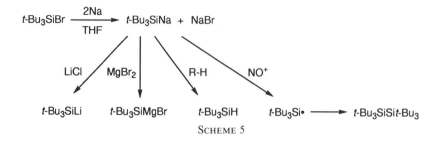

SCHEME 5

Similarly, $t\text{-Bu}_3\text{SiK}$ is obtained from $t\text{-Bu}_3\text{SiI}$ and K in refluxing heptane. Whereas $t\text{-Bu}_3\text{SiM}$ (M = Na, K) does not couple with $t\text{-Bu}_3\text{SiX}$ (X = Cl, Br), the disilane $t\text{-Bu}_3\text{SiSi}t\text{-Bu}_3$ can be obtained by coupling of the silyl radical $t\text{-Bu}_3\text{Si}\cdot$, which is generated by oxidation of $t\text{-Bu}_3\text{SiM}$ by the nitrosyl cation NO^+. $t\text{-Bu}_3\text{Si}^-$ acts as a strong base, and at higher temperatures it reacts with THF. The cesium derivative, $t\text{-Bu}_3\text{SiCs}$, deprotonates even hydrocarbons. $t\text{-Bu}_3\text{SiNa}$ can be converted into $t\text{-Bu}_3\text{SiLi}$ and $(t\text{-Bu}_3\text{Si})_2\text{Mg}$, respectively, by exposure to LiCl and MgBr_2.

The high electron-donating ability of Me_3SiNa causes reduction of naphthalene [Eq. (22)] (30a).

$$\text{Me}_3\text{SiNa} + \text{(naphthalene)} \longrightarrow \text{Me}_3\text{Si}\cdot + \left[\text{(naphthalene)}\right]^{\cdot-} \text{Na}^+ \tag{22}$$

$$\xrightarrow{\text{H}_2\text{O}} \text{(dihydronaphthalene)} + \text{(dihydronaphthalene)}$$

13% 11%

Nucleophilic substitition of aromatic halides with trimethylsilyl anions competes with reduction. The ratio of substitution/reduction increases in the direction of I < Br < Cl, and appears relatively insensitive to the metal cations (30a,b). When an electrophile is a poorer electron acceptor, such as Ph_3GeBr, nucleophilic substitution proceeds smoothly to form the coupling product in good yield (8).

B. Structural Study

Crystal structures of Me_3SiLi and its N,N,N',N'-tetramethylethylenediamine (TMEDA) complex have been investigated by X-ray analysis (41,42). Me_3SiLi exists as a hexamer in the solid state (Fig. 3) (41). The

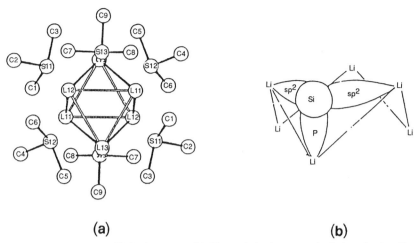

(a)   **(b)**

FIG. 3. (a) The $(Me_3SiLi)_6$ hexamer. (b) View of the hexamer depicting the bonding between the bridging silyl group and the lithium atoms (41). Reprinted with permission from *J. Am. Chem. Soc.* **1974,** *96,* 7593; **1980,** *102,* 3769. Copyright (1974 and 1980) American Chemical Society.

Me_3SiLi framework could best be represented as a "folded chair" six-membered lithium ring of approximate D_{3d} symmetry. The Li_6 chair conformation is much closer to the octahedron than to a cyclohexane structure. Each of the trimethylsilyl groups is located centrally above a triangular face in the periphery of the Li_6 ring. The averaged Si–C(methyl) bond distance is 1.89 Å, which is similar to the normal Si–C single bond (1.86–1.91 Å). The averaged Si–Li distance is 2.68 Å. The bonding is explained in terms of four-centered electron-deficient Si–Li bonds with minimal Li–Li interactions. This might be viewed as the mutual coordination of two trimers with essentially sp^2-hybridized lithium atoms, leaving a p orbital available that is properly oriented to interact with the carbon or silicon atoms of the trimethylsilyl groups on the second trimeric unit to provide the necessary bonding to lead to the observed hexameric structure.

The crystal structure of $(Me_3SiLi)_2(TMEDA)_3$ has also been investigated (Fig. 4) (42). The results reveal that (1) there is no Li–Si–Li bridge bonding, indicative of the complete destruction of the hexameric aggregate, $(LiSiMe_3)_6$, by complexation with TMEDA, and (2) the Si–Li bond distance is relatively long, 2.70 Å, compared to the averaged Si–Li distance of 2.68 Å in the uncomplexed hexamer.

Only one structural study on an Si–Mg compound has been performed. The X-ray analysis of $(Me_3Si)_2Mg\cdot DME$ (36) shows that the compound

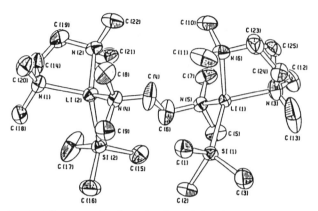

FIG. 4. An ORTEP diagram for the molecular unit (Me$_3$SiLi)$_2$(TMEDA)$_3$ (42). Reprinted with permission from *Organometallics* **1982**, *1*, 875. Copyright (1982) American Chemical Society.

is monomeric in the crystal and that the magnesium atom has a distorted tetrahedron surrounded by the two silicon atoms and two oxygen atoms. The Si–Mg distance of 2.63 Å is somewhat longer than the sum of the covalent radii (2.41 Å); the averaged Si–C(methyl) distance is 1.88 Å. The Si–Mg–Si angle of 125° is greater than the tetrahedral angle, whereas the O–Mg–O angle is much smaller (76°). These angle distortions result from the steric repulsion of the bulky trimethylsilyl groups and the necessarily small O–Mg–O angle required by the bite size of the DME ligand and the Mg–O distances.

Infrared (IR) and Raman spectra of Zn(SiMe$_3$)$_2$, Cd(SiMe$_3$)$_2$, and Hg(SiMe$_3$)$_2$ indicate that these molecules are linear, because only one Si-metal stretching is observed in all cases (*38a,43*).

V

FUNCTIONALIZED SILYL ANIONS

In contrast to triorganosilyl anions, functionalized silyl anions have been studied less extensively. Functional groups reported so far are hydrogen, chlorine, amino, alkoxy, and carbonyl groups. In addition to these species, silacyclopentadienide anions are also described in this section. Theoretical studies on the effect of functional groups are described in Section VIII.

1. Hydrosilyl Anions

Silylpotassium, H_3SiK, is obtained by the reaction of silane, SiH_4, and K in DME (44–47). The ^{29}Si NMR signal in HMPA at ambient temperature is observed at -165.0 ppm as a quartet ($^1J[Si, H] = 75$ Hz) (46). The photoelectron spectra of SiH_3^- and SiD_3^- have been investigated (48).

Reaction of $SiCl_4$ with gaseous lithium (under 10^{-4} torr at 700–800°C) appears to result in the formation of tetralithiosilane, $SiLi_4$ (49), because quenching with excess MeCl in the presence of DME gives Me_4Si in 5–10% yields.

HPh_2SiLi is prepared from HPh_2SiCl or $HPh_2SiSiPh_2H$ with Li in THF (50). This species, however, is obtained only in low yield and readily polymerizes during prolonged reaction periods or at higher temperatures. Still, a bulky analog, $HMes_2SiLi$ (Mes = mesityl = 2,4,6-trimethylphenyl), is obtained as stable species in good yield [Eq. (23)] (51–53). Furthermore, the off-white crystals are isolated in 71% yield (52). The ^{29}Si NMR chemical shift (δ -71.78 ppm) is consistent with the presence of the electropositive lithium atom.

$$HMes_2SiCl + 2\,Li \xrightarrow[\text{THF, 0 °C}]{} HMes_2SiLi + LiCl$$

$$\xrightarrow[\text{toluene}]{Ph_2SiCl_2} HMes_2Si\text{-}SiPh_2Cl + LiCl \qquad (23)$$
$$53\%$$

2. α-Heteroatom–Substituted Silyl Anions

Three {(amino)phenylsilyl}lithiums, $(Et_2N)Ph_2SiLi$, $(Et_2N)_2PhSiLi$, and $(Et_2N)PhMeSiLi$, are obtained by (1) action of lithium metal on (amino)-phenylchlorosilanes in THF, and (2) transmetallation of the corresponding (aminosilyl)stannanes with BuLi in THF (Scheme 6) (54a–d). In method

$$(Et_2N)_nPh_{3\text{-}n}SiCl \xrightarrow[\text{THF, 0 °C, 4 h}]{Li} (Et_2N)_nPh_{3\text{-}n}SiLi$$
$$n = 1, 2$$
$$97\text{-}98\%$$

$$(Et_2N)MePhSiCl \xrightarrow[\text{THF, 0 °C, 24 h}]{Li} (Et_2N)MePhSiLi$$
$$80\%$$

$$(Et_2N)_nPh_{3\text{-}n}SiSnMe_3 \xrightarrow{RLi} (Et_2N)_nPh_{3\text{-}n}SiLi$$

$n = 1$; R = n-Bu, THF, 0 °C, 30 min, 90%

$n = 2$; R = t-Bu, THF-Et_2O, -30 °C, 4 h, 77%

SCHEME 6

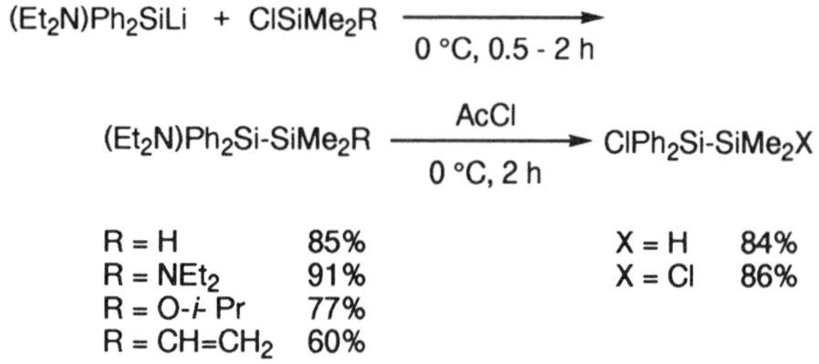

$(Et_2N)Ph_2SiLi + ClSiMe_2R \xrightarrow[0\ °C,\ 0.5\ -\ 2\ h]{}$

$(Et_2N)Ph_2Si\text{-}SiMe_2R \xrightarrow[0\ °C,\ 2\ h]{AcCl} ClPh_2Si\text{-}SiMe_2X$

R = H	85%	X = H	84%
R = NEt$_2$	91%	X = Cl	86%
R = O-i-Pr	77%		
R = CH=CH$_2$	60%		

SCHEME 7

(1), a highly reactive lithium dispersion is required for preparation of the last two species.

These three (aminosilyl)lithiums are stable at 0°C in THF for a few days under a nitrogen or argon atmosphere. The ^{29}Si NMR signal of $(Et_2N)Ph_2SiLi$ appears at 19.34 ppm in THF at 0°C. Its ^{13}C chemical shifts of aromatic carbons are quite similar to those for MePh$_2$SiLi ($19a$) (data are listed in Table I in Section III-B). The result implies that the Et$_2$N group exhibits essentially the same effect as the Me group on the charge distribution in silyl anions (see also Section III-B).

By use of these (aminosilyl)lithiums, a variety of functional disilanes and oligosilanes are obtained. The Et$_2$N group on silicon can be converted into an alkoxy group or chlorine (Scheme 7). The action of lithium on (amino)alkylchlorosilanes gives not silyllithiums but instead the disilane coupling products ($54c$).

(Aminosilyl)magnesium species are obtained by reaction of the silyllithiums with i-PrMgBr (Scheme 8). The silylmagnesium reagents are milder

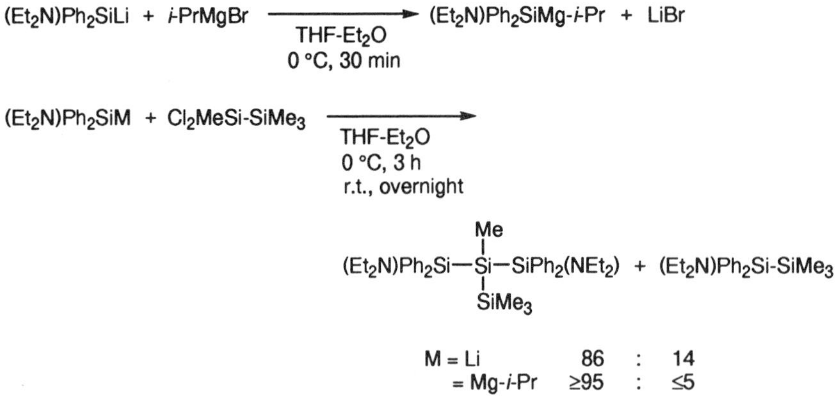

$(Et_2N)Ph_2SiLi + i\text{-}PrMgBr \xrightarrow[\substack{THF\text{-}Et_2O \\ 0\ °C,\ 30\ min}]{} (Et_2N)Ph_2SiMg\text{-}i\text{-}Pr + LiBr$

$(Et_2N)Ph_2SiM + Cl_2MeSi\text{-}SiMe_3 \xrightarrow[\substack{THF\text{-}Et_2O \\ 0\ °C,\ 3\ h \\ r.t.,\ overnight}]{}$

$$(Et_2N)Ph_2Si\overset{\overset{\displaystyle Me}{|}}{\underset{\underset{\displaystyle SiMe_3}{|}}{Si}}SiPh_2(NEt_2) + (Et_2N)Ph_2Si\text{-}SiMe_3$$

M = Li	86	:	14
= Mg-i-Pr	≥95	:	≤5

SCHEME 8

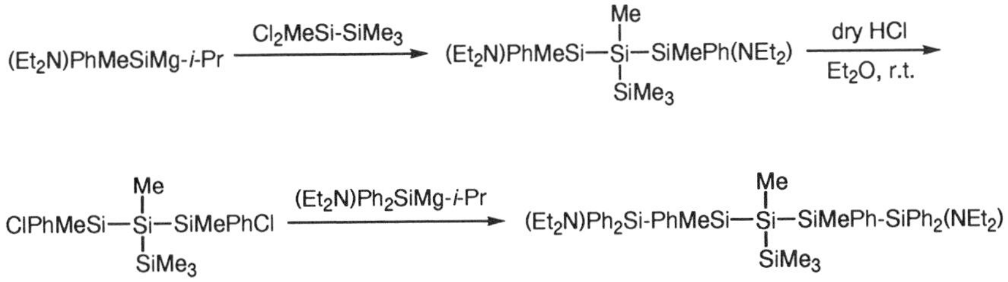

SCHEME 9

nucleophiles than the silyllithiums, so Si–Si bond cleavage is avoided in oligosilane synthesis (Schemes 8 and 9) (54b).

An (alkoxysilyl)lithium, $(t\text{-BuO})Ph_2SiLi$, can be prepared from $(t\text{-BuO})Ph_2SiSnMe_3$ by treatment with $n\text{-BuLi}$ in THF (Scheme 10). This species is stable at $-78°C$ for a few hours and behaves as an ordinary nucleophilic silyl anion (54d). $(t\text{-BuO})Ph_2SiLi$ also behaves as an electrophile under certain conditions: It undergoes self-condensation smoothly or butylation in the presence of an excess amount of $n\text{-BuLi}$ and TMEDA (Scheme 10). $(t\text{-BuO})Ph_2SiLi$ thus exhibits an ambiphilic nature, that is, silylenoid character. The ambiphilic reactivity of $(t\text{-BuO})Ph_2SiLi$, how-

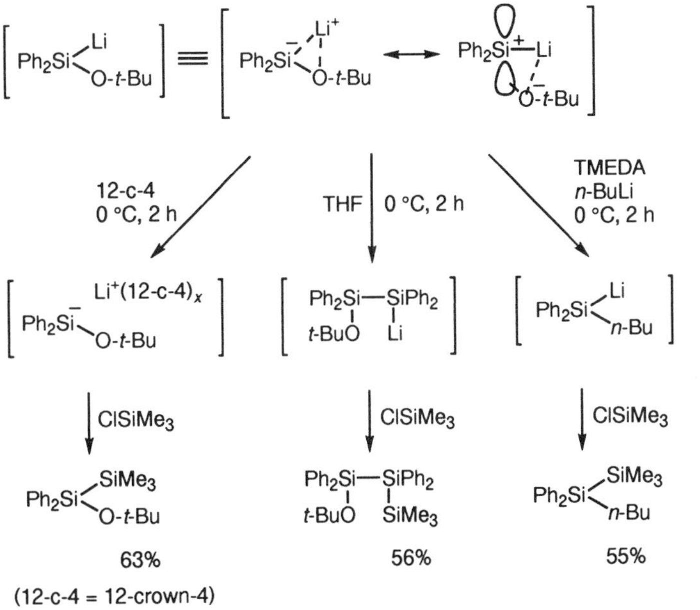

SCHEME 10

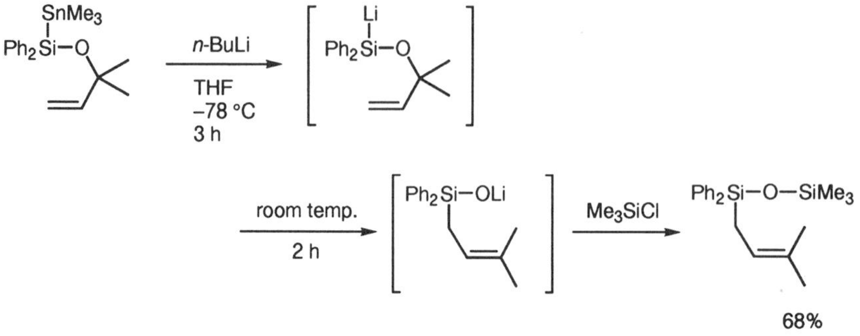

SCHEME 11

ever, disappears completely with the addition of a crown ether, which is expected to make a "solvent-separated ion pair." In this system, neither self-condensation nor butylation occurs at all (Scheme 10).

A variety of (alkoxysilyl)lithiums and other functionalized silyllithiums $X_n Ph_{3-n}SiLi$ (X = t-BuO, i-PrO, MeO, Et_2N, H; n = 1 and X = t-BuO; n = 2) are obtained by reduction of the corresponding chlorosilanes with lithium 1-(dimethylamino)naphthalenide ($54e$).

A (dialkoxysilyl)lithium $(t$-BuO$)_2$PhSiLi is also obtained in 45% yield by action of lithium metal on $(t$-BuO$)_2$PhSiCl in THF at 0°C for 4 h ($54f$).

(Allyloxysilyl)lithiums undergo a [2,3]Wittig-type rearrangement smoothly to form allylsilanolate anions in an intramolecular fashion (Scheme 11). This is the first example of the sila-Wittig rearrangement ($54g$).

(Methoxysilyl)sodiums, $(MeO)_n Me_{3-n}SiNa$ (n = 1 or 2), are generated by cleavage of the corresponding disilanes, $(MeO)_n Me_{3-n}SiSiMe_{3-n}(OMe)_n$, with MeONa in the common solvents THF, benzene, and triethylamine, requiring no strongly coordinating aprotic solvents, such as HMPA and DMI [Eq. (24)] ($55a,b$). This reaction is carried out in the presence of organic halides as trapping agents.

$$(MeO)_n Me_{3\text{-}n}Si\text{-}SiMe_{3\text{-}n}(OMe)_n \xrightarrow{\text{NaOMe}}$$
$$n = 1, 2$$

$$(MeO)_n Me_{3\text{-}n}Si^- Na^+ \xrightarrow{\text{R-X}} (MeO)_n Me_{3\text{-}n}Si\text{-}R$$

(24)

Trichlorosilyllithium, Cl_3SiLi, appears to be formed directly by a halogen–lithium exchange reaction of Cl_3SiBr with 2,4,6-tri-t-butylphenyllithium [Eq. (25)] (56).

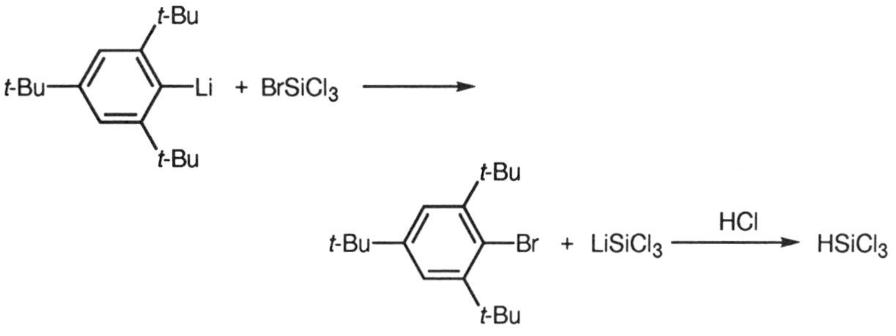

$$(25)$$

Trichlorosilane, Cl_3SiH, is deprotonated by Bu_3N in refluxing CH_3CN to give a salt $[Cl_3Si^-/Bu_3NH^+]$ [Eq. (26)] (57a–g). This species is a reductive silylating agent and is useful for organic synthesis (see Section VII). Similar species are prepared from Cl_3SiH and Cl_2RSiH (R = Me, Ph) with TMEDA (58) (see Section VII).

$$Cl_3SiH + Bu_3N \xrightarrow[\text{reflux}]{CH_3CN} \left[Cl_3Si^- / HNBu_3^+ \right] \qquad (26)$$

Evidence for the existence of the trichlorosilyl anion, Cl_3Si^-, has been obtained by ^1H NMR. The NMR spectrum of a solution of Cl_3SiH in CH_3CN shows a sharp singlet for the Cl_3SiH proton at δ 6.25. This singlet broadens and diminishes as $(n\text{-}C_3H_7)_3N$ is added. At the same time, a new singlet appears and grows at δ 11.03, which is attributed to the NH resonance of $(n\text{-}C_3H_7)_3NH^+$. This result indicates that proton exchanges are occurring between Cl_3SiH and $(n\text{-}C_3H_7)_3N$.

α-Halosilyl anions, $R_2Si(X)(M)$, are postulated intermediates in polysilane synthesis via reductive coupling of dihalosilanes with alkali metal. The α-halosilyl anion generated from dihalosilane and alkali metal rapidly undergoes coupling with dihalosilane to give dihalodisilane (Scheme 12) (59) (see also Section VI-B). A theoretical study on (fluorosilyl)lithium, H_2SiLiF, has been performed (see Section VIII).

3. Sila-Enolate

Silyl anions in which a carbonyl group is attached to silicon have been reported (60). An acyltris(trimethylsilyl)silane is treated with tris(trimeth-

$$R_2SiX_2 \xrightarrow[-MX]{2\,M} R_2Si{\overset{X}{\underset{M}{<}}} \xrightarrow[-MX]{R_2SiX_2} R_2Si\underset{X}{\overset{|}{-}}SiR_2\underset{X}{\overset{|}{}}$$

SCHEME 12

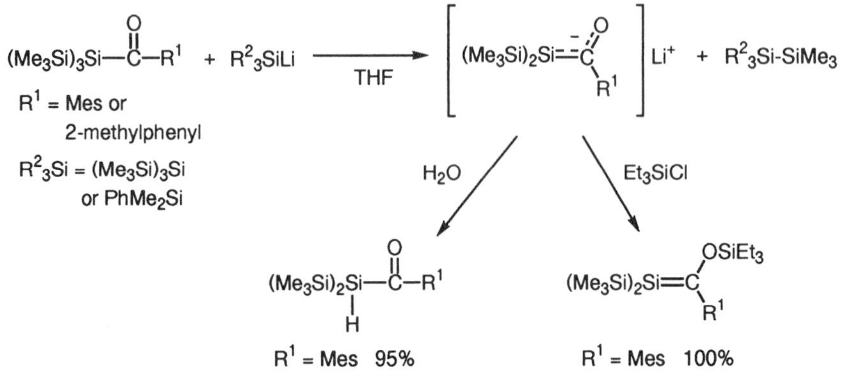

SCHEME 13

ylsilyl)silyllithium in THF at $-80°C$ for 2 hours to give a silicon analog of a lithium enolate, along with quantitative amount of tetrakis(trimethylsilyl)silane (Scheme 13). The lithium sila-enolate is trapped by H_2O to give the corresponding acylbis(trimethylsilyl)silane in 95% yield. In contrast, trapping by Et_3SiCl gives the corresponding sila-enol silyl ether (siloxysilene) in quantitative yield. The ^1H NMR spectrum of the enolate in THF at $-40°C$ shows two resonances, at -0.26 and 0.17 ppm, attributed to two nonequivalent trimethylsilyl protons. The ^{29}Si NMR spectrum at $-40°C$ shows three resonances, at -10.6, -12.8, and -59.9 ppm, attributed to two nonequivalent trimethylsilyl silicons and central silicon, respectively.

The formation of the sila-enolate has been supported by theoretical calculations (6-31G*) for a model reaction as shown in Eq. (27). The enthalpy for this reaction has been calculated to be -13.81 kcal/mol. The formation of the sila-enolate is thus thermodynamically favorable.

$$(H_3Si)_3SiCOCH_3 + H_3SiLi \longrightarrow (H_3Si)_2SiLi(COCH_3) + H_3SiSiH_3 \qquad (27)$$

4. *Silacyclopentadienide Anions*

Studies on silacyclopentadienide anions have been developing in recent years (*61*). The term *silole* means the same thing as *silacyclopentadiene*.

Theoretical calculations on a model system have predicted that the ground-state conformation of the anion is the pyramidal C_s structure (Fig. 5) (*62*). It is noteworthy that the C_s structure has unusually long C–Si bonds and a butadiene-like geometry of the carbon skeleton. The planar C_{2v} form is the transition state for pyramidal inversion between equivalent C_s structures, the inversion barrier being 16.2 kcal/mol (6-31 G*).

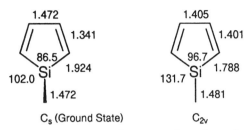

C_s (Ground State) C_{2v}

FIG. 5. Calculated bonding parameters for two conformations of a cyclopentadienide anion. Bond lengths are in angstroms and bond angles in degrees.

An energy difference for the isodesmic reaction [Eq. (28)] indicates a quantitative measure of the extent of electron delocalization (6-31G*) (62). The energy difference for the C_s structure of the silacyclopentadienide anion is only 2.2 kcal/mol, which is much smaller than 73.4 kcal/mol for the carbon analog, a planar cyclopentadienide anion (C_{2v}). On the basis of these calculations, it is concluded that the ground-state structure of the silacyclopentadienide anion has only ca. 3% of the resonance stabilization exhibited by the cyclopentadienide anion. However, the previous calculation (3-21G//STO-2G) on the C_{2v} planar structure shows that the resonance energy is 23 kcal/mol, which means ca. 25% as aromatic as the carbon analog (63).

$$\underset{\substack{\text{Si}\\\text{H}}}{\bigcirc}^{-} + H_4Si \longrightarrow \underset{\substack{\text{Si}\\\text{H}_2}}{\bigcirc} + H_3Si^- \qquad (28)$$

1-Lithio- and 1-sodio-1-t-butyl-2,3,4,5-tetraphenylsilacyclopentadienes are prepared by cleavage of the Si–Si bond of di(silacyclopentadienyl) with Li and Na, respectively, in THF at room temperature [Eq. (29)] (64a).

$$\text{(structure)} \xrightarrow[\substack{M = \text{Li, Na}}]{2\,M\,/\,\text{THF}} \left[\text{(structure)} \right] \xrightarrow{Me_2SiHCl} \text{(structure)}$$

M = Li (purple)
M = Na (dark purple) $\qquad (29)$

The ^{29}Si NMR chemical shifts are observed at 25.10 ppm for the lithio derivative and at 26.12 ppm for the sodio derivative, a large downfield shift compared to the starting material (3.62 ppm). The downfield shifts strongly suggest the incorporation of silicon p-orbitals to diffuse and/or

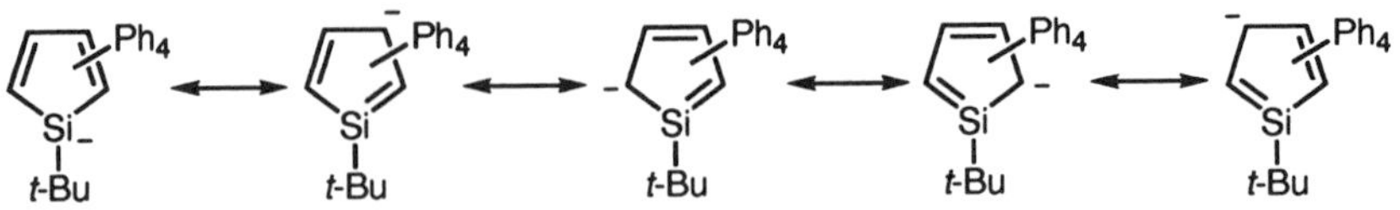

SCHEME 14

delocalize the negative charge into the butadiene moiety (Scheme 14). In the ^{13}C NMR spectrum of the lithio derivative, the C_α and C_β are shifted upfield ($\Delta\delta = -5.26$ and -2.28 ppm), the ipso and ortho carbons, C_i and C_o, of the four phenyl groups are shifted downfield ($\Delta\delta = 1.10$ and 7.03 ppm), and the meta and para carbons, C_m and C_p, are shifted upfield ($\Delta\delta = -1.19$ and -4.92 ppm). The upfield shielding of the C_α and C_β in the anion indicates that there is an increase in π-electron density at these positions, as shown in Scheme 14. The four phenyl groups on the ring provide additional stabilization of the silacyclopentadienide anion through polarization (cf. Fig. 2).

1-Lithio-1-methyl-2,3,4,5-tetraphenylsilacyclopentadiene is also prepared and characterized by X-ray analysis which shows a dimeric structure containing a 1,3-disilacyclobutane skeleton formed in such a fashion that each silyl anion adds to the C2-C3 double bond in the other ring (64b).

1,1-Disodio-2,3,4,5-tetraphenyl-1-silacyclopentadiene is obtained as a reddish-brown solid from the 1,1-dichloro derivative, with more than four equivalents of Na in refluxing dioxane for 7 hours in 80% yield (Scheme 15) (65a). Reaction of the dianion with MeI and Me$_3$SiCl gives the expected trapping products. Trapping with H$_2$O, however, causes ring opening to give 1,2,3,4-tetraphenyl-2-butene and silicate.

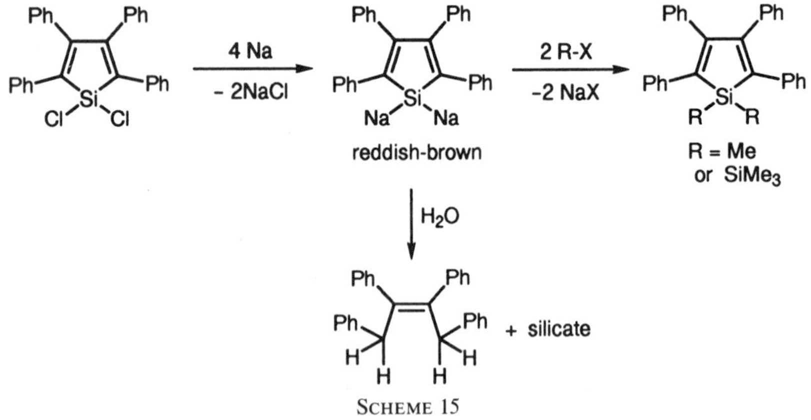

SCHEME 15

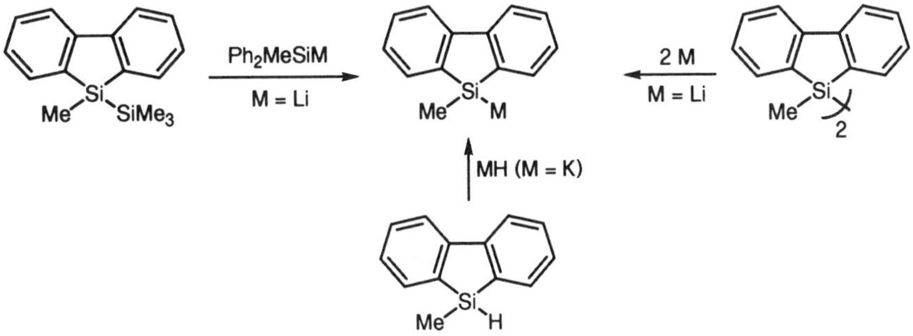

SCHEME 16

1,1-Dilithio-silacyclopentadienes are also prepared from the 1,1-dichloroderivatives with four equivalents of lithium in THF ($65b,c,d$). X-ray analysis indicates a monomeric structure in which one lithium atom lies in the plane and the other over the plane of the silacyclopentadiene ring, thus represented as η^5-(1-lithio-silacyclopentadienyl)lithium ($65d$).

Benzosilole anions have been obtained in three ways, as summarized in Scheme 16. 5-Lithio-5-methyldibenzosilole is thus obtained from bis(dibenzosilole) with lithium in THF, which affords 5,5-dimethyldibenzosilole (54% yield) after trapping by dimethyl sulfate (66). The same species is also formed from 5-methyl-5-(trimethylsilyl)dibenzosilole via the Si–Si bond cleavage on treatment with $Ph_2MeSiLi$ in THF at $-78°C$, together with tetramethyldiphenyldisilane (67). The potassium analog can be prepared from the 1-hydridobenzosilole by treatment with strong nonnucleophilic bases such as KH (14).

The ^1H NMR spectrum of the lithium species reveals a sharp resonance at δ 0.24 ppm due to the CH_3–Si protons. The UV spectrum of the anion in THF exhibits characteristic absorptions in the longer-wavelength region at 377 nm (ε 5700) and 546 nm (ε 1200), in comparison with the neutral dibenzosilole, which has several absorption maxima up to 318 nm (ε 170).

VI

POLYSILANYL ANIONS

These species may fall into two classes: isolable or spectroscopically detectable polysilanyl anions, and polysilanyl anions postulated as reaction intermediates.

$$Me_3SiSiMe_3 \;+\; MeLi \xrightarrow[\text{HMPA-THF}]{} Me_3SiLi \;+\; Me_4Si$$

$$Me_3SiSiMe_3 \;+\; Me_3SiLi \xrightarrow{} Me_3SiSiMe_2Li \;+\; Me_4Si$$

SCHEME 17

A. Preparation, Reaction, and Structure

1. Disilanyl Anions

Polysilanyl anions, H_3SiH_2SiK, $(H_3Si)_2HSiK$, and $(H_3Si)_3SiK$, are prepared, respectively, (a) from H_3SiK with SiH_4, (b) from H_3SiK and Si_3H_8, and (c) from Si_2H_6 or Si_3H_8 and K in HMPA, and identified by ^1H NMR and IR spectroscopy (46) (see also Section V).

As mentioned in Section IV-A, Me_3SiMe_2SiLi is prepared by the action of MeLi on $Me_3SiSiMe_3$ on a large scale (Scheme 17) (68a–c). It is proposed that in this reaction, Me_3SiLi, resulting from Si–Si bond cleavage by MeLi in the first stage, cleaves the Si–C(methyl) bond of the unreacted disilane to give Me_3SiMe_2SiLi and Me_4Si (68a). This reaction occurs also with n-BuLi instead of MeLi. A similar Si–C(methyl) bond cleavage is also observed in the case of dodecamethylcyclohexasilane (see Section VII-A-4).

The Si–Si bond cleavage of trisilanes by MeLi in THF is applicable to the preparation of disilanyllithiums, in which a phenyl or mesityl group is attached to α-silicon [Eq. (30)] (69). A similar cleavage reaction occurs with a Na/K alloy to afford disilanylpotassium. These species couple with AdCOCl (Ad = 1-adamanthyl) to give acyldisilane in modest yields.

$$Me_3Si\text{-}SiR_2\text{-}SiMe_3 \xrightarrow[\text{THF}]{MeLi} Me_3Si\text{-}SiR_2Li \xrightarrow[-78\,°C,\,THF]{AdCOCl} Me_3Si\text{-}SiR_2\text{-}COAd \qquad (30)$$

$$R = Ph \qquad \text{or Na/K}$$
$$R = Mes$$

A mercury–lithium exchange reaction of $(Me_3SiMe_2Si)_2Hg$ yields Me_3SiMe_2SiLi in toluene quantitatively (Scheme 18) (70). Similarly, a variety of polysilanyllithiums is also obtained (see Section VI-A-3). This method can avoid polar solvents, so the reagents undergo coupling reactions with chlorosilanes, without undesirable Si–Si bond cleavage, to afford various oligosilanes in high yields (see Section VI-A-3).

The ^{29}Si NMR signals of Me_3SiMe_2SiLi are observed at -74.9 ppm for α-silicon (bonded to Li) and at -12.4 ppm for β-silicon (in trimethysilyl) in THF-d_8 solution. These data indicate that the negative charge is highly localized on the α-silicon atom. At 180 K, the α-silicon is observed as a

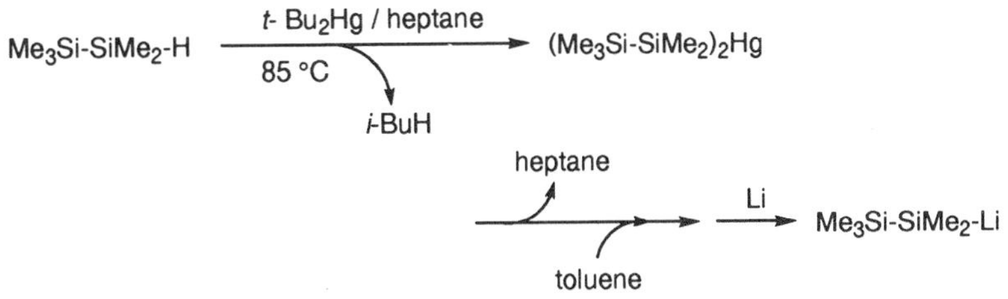

$$Me_3Si\text{-}SiMe_2\text{-}H \xrightarrow[85\,°C]{t\text{-}Bu_2Hg\,/\,heptane} (Me_3Si\text{-}SiMe_2)_2Hg$$

$i\text{-}BuH$

$$\xrightarrow[toluene]{heptane} \xrightarrow{Li} Me_3Si\text{-}SiMe_2\text{-}Li$$

SCHEME 18

triplet due to coupling with one ^6Li (I = 1, $^1J[^{29}Si, \,^6Li]$ = 18.8 Hz); Me_3SiMe_2SiLi thus exists as a monomer in THF (see Section III-B).

The crystal structure of Me_3SiMe_2SiLi has been determined by X-ray analysis. The disilanyllithium exists as a tetramer in the solid state. The lithium atoms are so arranged that a tetrahedron with an average Li–Li distance of 2.78 Å is defined. The pentamethyldisilanyl group caps each face of the tetrahedron with the three nearly equal Li–Si distances of 2.68 Å in average. While the Si–Si bond length (2.35 Å) is normal, the averaged Si–C(methyl) bond on the α-silicon (1.93 Å) is longer than a normal Si–C single bond (1.86–1.91 Å) (see Section IV-B).

A disilanylene 1,2-dianion is formed in the reaction of 1,2-dichloro-1,2-disilaacenaphthene with lithium in THF under ultrasonication (Scheme 19) (71). Trapping with MeI and CH_3OD gives the corresponding dimethyl- and dideuterio-disilaacenaphthene, respectively. It is emphasized that the action of lithium on the dichlorodisilane gives not disilene but disilanylene dianion. This may result because the distance between two silicon atoms of the starting material is fixed at Si–Si single bond length by the naphthalene skeleton, inhibiting Si–Si double bond formation.

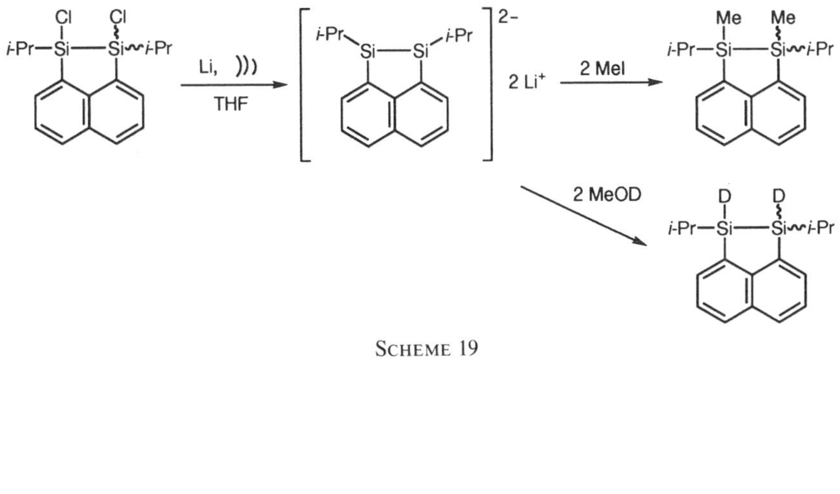

SCHEME 19

The resulting dianion decomposes at room temperature with a half-life of 13.5 hours at 30°C. The ^{29}Si resonance is observed at -1.24 ppm, shifted upfield relative to that of the starting dichlorodisilane (8.99 ppm). The ^{13}C NMR spectrum shows that the ipso carbon on the naphthalene ring (172.94 ppm) shifts remarkably downfield relative to the starting material (136.23 ppm). This may be caused by the π-polarization effect of the naphthyl ring as observed in the case of (phenylsilyl)lithiums (see section III-B).

Reduction of tetrakis(trialkylsilyl) disilene with Li in THF or K in DME gives the corresponding 1,2-dianions, together with the anion radicals [Eq. (31)] (72).

$$
\begin{array}{c}
\text{R}_3\text{Si} \quad \text{SiR}_3 \\
\text{Si=Si} \\
\text{R}_3\text{Si} \quad \text{SiR}_3
\end{array}
\xrightarrow[\substack{\text{THF or DME} \\ \text{M = Li or K}}]{2\,\text{M}}
\begin{array}{c}
\text{R}_3\text{Si} \overset{\text{M}}{\underset{}{|}} \overset{\text{M}}{\underset{}{|}} \text{SiR}_3 \\
\text{Si-Si} \\
\text{R}_3\text{Si} \quad \text{SiR}_3
\end{array}
\tag{31}
$$

$$
\begin{aligned}
\text{R}_3\text{Si} &= (i\text{-Pr})_2\text{MeSi} \\
&= t\text{-BuMe}_2\text{Si} \\
&= (i\text{-Pr})_3\text{Si}
\end{aligned}
$$

2. Tris(trimethylsilyl)silyllithium, (Me₃Si)₃SiLi, and Its Analogues

Tris(trimethylsilyl)silyllithium, $(\text{Me}_3\text{Si})_3\text{SiLi}$, and its derivatives that contain branched silicon chains are the most extensively investigated species among polysilanyllithiums.

Tetrakis(trimethylsilyl)silane, $(\text{Me}_3\text{Si})_4\text{Si}$, undergoes cleavage of the Si–Si bond with MeLi in THF/Et$_2$O (4/1) to give a pale greenish-yellow solution of $(\text{Me}_3\text{Si})_3\text{SiLi}$ [Eq. (32)] (73a). Tris(dimethylsilyl)silyllithium, $(\text{HMe}_2\text{Si})_3\text{SiLi}$, is obtained from $(\text{HMe}_2\text{Si})_4\text{Si}$ by a similar method (73a). Ph$_3$SiLi is applicable to the cleavage, indicative of the higher stability of $(\text{Me}_3\text{Si})_3\text{SiLi}$ than Ph$_3$SiLi; the reverse reaction results in the recovery of the starting materials [Eq. (33)] (73b). $(\text{Me}_3\text{Si})_3\text{SiLi}$ cannot be prepared by the reaction of $(\text{Me}_3\text{Si})_4\text{Si}$ and lithium metal; this may be ascribed to the formation of the thermodynamically less stable Me$_3$SiLi [Eq. (34)] (73b).

$$(\text{Me}_3\text{Si})_4\text{Si} + \text{MeLi} \xrightarrow[\substack{\text{THF - Et}_2\text{O} \\ (4/1)}]{} (\text{Me}_3\text{Si})_3\text{SiLi} + \text{Me}_4\text{Si} \tag{32}$$

$$(\text{Me}_3\text{Si})_4\text{Si} + \text{Ph}_3\text{SiLi} \rightleftharpoons (\text{Me}_3\text{Si})_3\text{SiLi} + \text{Me}_3\text{Si-SiPh}_3 \tag{33}$$

$$(\text{Me}_3\text{Si})_3\text{Si-SiMe}_3 + 2\,\text{Li} \xrightarrow{\not} (\text{Me}_3\text{Si})_3\text{SiLi} + \text{Me}_3\text{SiLi} \tag{34}$$

In contrast, hexakis(trimethylsilyl)disilane, $(\text{Me}_3\text{Si})_3\text{SiSi(SiMe}_3)_3$, is cleaved with lithium metal in THF [Eq. (35)] (73c), as well as with MeLi

[Eq. (36)] (74) regioselectively, to give $(Me_3Si)_3SiLi$. Exposure of $(Me_3Si)_3$-SiLi to the disilane yields $(Me_3Si)_3SiSi(Me_3Si)_2Li$ and $Li(Me_3Si)_2SiSi$-$(Me_3Si)_2Li$ [Eq. (37)] $(73c)$.

$$(Me_3Si)_3Si\text{-}Si(SiMe_3)_3 + 2\,Li \longrightarrow 2\,(Me_3Si)_3SiLi \tag{35}$$

$$(Me_3Si)_3Si\text{-}Si(SiMe_3)_3 + MeLi \longrightarrow (Me_3Si)_3SiLi + MeSi(SiMe_3)_3 \tag{36}$$

$$(Me_3Si)_3Si\text{-}Si(SiMe_3)_3 + (Me_3Si)_3SiLi \longrightarrow \tag{37}$$
$$(Me_3Si)_3Si\text{-}Si(SiMe_3)_2Li + Li(Me_3Si)_2Si\text{-}Si(SiMe_3)_2Li + (Me_3Si)_4Si$$

These results imply that the stability increases in the following order: Me_3SiLi, $Ph_3SiLi < Me_3SiMe_2SiLi < (Me_3Si)_3SiLi < (Me_3Si)_3SiSi(Me_3Si)_2$-Li. In terms of inductive effect, one would expect Me_3SiLi to be more stable than $(Me_3Si)_3SiLi$, since a trimethylsilyl group is a better electron donor than is a methyl group. The facts, however, are opposite. In the original paper, the stability of $(Me_3Si)_3SiLi$ was explained by the conjugation effect between the vacant $3d$ orbitals of peripheral silicon atoms (the electron acceptors) and the filled $3p$ orbital of the lithium-bearing silicon atom (the electron donor) (dative $p\pi$-$d\pi$ bonding) $(73b)$. However, the stabilization now may be explained by hyperconjugative mixing of the anion $3p$ (Si_α) with σ^* $(Si_\beta$-C) orbitals.

$(Me_3Si)_3SiLi$ decomposes slowly in THF. This problem is avoided by a solvent-exchange technique. The silyllithium is isolated as a colorless crystalline solid containing three mole-equivalents of THF of solvation, that is, $(Me_3Si)_3SiLi\cdot3THF$ (75). The compound is soluble in hydrocarbon solvents, giving a colorless solution, and coupling reactions in such nonpolar media are generally much cleaner and higher in yield.

The UV spectra in THF show absorption maxima at 236 nm (ε 6000, sh), 295 nm (ε 22000), and 370 nm (ε 10000), whereas its neutral derivatives, $(Me_3Si)_4Si$ and $(Me_3Si)_3SiH$, show no absorption above 210 nm (cyclohexane) $(73b)$. These marked absorptions of long wavelengths imply considerable resonance stabilization, as mentioned earlier. The ^{29}Si NMR signal (C_6D_6) is observed at much higher field (-185.4 ppm) for the central silicon in the anion than that for the central silicon of $(Me_3Si)_4Si$ (-135.5 ppm) (75). The ^{29}Si resonance is observed at -189.4 ppm in C_7H_8-C_6D_6 at room temperature and resolves into a quartet due to coupling with one ^7Li atom $(I = 3/2, ^1J[^{29}Si, ^7Li] = 38.6$ Hz). This result implies a partial covalent nature of the Si–Li bond $(76a)$.

The crystal structure of $(Me_3Si)_3SiLi\cdot3THF$ has been determined by X-ray analysis (Fig. 6) $(22,76)$. There is distorted tetrahedral coordination at both lithium and silicon. The Si–Li bond length is 2.67 Å, which is similar to the average Li–Si bond length (2.68 Å) observed in $(Me_3SiLi)_6$

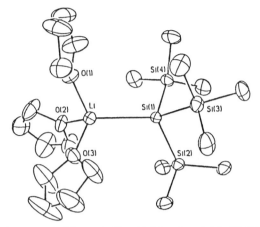

FIG. 6. Computer-generated thermal ellipsoidal drawing of $(Me_3Si)_3SiLi\cdot3THF$ (22). Reprinted with permission from *J. Organomet. Chem.* **1993,** *462,* 1. Copyright (1993) Elsevier Science S. A.

(see section IV-B). In addition, it is slightly longer than the Li–Si bonds (2.63 Å) observed in $(Me_3Si)_3SiLi\cdot1.5DME$ (77). The average Si–Si distance is 2.33 Å. For the central silicon atom, the average Si–Si–Si and Si–Si–Li bond angles are 103° and 116°, respectively. The geometry of the $(Me_3Si)_3Si$-moiety thus indicates increased s-character in the silicon orbital directed toward lithium.

$(Me_3Si)_3SiM$ (M = K, Rb, and Cs) can be obtained by the metal–metal exchange reaction of $[(Me_3Si)_3Si]_2M$ (M = Zn, Cd, and Hg) with the corresponding alkali metals in n-pentane (78). These compounds are colorless powders in their pure solvent-free state. The X-ray structure analyses reveal their dimeric molecular structures.

Reaction of $(Me_3Si)_3SiLi$ with MX_2 (M = Zn, Cd, Hg; X = halide) in Et_2O yields $[(Me_3Si)_3Si]_2M$ (M = Zn, Cd, Hg) [Eq. (38)] (79a). The Si–Tl and Si–Pb compounds cannot be obtained from $(Me_3Si)_3SiLi$ with the metal halides, the reaction affording the elemental metals and $(Me_3Si)_3SiSi$-$(SiMe_3)_3$ (76,79b,80a).

$$2\ (Me_3Si)_3SiLi + MX_2 \xrightarrow{Et_2O} [(Me_3Si)_3Si]_2M + 2\ LiX \qquad (38)$$

$$M = Zn,\ X = Cl$$
$$M = Cd,\ X = Cl$$
$$M = Hg,\ X = Br$$

While the zinc and mercury derivatives are stable in benzene-d_6 solution for at least 3 days, the cadmium derivative is much less stable. The ^{29}Si NMR shifts for these compounds move to lower field on descending the group from zinc to mercury $(79a)$. X-ray crystal structure analysis of the zinc derivative reveals that the compound adopts a D_{3d} structure in the solid state, with the central zinc atom lying at the inversion center $(79a)$. The two $(Me_3Si)_3Si$-ligands are bonded in a linear fashion to the metal (Si–Zn–Si angle = 180°). The Zn–Si bond distance, 2.34 Å, is close to that expected on the basis of covalent radii [2.35 Å, from covalent radii for Zn (1.18 Å) and for Si (1.17 Å)]. The average Si–Si–Si angle is 112°, which is larger than that of $(Me_3Si)_3SiLi$ (103°). The average Si–Si–Zn angle is 107°.

Silylcuprates $[\{(Me_3Si)_3Si\}_2Cu_5Cl_4](Li\cdot4THF)$ and $[\{(Me_3Si)_3Si\}_2Cu_2Br](Li\cdot3THF)$ are prepared by reaction of $(Me_3Si)_3SiLi$ with, respectively, CuCl in pentane-THF and CuBr in hexane-THF $(76,80)$. X-ray analysis of these silylcuprates shows the presence of a bridging $(Me_3Si)_3Si$ group on a Cu–Cu bond in both cases $(80b,c)$.

A tin-substituted silyllithium $(Me_3Sn)_3SiLi$ is prepared by reaction of $(Me_3Sn)_4Si$ with MeLi [Eq. (39)] $(81a,b)$. The compound is obtained in 81% yield as light yellow crystals from pentane. The ^{29}Si NMR shift for $(Me_3Sn)_3SiLi$, -122.0 ppm, is upfield from that of $(Me_3Sn)_4Si$, -69.2 ppm, and is downfield from the analogous $(Me_3Si)_3SiLi$ (-184.5 ppm).

$$(Me_3Sn)_4Si + MeLi \xrightarrow[THF]{} (Me_3Sn)_3SiLi + SnMe_4 \qquad (39)$$

A gem-dilithiooligosilane, $(Me_3Si)_2SiLi_2$, is prepared by thermal redistribution of $(Me_3Si)_3SiLi$ at 140–150°C [Eq. (40)] (82). If the reaction is carried out for a longer time, the coupling product $(Me_3Si)_3SiSi(SiMe_3)_3$ is produced at the expense of the dilithio compound. Flash-vaporization mass spectroscopy of a solid sample of $(Me_3Si)_2SiLi_2$ displays parent ions for both the monomer and the dimer.

$$(40)$$

3. Linear Oligosilanyllithiums

Action of lithium metal on linear phenyloligosilanes gives rise to the Si–Si bond cleavage with no regioselectivity to yield a mixture of various mono- and oligosilanyllithiums (83). In contrast, action of lithium metal on a cyclic oligosilane sometimes results in selective cleavage of the Si–Si bond, giving one species of linear oligosilanyllithium. 1,4-Dilithiooctaphenyltetrasilane and 1,5-dilithiopentasilane are thus prepared by the lithium cleavage of octaphenylcyclotetrasilane and decaphenylcyclopentasilane, respectively, in THF (84–85b). These reagents are useful for preparation of linear oligosilanes such as a nonasilane [Eq. (41)]. Octaphenylcyclotetrasilane is also cleaved by phenyllithium to give nonaphenyltetrasilanyllithium.

$$2 \text{ PhMe}_2\text{SiPh}_2\text{SiCl} + \text{LiPh}_2\text{Si(Ph}_2\text{Si)}_3\text{Ph}_2\text{SiLi} \xrightarrow[\text{THF}]{} \text{PhMe}_2\text{Si(Ph}_2\text{Si)}_7\text{SiMe}_2\text{Ph} \quad (41)$$

26 %

A variety of linear and branched oligosilanyllithiums can also be obtained by mercury–lithium exchange reactions in good yields (Scheme 20) (70).

1,4-Dilithiooctaphenyltetrasilane can be isolated as orange-red crystals (85b). X-ray structure determination shows that the molecule is monomeric and centrosymmetric, with an extended planar Li–Si_4–Li unit. The bond angle of Li–Si–Si is 126° and that of Si–Si–Si is 127°. The average Si–Li bond length of 2.68 Å is comparable to that of Ph_3SiLi (2.67 Å). While the Si–Si bond lengths, 2.41 Å and 2.43 Å, are longer than the normal Si–Si single bond (2.33–2.37 Å), the Si–C bond lengths are 1.90–1.92 Å, which are shorter than that of Ph_3SiLi (1.94 Å) (see Section III-B).

4. Cyclic Oligosilanyl Anions

Reaction of dodecamethylcyclohexasilane, $(\text{Me}_2\text{Si})_6$, with lithium metal (2 equivalents) in $\text{HMPA/Et}_2\text{O}$ solution gives no Si–Si bond cleavage product, $\text{Li(SiMe}_2)_6\text{Li}$, but rather cyclohexasilanyllithium, $\text{Me}_{11}\text{Si}_6\text{Li}$, the product arising from the Si–C bond cleavage (Scheme 21) (86). Even in

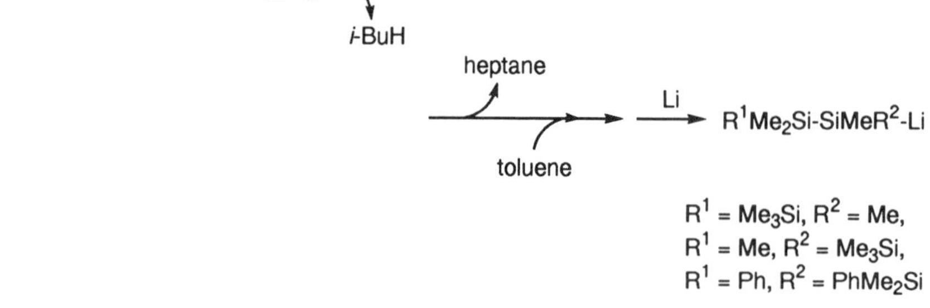

$$R^1\text{Me}_2\text{Si-SiMeR}^2\text{-H} \xrightarrow[85\,°C]{t\text{-Bu}_2\text{Hg / heptane}} (R^1\text{Me}_2\text{Si-SiMeR}^2)_2\text{Hg}$$

i-BuH

$$\xrightarrow[\text{toluene}]{\text{heptane}} \xrightarrow{\text{Li}} R^1\text{Me}_2\text{Si-SiMeR}^2\text{-Li}$$

$R^1 = \text{Me}_3\text{Si}, R^2 = \text{Me},$
$R^1 = \text{Me}, R^2 = \text{Me}_3\text{Si},$
$R^1 = \text{Ph}, R^2 = \text{PhMe}_2\text{Si}$

SCHEME 20

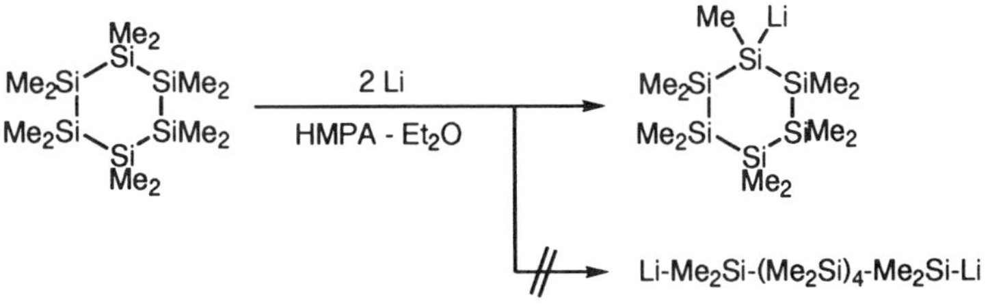

SCHEME 21

the presence of excess amounts of Li (4 equivalents), $Me_{11}Si_6Li$ is inert to further Si–C bond cleavage. A sealed solution of $Me_{11}Si_6Li$ in HMPA is stable for weeks at room temperature in the dark. The anion can be trapped with various agents, such as bromoethane and trimethylchlorosilane, to give the expected products in nearly 50% yield.

Treatment of the cyclohexasilane with MeLi, PhLi, or metal alkoxides ROM (M = Na, K; R = Me, Et, t-Bu) in HMPA/Et_2O offers additional routes to the anion (86). Some linear oligosilanes are also formed, due to concomitant ring cleavage. Reaction of decamethylcyclopentasilane, $(Me_2Si)_5$, with MeLi in HMPA does not give cyclopentasilanyllithium but rather species resulting from ring cleavage.

The thermodynamic stability of $Me_{11}Si_6Li$ may be due to the electron delocalization onto the contiguous silicon framework, consistent with the increasing order of thermodynamic stability: $Me_3SiLi < (Me_3Si)Me_2SiLi < (Me_3Si)_3SiLi$ (73b). The ^{29}Si NMR spectrum of $Me_{11}Si_6Li$ shows four sharp singlets, at δ −108.1, −33.8, −40.7, and −41.6 ppm, with a 1:2:2:1 ratio: The highest field peak can be assigned to the negatively charged silicon.

Nonamethylcyclopentasilanyl-potassium, Me_9Si_5K, can be prepared via a mercury–potassium exchange reaction from bis(cyclopentasilanyl)mercury, $(Me_9Si_5)_2Hg$, which is obtained by the reaction of hydro-cyclopentasilane with $(t$-Bu$)_2Hg$ (Scheme 22) (87). This potassium reagent offers a new route to permethylated polycyclic polysilanes.

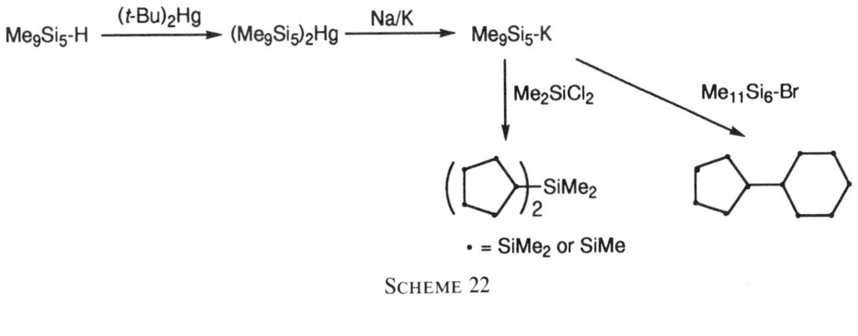

SCHEME 22

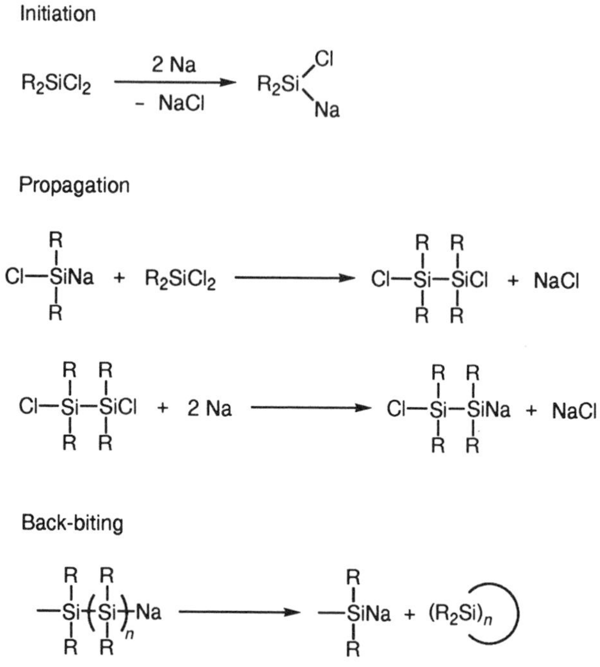

Initiation

Propagation

Back-biting

SCHEME 23

B. Intermediates in Polysilane Syntheses

Polysilanyl anions are reactive intermediates in various synthetic routes to polysilanes.

1. Reductive Coupling of Dichlorosilanes and Related Reactions

In the Wurtz coupling reaction of dichlorosilanes with sodium to polysilanes [Eq. (42)], polysilanyl anions are considered to be crucial reaction intermediates in the propagation step and in the cyclization via back-biting (Scheme 23) (88).

$$\text{Cl}-\underset{R}{\overset{R}{\text{Si}}}-\text{Cl} \ + \ 2n \text{ Na} \ \longrightarrow \ \left(\underset{R}{\overset{R}{\text{Si}}}\right)_n \ + \ 2n \text{ NaCl} \tag{42}$$

The reduction of Me_2SiCl_2 with sodium–potassium alloy in THF forms dodecamethylcyclohexasilane, $(Me_2Si)_6$, as the major product, in two stages. In the first stage, the main product is $(Me_2Si)_n$ polymer. After all the Si–Cl bonds disappear, depolymerization occurs via polysilanyl anions to form the equilibrium mixture of products enriched in $(Me_2Si)_6$ (89). Reaction of Me_2SiCl_2 with lithium in the presence of an equilibrating catalyst such as $Ph_3SiSiMe_3$ also affords a convenient route to $(Me_2Si)_6$ [Eq. (43)] (90). This reaction involves the thermodynamic redistribution of polysilanes with silyllithium.

$$Me_2SiCl_2 \xrightarrow[\text{THF}]{\text{Li, Ph}_3\text{SiSiMe}_3} \underset{90\%}{(Me_2Si)_6} + \underset{9\%}{(Me_2Si)_5} + \underset{1\%}{(Me_2Si)_7} \tag{43}$$

It has been demonstrated that α,ω-diphenylpermethylpolysilanes, $PhMe_2Si(Me_2Si)_nSiMe_2Ph$ ($n = 1$–3), undergo disproportionation readily in the presence of $PhMe_2SiLi$ as catalyst in THF to produce an equilibrium mixture consisting of linear oligosilanes and $(Me_2Si)_6$ [Eq. (44)] (91). The composition of the mixture depends on the chain length of the starting oligosilanes.

$$PhMe_2Si(Me_2Si)_nSiMe_2Ph \xrightarrow[\text{THF}]{\text{PhMe}_2\text{SiLi}} (Me_2Si)_6 +$$
$$\underset{n = 1\text{-}3}{} \tag{44}$$
$$PhMe_2Si(Me_2Si)_xSiMe_2Ph$$
$$x = 0\text{-}3$$

A kinetic study on the reaction of a chlorodisilane with Li reveals that oligosilanyl anions participate in the processes (28). The overall reaction of 1-chloro-2-phenyltetramethyldisilane with Li in THF consists of four stages (Schemes 24 and 25). In stage 1, a slow electron transfer from lithium to the disilanyl chloride produces a transient disilanyl anion, which immediately couples with the remaining disilanyl chloride to yield a linear tetrasilane. Stage 2 starts with the resulting tetrasilane having terminal phenyl groups, which makes for a much better electron acceptor than the disilanyl chloride. An electron transfer from lithium to the tetrasilane gives the corresponding radical anion, which decomposes unimolecularly to radicals and anions. The radicals may recombine to form oligosilanes or may take a second electron from lithium to form anions. The anions also directly attack any remaining disilanyl chloride to produce linear oligosilanes. The lifetime of the anions appears to be very short when the disilanyl chloride is present. No dodecamethylcyclohexasilane is formed

$$Ph(Me_2Si)_2Cl + 2\,Li \xrightarrow{-\text{LiCl}} Ph(Me_2Si)_xLi \rightleftharpoons 1/6\,(Me_2Si)_6 + PhMe_2SiLi$$

SCHEME 24

stage 1

$$Ph(Me_2Si)_2Cl + 2\ Li \longrightarrow Ph(Me_2Si)_2Li + LiCl$$

$$Ph(Me_2Si)_2Cl + Ph(Me_2Si)_2Li \longrightarrow Ph(Me_2Si)_4Ph + LiCl$$

stage 2

$$Ph(Me_2Si)_4Ph + Li \longrightarrow [Ph(Me_2Si)_4Ph]^{\bullet-}Li^+$$

$$[Ph(Me_2Si)_4Ph]^{\bullet-}Li^+ \longrightarrow Ph(Me_2Si)_nLi + Ph(Me_2Si)_{4-n}{}^{\bullet}$$

$$Ph(Me_2Si)_nLi + Ph(Me_2Si)_2Cl \longrightarrow Ph(Me_2Si)_{n+2}Ph$$

$$Ph(Me_2Si)_{4-n}{}^{\bullet} + Li \longrightarrow Ph(Me_2Si)_{4-n}Li$$

$$Ph(Me_2Si)_{4-n}{}^{\bullet} \longrightarrow Ph(Me_2Si)_xPh$$

stage 3

$$Ph(Me_2Si)_nLi + Ph(Me_2Si)_mPh \longrightarrow Ph(Me_2Si)_xLi + Ph(Me_2Si)_yPh$$
$$x+y = m+n$$

stage 4

$$Ph(Me_2Si)_zLi \rightleftharpoons Ph(Me_2Si)_{z-6}Li + (SiMe_2)_6$$

SCHEME 25

at this stage. Stage 3 starts when the disilanyl chloride is completely consumed. As the concentration of the linear oligosilanes decreases, the cyclohexasilane appears. The consumption of lithium continues throughout stage 3. The silyl anions, which have a longer lifetime in this stage, cleave the Si–Si bonds, resulting in the scrambling processes. The intermediate silyl anions are generated continuously and provide routes to cyclic oligosilanes. The relative concentrations of cyclic to linear oligosilanes are determined by their relative stabilities, i.e., by thermodynamics. In stage 4, the oligosilanes are cleaved by the continuous donation of electrons from lithium to lead to the formation of radical anions and anions. A thermodynamic mixture of monosilyl anions, disilanyl anions, and the cyclohexasilane is produced.

Electron-donating substituents on the aromatic rings decrease the electron affinity of the oligosilanes. 1-Chloro-2-[p -(dimethylamino)phenyl]tetramethyldisilane thus reacts with lithium to form exclusively the tetrasilane that is stable in the presence of excess Li metal [Eq. (45)].

$$2\ (p\text{-}Me_2NC_6H_4)Me_2SiMe_2SiCl\ +\ 2\ Li$$

$$\longrightarrow\ (p\text{-}Me_2NC_6H_4)Me_2Si(Me_2Si)_2SiMe_2(C_6H_4NMe_2\text{-}p)\ +\ 2\ LiCl \tag{45}$$

2. Oligomerization

The action of a catalytic amount of MeONa on methoxymethyldisilane gives α,ω-dimethoxypermethylpolysilanes or cyclic polysilanes, depending on the reaction time ($55b$). It was concluded that the mechanism involves a concerted nucleophilic substitution with base-assistance or stepwise substitution by silyl anion; a silylene mechanism is ruled out.

3. Ring-Opening Polymerization of Cyclotetrasilanes and Cyclopentasilanes

Cyclotetrasilanes undergo ring-opening by the action of a variety of nucleophiles, such as silyl anions and fluoride anion, to form the corresponding oligosilanyl anions. In the presence of 1,2-diphenyltetramethyldisilane, the oligosilanyl anions are trapped efficiently, resulting in the ring-opening disilylation of cyclotetrasilanes [Eq. (46)] (92): The reaction proceeds catalytically with respect to $PhMe_2SiLi$ (Scheme 26).

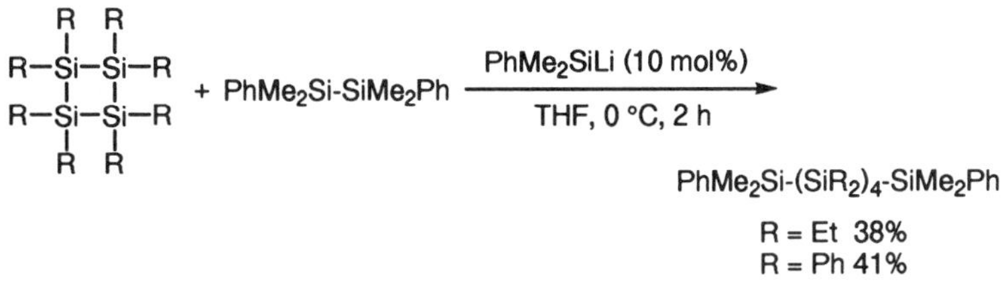

$$\tag{46}$$

$PhMe_2Si\text{-}(SiR_2)_4\text{-}SiMe_2Ph$

R = Et 38%
R = Ph 41%

Reaction of tetramethyltetraphenylcyclotetrasilanes $(PhMeSi)_4$ with a catalytic amount of butyllithium, silylpotassium or silylcuprate, $(PhMe_2Si)_2CuLi$, brings about anionic ring-opening polymerization (Scheme 27) (93); the ring strain of the tetrasilane is sufficient for subsequent ring-opening to cause completion of polymerization before the depolymerization due to back-biting becomes significant. Phenylnonamethylcyclopen-

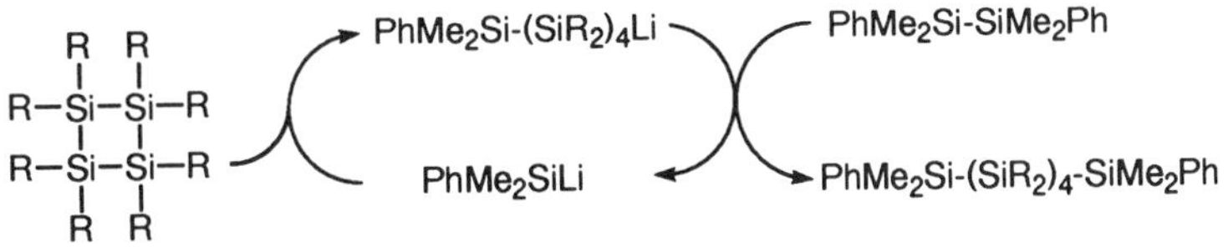

SCHEME 26

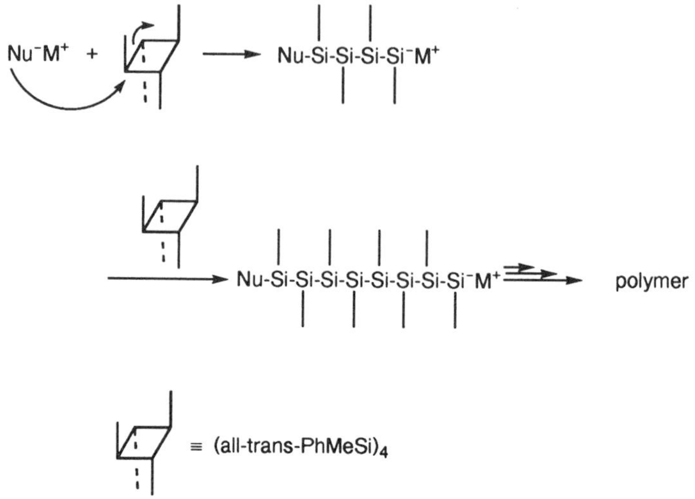

SCHEME 27

tasilane undergoes anionic ring-opening polymerization induced by a catalytic amount of Bu_4NF, Me_3SiK, or $PhMe_2SiK$ in polar solvents at low temperatures ($<-50°C$) [Eq. (47)] (*94*). The ring-opening is regioselective in such a way that the phenyl-bearing silyl anion is formed in every step. At higher temperatures, only a mixture of cyclic oligomers results.

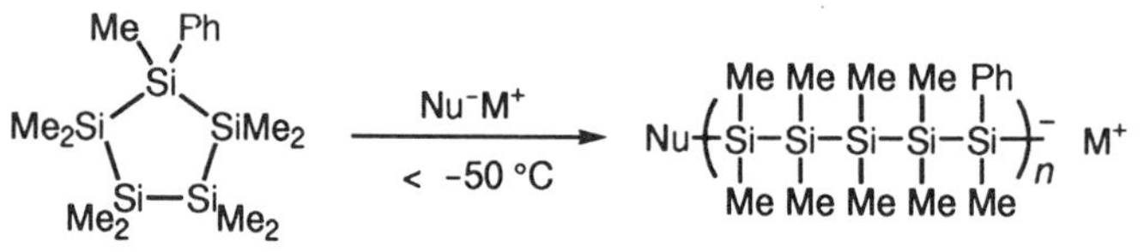

$$(47)$$

4. *Anionic Polymerization of Masked Disilenes*

Anionic polymerization of masked disilenes has opened up a novel route to polysilanes (*95*). 1-Phenyl-7,8-disilabicyclo[2.2.2]octa-2,5-dienes can be used as masked disilenes. *n*-BuLi works as an initiator. The polymerization may involve the attack of the polysilanyl anions on a silicon atom of the monomer, resulting in the formation of the new propagating polymer anion and biphenyl. This method is applicable to aminopolysilane synthesis (Scheme 28).

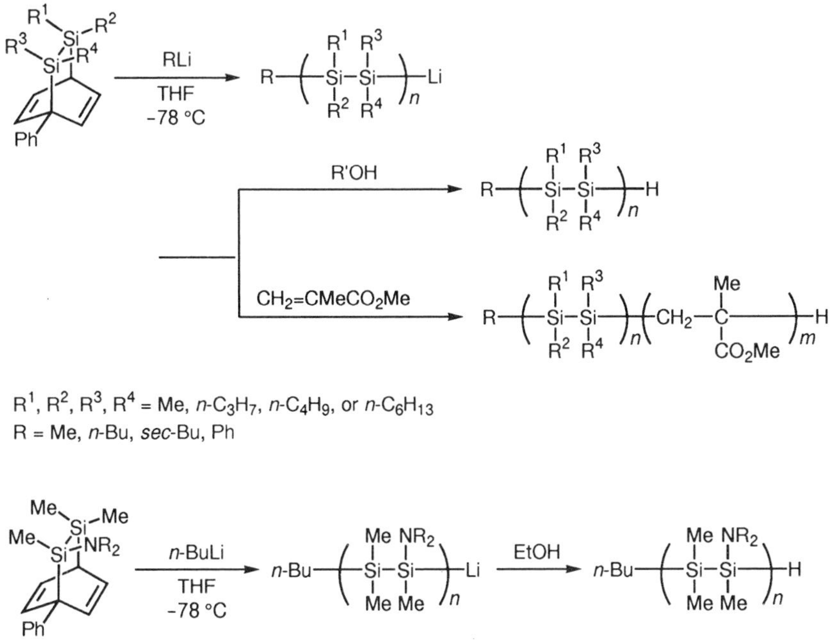

R^1, R^2, R^3, R^4 = Me, n-C$_3$H$_7$, n-C$_4$H$_9$, or n-C$_6$H$_{13}$
R = Me, n-Bu, sec-Bu, Ph

SCHEME 28

VII

SYNTHETIC APPLICATION OF SILYL ANIONS

Certain silyl anions are useful silylating reagents in organic synthesis. Since some leading reviews have covered the synthetic applications $(1d,e,f,2)$, only a few typical examples are mentioned here.

Me$_3$SiLi undergoes smoothly conjugate addition to α,β-unsaturated ketones in the absence of copper salt [Eq. (48)] (29).

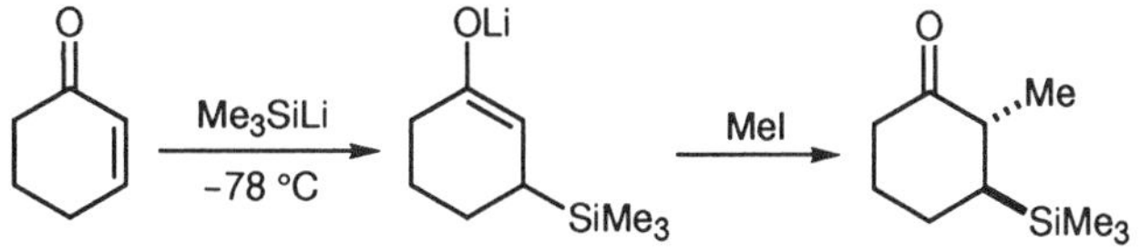

(48)

By use of Me$_3$SiK, some mono-, di-, and trisubstituted epoxides can be deoxygenated stereospecifically, with inversion of stereochemistry [Eq. (49)] $(96, 97)$. Thus, reaction of cis and trans epoxides with Me$_3$SiK in

HMPA at 65°C for 3 hours affords the corresponding *trans* and *cis* olefins, respectively. Backside attack of Me_3SiK on the cis (and trans) epoxides generates the threo- (and erythro-) β-alkoxysilanes, respectively, followed by syn elimination (Peterson elimination) to give the olefin with inverted stereochemistry.

$$(49)$$

Disilanyllithium Me_3SiMe_2SiLi also undergoes conjugate addition to an α,β-unsaturated ketone just like Me_3SiLi (*68a–c*). In comparison with Me_3SiLi, Me_3SiMe_2SiLi shows a better stereoselectivity in the addition [Eq. (50)].

$$(50)$$

R = SiMe₂SiMe₃ (63 %)
R = SiMe₃ (64 %; stereoselectivity 19 : 1)

Reaction of adamantanone with $(Me_3Si)_3SiLi$ in hexane leads, via a Peterson-type elimination, to the silene, which dimerizes spontaneously to 1,2-disilacyclobutane in 85% yield [Eq. (51)] (*98a,b*).

$$(51)$$

Silyl–copper and silylcuprate reagents are readily prepared by reaction of silyllithiums with copper salts such as CuCN, CuI, and $CuBr/Me_2S$ in ethereal solvent. The silylcuprates $(PhMe_2Si)_2CuLi$, $PhMe_2SiCu(CN)Li$, $(PhMe_2Si)_2Cu(CN)Li_2$, $Me(PhMe_2Si)Cu(CN)Li_2$, and $(PhMe_2Si)_3CuLi_2$

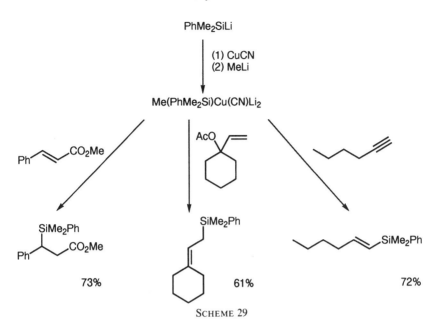

SCHEME 29

have been studied by NMR spectroscopy (see Table I in Section III and Table II in Section IX) ($99a$–d). Reactions of silylcuprate $(PhMe_2Si)_2Cu$-$Li \cdot LiI$ or $(PhMe_2Si)_2Cu(CN)Li_2$ with various electrophiles have been well investigated ($100a$–f). Silyl–copper reagents prefer S_N2' displacement to S_N2 in the allylic substitution [Eq. (52)] ($100a$–f). A mixed cuprate, which is prepared from $PhMe_2SiLi$, MeLi, and CuCN in an equimolar ratio, transfers specifically the silyl group, not the methyl group, to various electrophiles (Scheme 29).

$$C_6H_{13} \diagup\diagdown X \xrightarrow[\substack{Et_2O\text{-HMPA} \\ -60\ °C}]{Me_3SiCu} \quad C_6H_{13}\diagup\underset{SiMe_3}{\diagdown} + C_6H_{13}\diagup\diagdown SiMe_3 \tag{52}$$

87%   98 : 2

The phenylsilyl group introduced can be converted into the hydroxy group by oxidative cleavage of the Si–C bond, after changing the phenyl group on silicon into fluorine with $HBF_4 \cdot Et_2O$. Thus, the $PhMe_2Si^-$ anion works as the hydroxy anion HO^- equivalent [Eq. (53)] ($100a$–f). A one-pot procedure is also successful by use of bromine or mercuric acetate in the presence of peracetic acid [Eq. (54)] (*100a–f*).

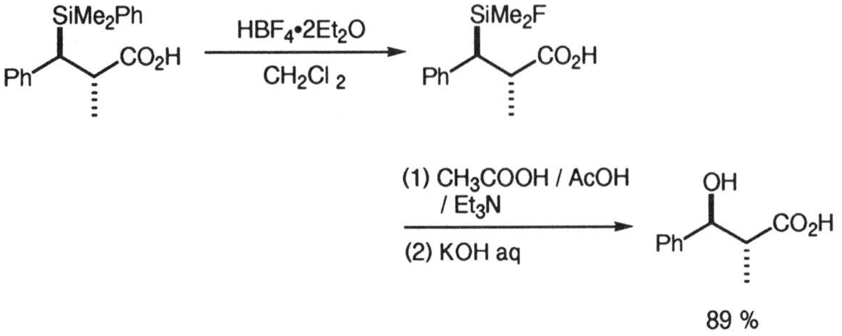

$$(53)$$

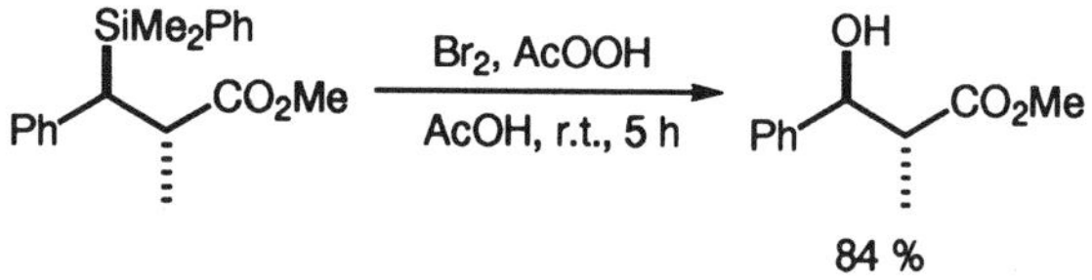

$$(54)$$

84 %

(Aminosilyl)lithium, $(Et_2N)Ph_2SiLi$ (see also Section V-2), can also be converted into silylcuprate just by mixing with 1 equivalent of CuCN in THF (*54a–d*). The silyllithium undergoes S_N2 replacement and the silylcuprate does S_N2' replacement, with high regio- and stereoselectivity in the allylic substitution reactions (Scheme 30) (*54b*). The aminosilyl anions serve as a hydroxy anion equivalent in combination with the H_2O_2 oxidation of the Si–C bond. These reagents are superior to the phenylsilyl

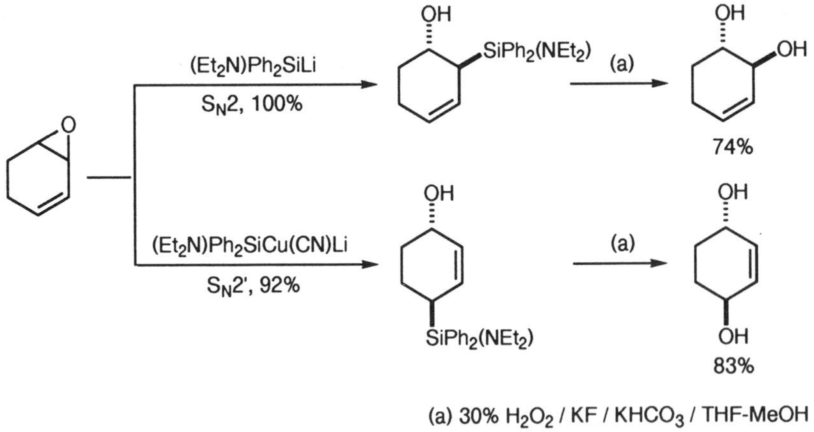

(a) 30% H_2O_2 / KF / $KHCO_3$ / THF-MeOH

SCHEME 30

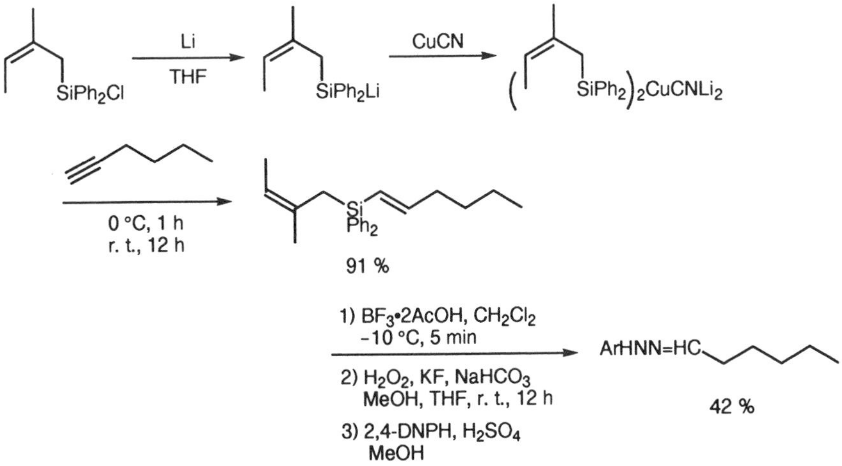

SCHEME 31

analogs, $PhMe_2SiLi$ and $(PhMe_2Si)_2Cu(CN)Li_2$, since no acid treatment is required before the oxidation.

An (allyl)silylcuprate, [(2-methyl-2-butenyl)diphenylsilyl]cuprate, is also a hydroxy anion equivalent. The 2-methyl-2-butenyl group can easily be removed selectively in the presence of other allyl or vinyl groups by protodesilylation under mild conditions (Scheme 31) (13).

Trialkylsilyl higher-order cyanocuprates are prepared directly by transmetallation of silylstannanes with higher-order dibutylcyanocuprates (Scheme 32) (101). In this method, steric bulkiness around silicion is essential for formation of the silylcuprates, so that the t-butyl or thexyl group is present on silicon; otherwise, stannyl anion is formed. These reagents undergo typical silyl-cupration reactions, as mentioned earlier.

The silylmagnesium reagent $Ph_3SiMgBr$, which is formed by the addition

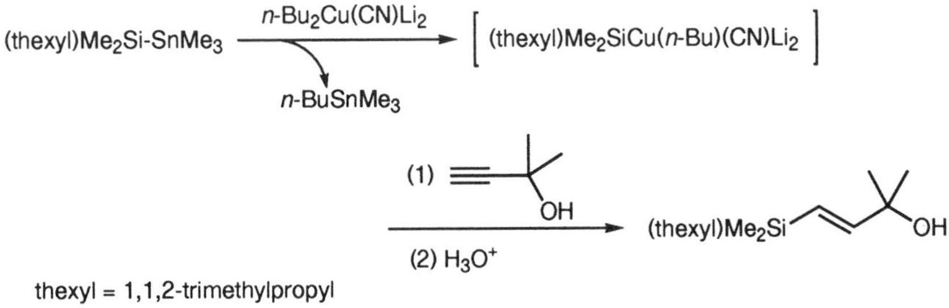

SCHEME 32

of $MgBr_2$ to Ph_3SiLi, undergoes stereoselective addition to chiral 1-acyl-pyridinium salt to give dihydropyridone [Eq. (55)] (*11*). The silylmagnesium reagent $PhMe_2SiMgMe$, prepared from $PhMe_2SiLi$ with Grignard reagent MeMgI, undergoes addition smoothly to acetylenes in the presence of a transition-metal catalyst with high regio- and stereoselectivities [Eq. (56)] (*102*); the silylmagnesium reagent does not cause acetylenic proton–metal exchange with terminal acetylenes.

$$(55)$$

$$(56)$$

$$> 99 : < 1$$

Me_3SiLi or $PhMe_2SiLi$ reacts with MeMgI and $MnCl_2$ to afford a silyl-manganese reagent. These reagents react with acetylenes to form disilylated products (Scheme 33) (*103*); even the highly crowded tetrasilylethylene can also be prepared. MeMgI is essential for the formation of disilylated products; without MeMgI, monosilylated products are obtained predominantly.

The trichlorosilyl anion $[Cl_3Si^-/Bu_3NH^+]$ (see also Section V-2) is a reductive silylating agent and is useful for some transformations (Scheme 34) (*57a–g*). A similar species prepared from Cl_3SiH or Cl_2RSiH (R = Me, Ph) with TMEDA undergoes β-hydrosilylation of acrylates in the presence of a copper salt [Eq. (57)] (*58*).

$$(57)$$

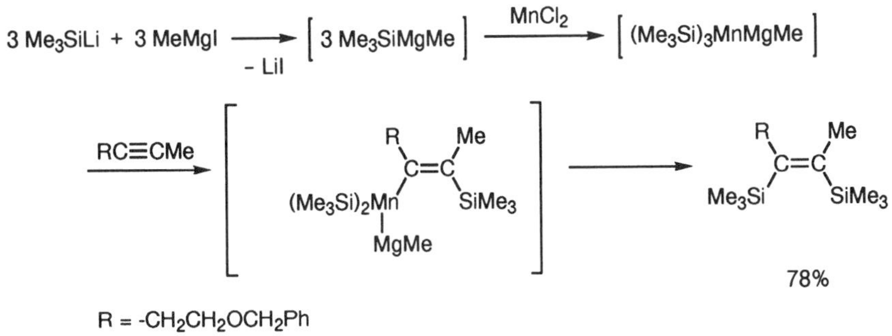

SCHEME 33

The synthetic applications of Me_3Si^-/NBu_4^+ are illustrated by the reaction with aldehydes and 1,3-dienes (Scheme 35) (39).

VIII

THEORETICAL STUDY

Ab initio calculations for the parent H_3Si^- have been carried out (104), and the effect of second-row substituents on free anions XH_2Si^- (X = BH_2, CH_3, NH_2, OH, and F) have been studied (104d,105,106).

The anions were treated using the larger double-zeta basis set for geometry optimizations on both the silyl anions, XH_2Si^-, and their patent silanes, XH_3Si. The following three points may be noted. (1) The stability order has been estimated as follows: $(CH_3)H_2Si^- \approx (H_2N)H_2Si^- < (HO)H_2Si^- = H_3Si^- < FH_2Si^- < (H_2B)H_2Si^-$. (2) The silyl anions, XH_2Si^-, are 50–60

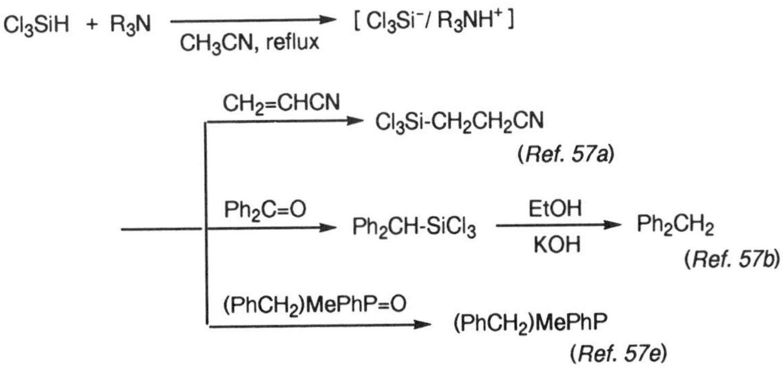

SCHEME 34

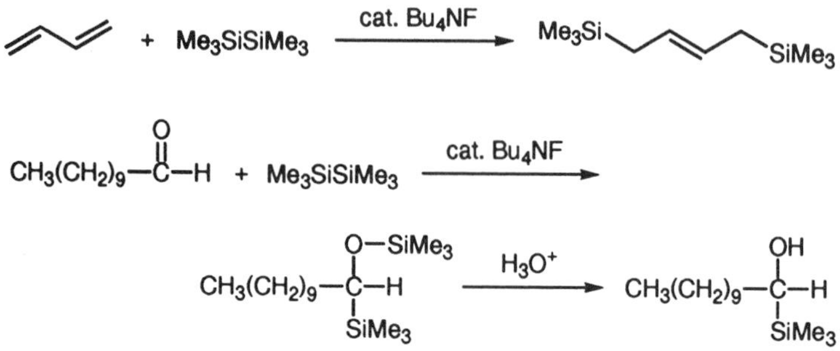

SCHEME 35

kcal/mol more stable than the corresponding carbanions, XH_2C^-. (3) The anions are pyramidal, with larger out-of-plane angles than their carbanion analogs, and inversion barriers are high, varying from 34.3 kcal/mol for $X = H$ up to 57.3 kcal/mol for $X = F$. The calculations agree with the experimental observation that inversion in silyl anions has a lower limit of 24 kcal/mol (24). The out-of-plane angles increase with the electronegativity of the substituent. This is due to π-donation from the substituent (which interacts with the lone pair at the inversion center to produce 4π electron destabilization) and σ-withdrawal to the substituent. For the same reason, elongation of the Si–X bond is observed. However, the borylsilyl anion, $(H_2B)H_2Si^-$, is planar at both boron and silicon and has a Si–B bond length shorter than in borylsilane, H_2BSiH_3.

Other potentially conjugated anions, $NCSiH_2^-$, $O{=}CH\text{-}SiH_2^-$, $HC{\equiv}CSiH_2^-$, have also been studied (104d,107a,b). They are also pyramidal at silicon, and their inversion barriers are large, 35.2, 19.9, and 34.3 kcal/mol, respectively.

Ab initio calculations on metal-containing silyllithiums and silylsodiums, $SiLi_4$, H_2SiLi_2, H_3SiLi and H_2SiNa_2, have also been performed (108a–c). The H_3SiLi species has been calculated with 6-31G* geometries at the MP4STDQ/6-31G** level, with a correction for zero-point energy (Fig. 7) (108a). The inverted SiH_3Li C_{3v} structure **A** is 2.4 kcal/mol more stable than the tetrahedral C_{3v} isomer **B** because of more favorable electrostatic interactions between Li and H. The best interconversion pathway involves transition structure **C**, which lies 13.5 kcal/mol higher in energy than **A**. The pyramidal inversion barrier of the silyl anion is quite high (24,108a–c), and the lithium counter-ion wanders over the surface like **C**. Furthermore, the nature of Si–Li bonding has been studied by natural population analysis using natural, localized molecular orbitals. The result indicates that the Si–Li bond is essentially ionic, even in the regular structures like **B**,

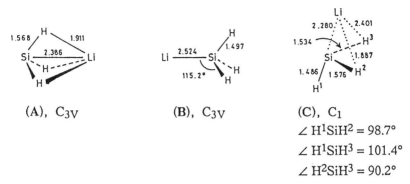

(A), C_{3v} (B), C_{3v} (C), C_1

$\angle$ H^{1}SiH^2 = 98.7°

$\angle$ H^{1}SiH^3 = 101.4°

$\angle$ H^{2}SiH^3 = 90.2°

FIG. 7. Three possible geometries for H$_3$SiLi (*108a*). Reprinted with permission from *J. Chem. Soc., Chem. Commun.* **1986**, 1371. Copyright (1986) The Royal Society of Chemistry.

although the partial covalent nature has been confirmed experimentally by NMR spectroscopy (see Section III-B).

Fluorosilyllithium, FH$_2$SiLi, is an interesting species, since it has been thought to be a model for silylenoids, analogous to carbenoids. The *ab initio* calculation has been performed with STO-3G and 3-21G (*109*). The two most stable forms are suggested to be the H$_2$SiLi$^+$F$^-$ ion pair and the H$_2$Si:FLi complex (Fig. 8). It is interesting that lithium interacts with fluorine in both forms. In H$_2$SiLi$^+$F$^-$, lithium substitution causes effective stabilization of SiH$_2^+$ (56.8 kcal/mol at STO-3G for H$_2$SiLi$^+$), and results in the weakening of the strong Si–F bond. These results suggest that SiH$_2$LiF exhibits electrophilicity due to the cationic silicon center. Thus, silylenoid nature appears in SiH$_2$LiF.

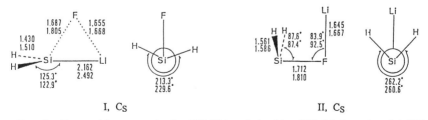

I, C_S II, C_S

FIG. 8. Two stable geometries for FH$_2$SiLi optimized by STO-3G (*upper*) and 3-21G (*lower*) levels. Bond lengths are in angstroms and bond angles in degrees (*110*). Reprinted with permission from *J. Organomet. Chem.* **1980**, *191*, 347. Copyright (1980) Elsevier Science S. A.

IX

^{29}Si NMR DATA OF SILYL ANIONS

^{29}Si NMR chemical shifts and coupling constants are compiled in Table II.

TABLE II

^{29}Si NMR Chemical Shifts and Coupling Constants of Silyl Anions

Compound	δ (ppm)		^1J (Hz)	Temp.	Solvent	Reference
Ph$_3$SiLi	-30.0			room temp.	THF	(20)
	-9.0			295 K	THF	(19b)
	-9.3	t	17[^{29}Si, ^{6}Li]	173 K	MTHF	(19b)
	9.28	q	45[^{29}Si, ^{7}Li]	193 K	C$_7$D$_8$	(22)
Ph$_2$MeSiLi	-20.5			room temp.	THF	(20)
	-23.1	t	15[^{29}Si, ^{6}Li]	158 K	THF	(19b)
PhMe$_2$SiLi	-22.1			room temp.	THF	(20)
	-29.1	t	18[^{29}Si, ^{6}Li]	173 K	THF	(19b)
(PhMe$_2$Si)$_2$CuLi	-24.4			223 K	THF	(99a–d)
PhMe$_2$SiCu(CN)Li	-25.5			223 K	THF	(99a–d)
(PhMe$_2$Si)$_2$Cu(CN)Li$_2$	-24.4			223 K	THF	(99a–d)
Me(PhMe$_2$Si)Cu(CN)Li$_2$	-20.6			223 K	THF	(99a–d)
(PhMe$_2$Si)$_3$CuLi$_2$	-18.9			223 K	THF	(99a–d)
Me$_3$SiLi	-32.8	t	21.1[^{29}Si, ^{6}Li]	223 K	THF-d$_8$	(70)
Me$_3$SiK	-34.4			room temp.	HMPA	(20)
H$_3$SiK	-165.0	q	75[^{29}Si, ^{1}H]	room temp.	HMPA	(46)
Cl$_3$SiLi	30.9			room temp.	THF	(56)
HMes$_2$SiLi	-71.78	d	101[^{29}Si, ^{1}H]	293 K	C$_6$D$_6$	(52)
(Et$_2$N)Ph$_2$SiLi	19.34			273 K	THF	(54a)
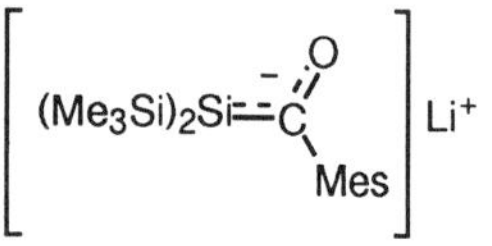 $\left[(Me_3Si)_2Si=C{\overset{O}{\underset{Mes}{}}} \right] Li^+$	-59.9			233 K	THF	(60)

t-Bu, Si, M Ph$_4$ M=Li, Na

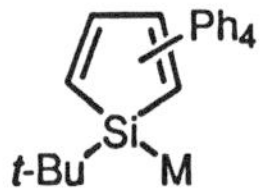

| M = Li | 25.10 | | | room temp. | THF-d$_8$ | (64a) |
| M = Na | 26.12 | | | room temp. | THF-d$_8$ | (64a) |

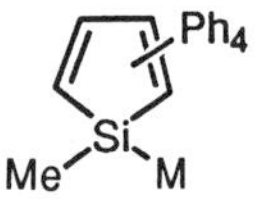

Me, Si, M Ph$_4$ M=Li, Na

TABLE II (continued)

Compound	δ (ppm)		^1J (Hz)	Temp.	Solvent	Reference
M = Li	-19.29			room temp.	THF-d_8	(64b)
M = Na	-33.51			room temp.	THF-d_8	(64b)
	68.54			room temp.	THF-d_8	(65b)
PhMe$_2$SiMe$_2$SiLi	-69.8		31[^{29}Si, ^{29}Si]	room temp.	THF-d_8	(28)
Me$_3$SiMe$_2$SiLi	-74.9	t	18.8[^{29}Si, ^{6}Li]	180 K	THF-d_8	(70)
			19.1[^{29}Si, ^{29}Si]			
	-1.24	q	43.5[^{29}Si, ^{7}Li]	173 K	THF-d_8	(71)
Me$_3$SiMe$_2$SiMe$_2$SiLi	-62.7	t	18.6[^{29}Si, ^{6}Li]	180 K	THF-d_8	(70)
(Me$_3$Si)$_2$MeSiLi	-133.8			180 K	THF-d_8	(70)
(Me$_3$Si)$_3$SiLi	-185.4			room temp.	C$_6$D$_6$	(75)
[{(Me$_3$Si)$_3$Si}$_2$Cu$_2$Br](Li·3THF)	-129.88			room temp.	C$_6$D$_6$	(76c)
[(Me$_3$Si)$_3$Si]$_2$M						
M = Zn	-123.9			296 K	C$_6$D$_6$	(79a)
M = Cd	-109.2			296 K	C$_6$D$_6$	(79a)
M = Hg	-54.5			296 K	C$_6$D$_6$	(79a)
(Me$_3$Sn)$_3$SiLi	-122.0		239[^{29}Si, Sn]	296 K	C$_6$D$_6$	(81b)
Li(Ph$_2$Si)$_4$Li	-28.54			303 K	THF	(85b)
Me$_{11}$Si$_6$Li	-108.1			310 K	HMPA	(86)

X

LIST OF SILYL ANIONS

Known silyl anions are summarized, together with references.

Arylsilyl Anions

Ph_3SiLi (*3,6,7,9a*); Ph_3SiNa (*6*); Ph_3SiK (*5,8*); Ph_3SiRb (*3*); Ph_3SiCs (*3*)
$Ph_3SiMgBr$ (*9–11*); $(Ph_3Si)_2Hg$ (*12*); $(o\text{-}Tolyl)_3SiLi$ (*3*); $[Ph_3Si^-/n\text{-}Bu_4N^+]$ (*39*)
$Ph_2MeSiLi$ (*26*); Ph_2MeSiK (*19a*); $Ph_2(t\text{-}Bu)SiLi$ (*74*); $Ph_2(t\text{-}Bu)SiK$ (*110*); $Ph_2(n\text{-}Bu)SiLi$ (*54d*)

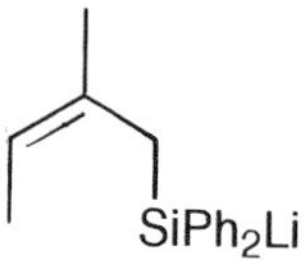

$$\text{(13)}$$

$$\text{(14)}$$

$(\alpha\text{-}Naph)PhMeSi*Li$ (*9a,b*); $(\alpha\text{-}Naph)PhMeSi*MgBr$ (*9a,b*); $[(\alpha\text{-}Naph)\text{-}PhMeSi*]_2Hg$ (*16*)
$PhMe_2SiLi$ (*26*); $PhMe_2SiK$ (*19a*); $[(PhMe_2Si)_4Hg]Li_2$ (*33b*)
$PhEt_2SiLi$ (*112*); $Ph(i\text{-}Pr)_2SiLi$ (*24*)
$Ph(PhCH_2)_2SiLi$ (*24*); $(neo\text{-}C_5H_{11})PhMeSi*Li$ (*15*)
$(PhMe_2Si)_2CuLi$ (*99a–d,100b*); $PhMe_2SiCu(CN)Li$ (*99a–d*)
$(PhMe_2Si)_2Cu(CN)Li_2$ (*99a–d,100b*); $Me(PhMe_2Si)Cu(CN)Li_2$ (*99a–d, 100b*); $(PhMe_2Si)_3CuLi_2$ (*99a–d*)
$PhMe_2SiLi/MeMgI$ (*102*); $PhMe_2SiLi/ZnBr_2$ (*102*)
$[PhMe_2Si^-/n\text{-}Bu_4N^+]$ (*39*)

Alkylsilyl Anions

Me_3SiLi (*29,34b,70*); Me_3SiNa (*8,30a,b*); Me_3SiK (*8,30a,b,31*); $(Me_3Si)_2Mg$ (*32,36*)
$(Me_3Si)_2Hg$ (*35*); $(Me_3Si)_2Zn$ (*38a–c*); $(Me_3Si)_2Cd$ (*38a–c*)
$(Me_3Si)_3HgLi$ (*33a*); $(Me_3Si)_4HgLi_2$ (*33a,b*); $(Me_3Si)_4HgNa_2$ (*33a*); $(Me_3Si)_4HgK_2$ (*33a*)
$(Me_2SiCH_2SiMe_2)_2Hg_2$ (*112*)
Et_3SiLi (*34b*); Et_3SiNa (*34b*); Et_3SiK (*8,34a,b*); $(Et_3Si)_2Hg$ (*34a,b*)
$t\text{-}Bu_3SiLi$ (*40*); $t\text{-}Bu_3SiNa$ (*40*); $t\text{-}Bu_3SiK$ (*40*); $t\text{-}Bu_3SiRb$ (*40*); $t\text{-}Bu_3SiCs$ (*40*)
$t\text{-}Bu_3SiMgBr$ (*40*); $(t\text{-}Bu_3Si)_2Cd$ (*38a–c*)
$(Me_3Si)_2CuLi$ (*99b*); $Me_3SiLi/CuI/Me_2S$ (*100f*); $Me_3SiLi/MeMgI/MnCl_2$ (*103*)
$t\text{-}BuMe_2SiCu(n\text{-}Bu)(CN)Li_2$ (*101*); $(Thexyl)Me_2SiCu(n\text{-}Bu)(CN)Li_2$ (*101*)
$[Me_3Si^-/n\text{-}Bu_4N^+]$ (*39*)

Functionalized Silyl Anions

$SiLi_4$ (*49*); H_3SiK (*44*)
HPh_2SiLi (*54e,56*); $HMes_2SiLi$ (*51–53*)
Cl_3SiLi (*50*); $Cl_3Si^-[R_3NH]^+$ (*57a–g*)
$Cl_3Si^-[TMEDA-H]^+$ (*58*); $Cl_2PhSi^-[TMEDA-H]^+$ (*58*);
  $Cl_2MeSi^-[TMEDA-H]^+$ (*58*)
$(Et_2N)Ph_2SiLi$ (*54a,e*); $(Et_2N)_2PhSiLi$ (*54a*); $(Et_2N)PhMeSiLi$ (*54a*)
$(Et_2N)Ph_2SiMg-i-Pr$ (*54b*); $(Et_2N)PhMeSiMg-i-Pr$ (*54b*)
$(Et_2N)Ph_2SiCu(CN)Li$ (*54a*); $(Et_2N)_2PhSiCu(CN)Li$ (*54b*)
$(t-BuO)Ph_2SiLi$ (*54d,e*); $(i-PrO)Ph_2SiLi$ (*54e*); $(MeO)Ph_2SiLi$ (*54e*);
  $(t-BuO)_2PhSiLi$ (*54e,f*); $(CH_2{=}CH-Me_2C-O-)Ph_2SiLi$ (*54g*);
  $(CH_2{=}CH-Et_2C-O-)(p-tolyl)_2SiLi$ (*54f*)
$(MeO)Me_2SiNa$ (*55a*); $(MeO)_2MeSiNa$ (*55a*)

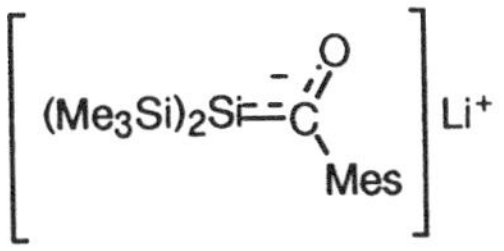

$$\left[(Me_3Si)_2Si{=}\overset{-}{\underset{Mes}{C}}{\overset{O}{}} \right] Li^+$$

(*60*)

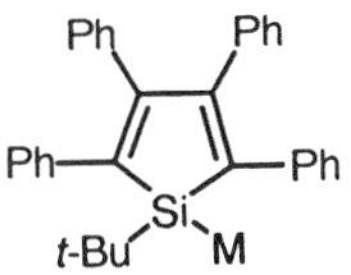

$M{=}Li, Na$ $M{=}Li, Na$

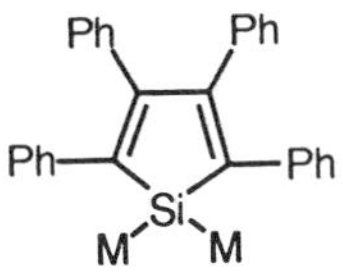

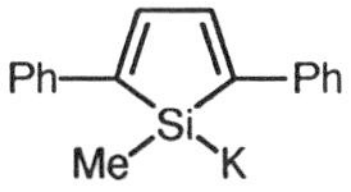

(*64*)   (*65*)

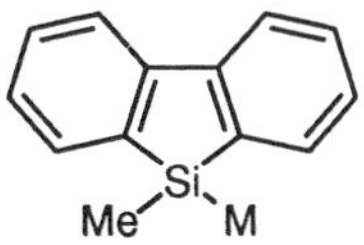

$M{=}Li, K$

(*14*)   (*14, 66, 67*)

Disilanyl Anions

H_3SiH_2SiK (*44*); $(H_3Si)_2HSiK$ (*44*); $(H_3Si)_3SiK$ (*44*)
Me_3SiMe_2SiLi (*68a–c,70*); $Me_3SiMe_2Si^-[Bu_4N]^+$ (*39*)
Me_3SiPh_2SiLi (*69*); Me_3SiMes_2SiLi (*69*); $PhMe_2SiMe_2SiLi$ (*28*)
Ph_3SiPh_2SiLi (*113*); Ph_3SiPh_2SiK (*114*); $(t-BuO)Ph_2SiPh_2SiLi$ (*54d*)

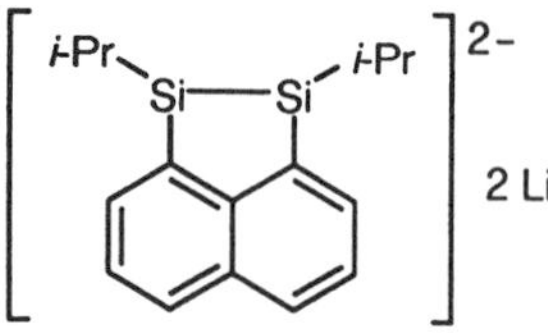

(72)

Polysilanyl Anions

$(Me_3Si)_3SiLi$ (73–77,80b,c,98b); $(HMe_2Si)_3SiLi$ (73a–c); $(Me_3Si)_2MeSiLi$
(70,73a–c)
$(Me_3Si)_2SiLi_2$ (82)
$(Me_3Si)_3SiK$ (79b); $(Me_3Si)_3SiRb$ (79b); $(Me_3Si)_3SiCs$ (79b)
$[(Me_3Si)_3Si]_2Zn$ (79a); $[(Me_3Si)_3Si]_2Cd$ (79a); $[(Me_3Si)_3Si]_2Hg$ (79a)
$[\{(Me_3Si)_3Si\}_2Cu_5Cl_4](Li\cdot4THF)$ (80b); $[\{(Me_3Si)_3Si\}_2Cu_2Br](Li\cdot3THF)$
(80c)
$(Me_3Sn)_3SiLi$ (81a,b)
$[(R_3Si)_3SiSi(SiR_3)_2]Li$ (74); $Li[(R_3Si)_2SiSi(SiR_3)_2]Li$ (72,74);
$\quad K[(R_3Si)_2SiSi(SiR_3)_2]K$ (72)
$LiPh_2Si(Ph_2Si)_2Ph_2SiLi$ (84b,85b); $LiPh_2Si(Ph_2Si)_3Ph_2SiLi$ (84c)
$Me_3SiMe_2SiMe_2SiLi$ (70); $(PhMe_2Si)_2MeSiLi$ (70)
cyclo-$Me_{11}Si_6$-M (M = Li, Na, K) (86)
cyclo-Me_9Si_5-K (87); (cyclo-Me_9Si_5-)$_2Hg$ (87)

Model for Calculation

H_3Si^- (104–106); $(H_2B)H_2Si^-$ (105,106); $(H_3C)H_2Si^-$ (105,106);
$\quad(H_2N)H_2Si^-$ (105,106)
$(HO)H_2Si^-$ (105,106); FH_2Si^- (105,106)
$(N{\equiv}C{-})H_2Si^-$ (107a); $(O{=}CH{-})H_2Si^-$ (107a); $(HC{\equiv}C{-})H_2Si^-$ (107b);
$\quad H_3SiH_2Si^-$ (104d)
$SiLi_4$ (108b); H_2SiLi_2 (108b); H_2SiNa_2 (108b); H_3SiLi (108b,108d)
FH_2SiLi (109); $(H_3Si)_2SiLi(COCH_3)$ (60)

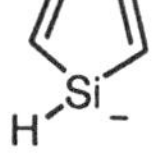

(62, 63)

XI

CONCLUDING REMARKS

The chemistry of silyl anions has been surveyed. While silyl anion chemistry has been developing steadily and rapidly in recent years, it still lags far behind carbanion chemistry in terms of the number and kind of anions so far studied. When compared to carbanion chemistry, however, silyl anion chemistry has intriguing and unique features, such as softer nucleophilicity, preference for monomeric structure in solution, high inversion barrier of pyramidal structure, generation of 1,1- and 1,ω-dianions, and redistribution reactions of oligo- and polysilanes with silyl anions. More extensive studies are expected to disclose other properties, reactivities, and utilities of silyl anions. Studies of silyl anions, however, sometimes encounter difficulty in preparation of silyl anions with various kinds of substituents, especially functional groups. Further development thus greatly depends on the exploitation of new methodologies for preparation of silyl anions.

REFERENCES

(1) Previous reviews include (a) Wittenberg, D.; Gilman, H. *Q. Rev.* **1969**, *13*, 116; (b) Davis, D. D.; Gray, C. E. *Organomet. Chem. Rev., Sect. A* **1970**, *6*, 283; (c) Wiberg, E.; Stecher, O.; Andrascheck, H. J.; Kreuzbichler, L.; Staude, E. *Angew. Chem., Int. Ed. Engl.* **1963**, *2*, 507; (d) Fujita, M.; Hiyama, T. *J. Synth. Org. Chem. Jpn.* **1984**, *42*, 293; (e) Colvin, E. W. "Silicon Organic Synthesis"; Butterworth: London, 1981; pp. 134–140; (f) Fleming, I. In "Comprehensive Organic Chemistry"; Barton, E.; Ollis, W. D., Eds.; Pergamon: Oxford, 1979; Vol. 3, pp. 664–669; for theoretical studies, see (g) Apeloig, Y. In "The Chemistry of Organic Silicon Compounds"; Patai, S.; Rappoport, Z., Eds.; Wiley: Chichester, 1989; p. 201; (h) Lambert, J. B.; Schultz, W. J., Jr. In "The Chemistry of Organic Silicon Compounds"; Patai, S.; Rappoport, Z., Eds.; Wiley: Chichester, 1989; Chapter 16, pp. 1007–1010; for structural studies, see (i) Sheldrick, W. S. In "The Chemistry of Organic Silicion Compounds"; Patai, S.; Rappoport, Z., Eds.; Wiley: Chichester, 1989; p. 268.

(2) Tilley, T. D. In "The Chemistry of Organosilicon Compounds"; Patai, S.; Rappoport, Z., Eds.; Wiley: Chichester, 1989; Chapter 24, pp. 1415–1477; Silylcuprates and silylcopper: Lipshutz, B. H.; Wilhelm, R. S.; Kozlowski, J. A. *Tetrahedron* **1984**, *40*, 5005; Transition metal catalyzed silylmetalation: Oshima, K. In "Advances in Metal-Organic Chemistry"; Liebeskind, L. S., Ed.; JAI Press: London, 1991, Vol. 2, pp. 101–141.

(3) George, M. V.; Peterson, D. J.; Gilman, H. *J. Am. Chem. Soc.* **1960**, *82*, 404.

(4) Steudel, W.; Gilman, H. *J. Am. Chem. Soc.* **1960**, *82*, 6129.

(5) Gilman, H.; Wu, T. C. *J. Am. Chem. Soc.* **1951**, *73*, 4031.

(6) Brook, A. G.; Gilman, H. *J. Am. Chem. Soc.* **1954**, *76*, 278.

(7) Benkeser, R. A.; Severson, R. G. *J. Am. Chem. Soc.* **1951**, *73*, 1424.

(8) Corriu, R. J. P.; Guérin, C. *J. Chem. Soc., Chem. Commun.* **1980**, 168.

(9a) Colomer, E.; Corriu, R. J. P. J. Chem. Soc., Chem. Commun. 1976, 176.
(9b) Colomer, E.; Corriu, R. J. P. J. Organomet. Chem. 1977, 133, 159.
(10) Selin, T. G.; West, R. Tetrahedron 1959, 5, 97.
(11) Comins, D. L.; Killpack, M. O. J. Am. Chem. Soc. 1992, 114, 10972.
(12) Jackson, R. A. Chem. Commun. 1966, 827.
(13) Fleming, I.; Winter, S. B. D. Tetrahedron Lett. 1993, 34, 7287.
(14) Corriu, R. J. P.; Guérin, C.; Kolani, B. Bull. Soc. Chim. Fr. 1985, 973.
(15) Sommer, L. H.; Mason, R. J. Am. Chem. Soc. 1965, 87, 1619.
(16) Eaborn, C.; Jackson, R. A.; Rahman, M. T. J. Organomet. Chem. 1972, 34, 7; Eaborn, C.; Jackson, R. A.; Tune, D. J.; Walton, D. R. M. ibid. 1973, 63, 85.
(17) Brefort, J. L.; Corriu, R. J. P.; Guérin, C.; Henner, B. J. Organomet. Chem. 1989, 370, 9.
(18) Hajdasz, D. J.; Squires, R. R. J. Am. Chem. Soc. 1986, 108, 3139.
(19a) Buncel, E.; Venkatachalam, T. K.; Eliasson, B.; Edlund, U. J. Am. Chem. Soc. 1985, 107, 303.
(19b) Edlund, U.; Lejon, T.; Venkatachalam, T. K.; Buncel, E. J. Am. Chem. Soc. 1985, 107, 6408.
(19c) Rakita, P. E.; Srebro, J. P.; Worsham, L. S. J. Organomet. Chem. 1976, 104, 27.
(20) Olah, G. A.; Hunadi, R. J. J. Am. Chem. Soc. 1980, 102, 6989.
(21) Waack, R.; Doran, M. A. Chem. Ind. (London) 1965, 563; Evans, A. G.; Hamid, M. A.; Rees, N. H. J. Chem. Soc. B 1971, 1110.
(22) Davis, H. V. R.; Olmstead, M. M.; Ruhlandt-Senge, K.; Power, P. P. J. Organomet. Chem. 1993, 462, 1.
(23) Bartlett, R. A.; Dias, H. V. R.; Power, P. P. J. Organomet. Chem. 1988, 341, 1.
(24) Lambert, J. B.; Urdaneta-Pérez, M. J. Am. Chem. Soc. 1978, 100, 157.
(25) Hayashi, T.; Okamoto, Y.; Kumada, M. J. Chem. Soc., Chem. Commun. 1982, 1072.
(26) Gilman, H.; Lichtenwalter, G. D. J. Am. Chem. Soc. 1958, 80, 608.
(27) Gilman, H.; Lichtenwalter, G. D.; Wittenberg, D. J. Am. Chem. Soc. 1959, 81, 5320.
(28) Ruehl, K. E.; Davis, M. E.; Matyjaszewski, K. Organometallics 1992, 11, 788.
(29) Still, W. C. J. Org. Chem. 1976, 41, 3063.
(30a) Sakurai, H.; Okada, A.; Kira, M.; Yonezawa, K. Tetrahedron Lett. 1971, 19, 1511.
(30b) Sakurai, H.; Kondo, F. J. Organomet. Chem. 1975, 92, C46.
(31) Shippey, M. A.; Dervan, P. B. J. Org. Chem. 1977, 42, 2654.
(32) Eaborn, C.; Jackson, R. A.; Walsingham, R. W. J. Chem. Soc. C 1967, 2188.
(33a) Schaaf, T. F.; Oliver, J. P. J. Am. Chem. Soc. 1969, 91, 4327.
(33b) Isley, W. H.; Albright, M. J.; Anderson, T. J.; Glick, M. D.; Oliver, J. P. Inorg. Chem. 1980, 19, 3577.
(34a) Vyazankin, N. S.; Razuvaev, G. A.; Gladyshev, E. N.; Korneva, S. P. J. Organomet. Chem. 1967, 7, 353.
(34b) Gladyshev, E. N.; Fedorova, E. A.; Yuntila, L. O.; Razuvaev, G. A.; Vyazankin, N. S. J. Organomet. Chem. 1975, 96, 169.
(35) Hengge, E.; Holtschmidt, N. J. Organomet. Chem. 1968, 12, P5.
(36) Glaggett, A. R.; Ilsley, W. H.; Anderson, T. J.; Glick, M. D.; Oliver, J. P. J. Am. Chem. Soc. 1977, 99, 1797.
(37) Rösch, L. J. Organomet. Chem. 1976, 121, C15.
(38a) Rösch, L.; Müller, H. Angew. Chem., Int. Ed. Engl. 1976, 15, 620.
(38b) Rösch, L.; Altnau, G. Angew. Chem., Int. Ed. Engl. 1979, 18, 60.
(38c) Rösch, L.; Altnau, G. Z. Naturforsch., B: Anorg. Chem., Org. Chem. 1980, 35, 195.
(39) Hiyama, T.; Obayashi, M.; Mori, I.; Nozaki, H. J. Org. Chem. 1983, 48, 912; Hiyama, T.; Obayashi, M. Tetrahedron Lett. 1983, 24, 4109; Hiyama, T.; Obayashi, M.; Sawahata, M. ibid., 4113.

(40) Wiberg, N. In "Frontiers of Organosilicon Chemistry"; Bassindale, A. R.; Gaspar, P. P., Eds.; Royal Society of Chemistry: London, 1991; pp. 263–270; Wiberg, N.; Fischer, G.; Karampatses, P. *Angew. Chem., Int. Ed. Engl.* **1984,** *23,* 59; Wiberg, N.; Schuster, H.; Simon, A.; Peters, K. *ibid.* **1986,** *25,* 79; Wiberg, N.; Schurz, K. *J. Organomet. Chem.* **1988,** *341,* 145.

(41) Schaaf, T. F.; Butler, W.; Glick, M. D.; Oliver, J. P. *J. Am. Chem. Soc.* **1974,** *96,* 7593; Ilsley, W. H.; Schaaf, T. F.; Glick, M. D.; Oliver, J. P. *ibid.* **1980,** *102,* 3769.

(42) Teclé, B.; Ilsley, W. H.; Oliver, J. P. *Organometallics* **1982,** *1,* 875.

(43) Bleckmann, P.; Soliman, M.; Reuter, K.; Neumann, W. P. *J. Organomet. Chem.* **1976,** *108,* C18.

(44) Ring, M. A.; Ritter, D. M. *J. Am. Chem. Soc.* **1961,** *83,* 802; *J. Phys. Chem.* **1961,** *65,* 182.

(45) Amberger, E.; Mühlhofer, E. *J. Organomet. Chem.* **1968,** *12,* 55.

(46) Bürger, H.; Eujen, R.; Marsmann, H. C. *Z. Naturforsch., B: Anorg. Chem., Org. Chem.* **1974,** *29B,* 149.

(47) Mundt, O.; Becker, G.; Hartmann, H. M.; Schwarz, W. *Z. Anorg. Allg. Chem.* **1989,** *572,* 75.

(48) Nimlos, M. R.; Ellison, G. B. *J. Am. Chem. Soc.* **1986,** *108,* 6522.

(49) Morrison, J. A.; Lagow, R. J. *Inorg. Chem.* **1977,** *16,* 2972.

(50) Gilman, H.; Steudel, W. *Chem. Ind. (London)* **1959,** 1094.

(51) Weidenbruch, M.; Kramer, K.; von Schnering, H. G. *Z. Naturforsch., B: Anorg. Chem., Org. Chem.* **1985,** *40B,* 601.

(52) Roddick, D. M.; Heyn, R. H.; Tilley, T. D. *Organometallics* **1989,** *8,* 324.

(53) Stüger, H. *J. Organomet. Chem.* **1993,** *458,* 1.

(54a) Tamao, K.; Kawachi, A.; Ito, Y. *J. Am. Chem. Soc.* **1992,** *114,* 3989. The ^{29}Si NMR data have been corrected by reinvestigation (1994).

(54b) Tamao, K.; Kawachi, A.; Ito, Y. "Proceedings of the 10th International Symposium on Organosilicon Chemistry, Poznan, Poland, **1993** (in press).

(54c) Tamao, K.; Kawachi, A.; Ito, Y. *Organometallics* **1993,** *12,* 580.

(54d) Tamao, K.; Kawachi, A. *Angew. Chem., Int. Ed. Engl.* **1995,** *34,* 818.

(54e) Tamao, K.; Kawachi, A. *Organometallics* **1995,** in press.

(54f) Tamao, K. "28th Organosilicon Symposium, 1995, Gainesville, Florida, USA," Abstracts, p. A-4.

(54g) Tamao, K.; Kawachi, A. "6th International Kyoto Conference on New Aspects of Organic Chemistry, 1994, Kyoto, Japan," Abstracts, p. 143.

(55a) Watanabe, H.; Higuchi, K.; Kobayashi, M.; Hara, M.; Koike, Y.; Kitahara, T.; Nagai, Y. *J. Chem Soc., Chem. Commun.* **1977,** 534.

(55b) Watanabe, H.; Higuchi, K.; Goto, T.; Muraoka, T.; Inose, J.; Kageyama, M.; Iizuka, Y.; Nozaki, M.; Nagai, Y. *J. Organomet. Chem.* **1981,** *218,* 27.

(56) Oehme, H.; Weiss, H. *J. Organomet. Chem.* **1987,** *319,* C16.

(57a) Nozakura, S.; Konotune, S. *Bull. Chem. Soc. Jpn.* **1956,** *29,* 322.

(57b) Benkeser, R. A.; Foley, K. M.; Gaul, J. M.; Li, G. S. *J. Am. Chem. Soc.* **1970,** *92,* 3232.

(57c) Li, G. S.; Ehler, D. F.; Benkeser, R. A. *Org. Synth., Collect. Vol.* **1988,** *6,* 747.

(57d) Benkeser, R. A. *Acc. Chem. Res.* **1971,** *4,* 94.

(57e) Naumann, K.; Zon, G.; Mislow, K. *J. Am. Chem. Soc.* **1969,** *91,* 7012.

(57f) Benkeser, R. A.; Foley, K. M.; Grutsner, J. B.; Smith, W. E. *J. Am. Chem. Soc.* **1970,** *92,* 697.

(57g) Bernstein, S. C. *J. Am. Chem. Soc.* **1970,** *92,* 699.

(58) Boudjouk, P.; Kloos, S.; Amirthini, B. *J. Organomet. Chem.* **1993,** *443,* C41.

(59) Boudjouk, P.; Samaraweera, U.; Sooriyakumaran, R.; Chrusciel, J.; Anderson,

K. R. *Angew. Chem., Int. Ed. Engl.* **1988**, *27*, 1355; Tsumuraya, T.; Batcheller, S. A.; Masamune, S. *ibid.* **1991**, *30*, 902; Corriu, R.; Lanneau, G.; Priou, C.; Soulairol, F.; Auner, N.; Probst, R.; Conlin, R.; Tan, C. J. *J. Organomet. Chem.* **1994**, *466*, 55.

(60) Ohshita, J.; Masaoka, Y.; Masaoka, S.; Ishikawa, M.; Tachibana, A.; Yano, T.; Yamabe, T. *J. Organomet. Chem.* (in press).

(61) Colomer, E.; Corriu, R. J. P.; Lheureux, M. *Chem. Rev.* **1990**, *90*, 265.

(62) Damewood, J. R., Jr. *J. Org. Chem.* **1986**, *51*, 5028.

(63) Gordon, M. S.; Boudjouk, P.; Anwari, F. *J. Am. Chem. Soc.* **1983**, *105*, 4972.

(64a) Hong, J. H.; Boudjouk, P. *J. Am. Chem. Soc.* **1993**, *115*, 5883.

(64b) Sohn, H.; Bankwitz, U.; Powell, D. R.; West, R. "28th Organosilicon Symposium, 1995, Gainesville, Florida, USA"; Abstracts, p. 68.

(65a) Joo, W. C.; Hong, J. H.; Choi, S. B.; Son, H. E. *J. Organomet. Chem.* **1990**, *391*, 27.

(65b) Hong, J. H.; Boudjouk, P.; Castellino, S. *Organometallics* **1994**, *13*, 3387.

(65c) Bankwitz, U.; Sohn, H.; Powell, D. R.; West, R. "28th Organosilicon Symposium, 1995, Gainesville, Florida, USA," Abstracts, p. 65.

(65d) Sohn, H.; Bankwitz, U.; Calabrese, J.; West, R. "28th Organosilicon Symposium, 1995, Gainesville, Florida, USA," Abstracts, p. 69.

(66) Gilman, H.; Gorsich, R. D. *J. Am. Chem. Soc.* **1958**, *80*, 3243.

(67) Ishikawa, M.; Tabohashi, T.; Ohashi, H.; Kumada, M.; Iyoda, J. *Organometallics* **1983**, *2*, 351.

(68a) Hudrlik, P. F.; Waugh, M. A.; Hudrlik, A. M. *J. Organomet. Chem.* **1984**, *271*, 69.

(68b) Hwu, J. R.; Wetzel, J. M.; Lee, J. S.; Butcher, R. J. *J. Organomet. Chem.* **1993**, *453*, 21.

(68c) Krohn, K.; Khanbabaee, K. *Angew. Chem., Int. Ed. Engl.* **1994**, *33*, 99.

(69) Brook, A. G.; Baumegger, A.; Lough, A. J. *Organometallics* **1992**, *11*, 310.

(70) Sekiguchi, A.; Nanjo, M.; Kabuto, C.; Sakurai, H. *J. Am. Chem. Soc.* **1995**, *117*, 4195.

(71) Wakahara, T.; Akasaka, T.; Ando, W. *Organometallics* **1994**, *13*, 4683.

(72) Kira, M.; Maruyama, T.; Kabuto, C.; Ebata, K.; Sakurai, H. "10th International Symposium on Organosilicon Chemistry, 1993, Poznán, Poland"; Abstracts, p. 189; *Angew. Chem., Int. Ed. Engl.* **1994**, *33*, 1489.

(73a) Gilman, H.; Holmes, J. M.; Smith, C. L. *Chem. Ind. (London)* **1965**, 848.

(73b) Gilman H.; Smith, C. L. *J. Organomet. Chem.* **1968**, *14*, 91.

(73c) Gilman, H.; Harrell, R. L., Jr. *J. Organomet. Chem.* **1967**, *9*, 67.

(74) Smith, C. M.; Lickiss, P. D. "10th International Symposium on Organosilicon Chemistry, 1993, Poznán, Poland"; Abstracts, p. 263.

(75) Gutekunst, G.; Brook, A. G. *J. Organomet. Chem.* **1982**, *225*, 1.

(76) Heine, A.; Herbst-Irmer, R.; Sheldrick, G. M.; Stalke, D. *Inorg. Chem.* **1993**, *32*, 2694.

(77) Becker, G.; Hartmann, H. M.; Münch, A.; Riffel, H. *Z. Anorg. Allg. Chem.* **1985**, *530*, 29.

(78) Klinkhammer, K. W.; Schwarz, W. "10th International Symposium on Organosilicon Chemistry, 1993, Poznán, Poland"; Abstracts, p. 191.

(79a) Arnold, J.; Tilley, T. D.; Rheingold, A. L.; Geib, S. J. *Inorg. Chem.* **1987**, *26*, 2106.

(79b) Henkel, S.; Klinkhammer, K. W.; Schwarz, W. *Angew. Chem., Int. Ed. Engl.* **1994**, *33*, 681.

(80a) Mallela, S. P.; Bernal, I.; Geanangel, R. A. *Inorg. Chem.* **1992**, *31*, 1626.

(80b) Heine, A.; Stalke, D. *Angew. Chem., Int. Ed. Engl.* **1993**, *32*, 121.

(80c) Heine, A.; Herbst-Irmer, R.; Stalke, D. *J. Chem. Soc., Chem. Commun.* **1993**, 1729.

(81a) Biffar, W.; Gasparis-Ebeling, T.; Nöth, H.; Storch, W.; Wrackmeyer, B. *J. Magn. Reson.* **1981**, *44*, 54.

(*81b*) Heyn, R. H.; Tilley, T. D. *Inorg. Chem.* **1990**, *29*, 4051.

(*82*) Mehrotra, S. K.; Kawa, H.; Baran, J. R., Jr.; Ludvig, M. M.; Lagow, R. J. *J. Am. Chem. Soc.* **1990**, *112*, 9003.

(*83*) Wittenberg, D.; George, M. V.; Gilman, H. *J. Am. Chem. Soc.* **1959**, *81*, 4812.

(*84a*) Jarvie, A. W. P.; Gilman, H. *J. Org. Chem.* **1961**, *26*, 1999.

(*84b*) Jarvie, A. W.; Winkler, H. J. S.; Peterson, D. J.; Gilman, H. *J. Am. Chem. Soc.* **1961**, *83*, 1921.

(*84c*) Gilman, H.; Tomasi, R. A. *Chem. Ind. (London)* **1963**, 954.

(*85a*) Hengge, E.; Wolfer, D. *J. Organomet. Chem.* **1974**, *66*, 413.

(*85b*) Becker, G.; Hartmann, H. M.; Hengge, E.; Schrank, F. *Z. Anorg. Allg. Chem.* **1989**, *572*, 63.

(*86*) Allred, A. L.; Smart, R. T.; Beek, D. A. V., Jr. *Organometallics* **1992**, *11*, 4225.

(*87*) Hengge, E.; Spielberger, A.; Gspaltl, P. "10th International Symposium on Organosilicon Chemistry, 1993, Poznán, Poland"; Abstracts, p. 265.

(*88*) West, R. *J. Organomet. Chem.* **1986**, *300*, 327; Worsfold, D. J. *ACS Symp. Ser.* **1988**, *360*, 101–111; Jones, R. G.; Benfield, R. E.; Cragg, R. H.; Swain, A. C.; Webb, S. J. *Macromolecules* **1993**, *26*, 4878.

(*89*) Brough, L. F.; West, R. *J. Organomet. Chem.* **1980**, *194*, 139.

(*90*) Carberry, E.; West, R. *J. Organomet. Chem.* **1966**, *6*, 582.

(*91*) Kumada, M.; Sakamoto, S.; Ishikawa, M. *J. Organomet. Chem.* **1969**, *17*, 231.

(*92*) Hatanaka, Y.; Hiyama, T. "40th Symposium on Organometallic Chemistry, 1993, Sapporo, Japan"; Abstracts, p. 136.

(*93*) Cypryk, M.; Gupta, Y.; Matyjaszewski, K. *J. Am. Chem. Soc.* **1991**, *113*, 1046.

(*94*) Suzuki, M.; Kotani, J.; Gyobu, S.; Kaneko, T.; Saegusa, J. *Macromolecules* **1994**, *27*, 2360.

(*95*) Sakamoto, K.; Obata, K.; Hirata, H.; Nakajima, M.; Sakurai, H. *J. Am. Chem. Soc.* **1989**, *111*, 7641; Sakamoto, K.; Yoshida, M.; Sakurai, H. *Macromolecules* **1990**, *23*, 4494; *Polym. Prepr., Am. Chem. Soc., Div. Polym. Chem.* **1993**, *34*, 218.

(*96*) Dervan, P. B.; Shippey, M. A. *J. Am. Chem. Soc.* **1976**, *98*, 1265.

(*97*) Reetz, M. T.; Plachky, M. *Synthesis* **1976**, *199*.

(*98a*) Zhivotovskii, D. B.; Braude, V.; Stanger, A.; Kapon, M.; Apeloig, Y. *Organometallics* **1992**, *11*, 2326; Zhivotovskii, D. B.; Zharov, I.; Apeloig, Y. "10th International Symposium on Organosilicon Chemistry, 1993, Poznán, Poland"; Abstracts, p. 295.

(*98b*) Oehme, H.; Wustrack, R.; Heine, A.; Sheldrick, G. M.; Stalke, D. *J. Organomet. Chem.* **1993**, *452*, 33.

(*99a*) Sharma, S.; Oehlschlager, A. C. *Tetrahedron* **1989**, *45*, 557.

(*99b*) Sharma, S.; Oehlschlager, A. C. *J. Org. Chem.* **1989**, *54*, 5383.

(*99c*) Sharma, S.; Oehlschlager, A. C. *J. Org. Chem.* **1991**, *56*, 770.

(*99d*) Singer, R. D.; Oehlschlager, A. C. *J. Org. Chem.* **1991**, *56*, 3510.

(*100a*) Colvin, E. W. "Silicon Reagents in Organic Synthesis"; Academic Press: San Diego, 1988; pp. 27, 60.

(*100b*) Fleming, I.; Newton, T. W. *J. Chem. Soc., Perkin. Trans. 1* **1984**, 1805.

(*100c*) Fleming, I. *Pure. Appl. Chem.* **1988**, *60*, 71.

(*100d*) Fleming, I. *J. Chem. Soc., Perkin Trans. 1*, **1992**, 3363, and references cited therein.

(*100e*) Fleming, I.; Sanderson, P. E. J. *Tetrahedron Lett.* **1987**, *28*, 4229.

(*100f*) Smith, J. G.; Drozda, S. E.; Petraglia, S. P.; Quinn, N. R.; Rice, E. M.; Taylor, B. S.; Viswanathan, M. *J. Org. Chem.* **1984**, *49*, 4112.

(*101*) Lipshutz, B. H.; Reuter, D. C.; Ellsworth, E. L. *J. Org. Chem.* **1989**, *54*, 4975.

(*102*) Hiyama, T.; Sato, M.; Kanemoto, S.; Morizawa, Y.; Oshima, K.; Nozaki, H. *J. Am. Chem. Soc.* **1985**, *105*, 4491.

(103) Hibino, J-I.; Nakatsukasa, S.; Fugami, K.; Matsubara, S.; Oshima, K.; Nozaki, H. *J. Am. Chem. Soc.* **1985,** *107,* 6416.

(104) Eades, R. A.; Dixon, D. A. *J. Chem. Phys.* **1980,** *72,* 3309; Sheldon, J. C.; Bowie, J. H.; DePuy, C. H.; Damrauer, R. *J. Am. Chem. Soc.* **1986,** *108,* 6794; Damewood, Jr., J. R.; Hadad, C. M. *J. Phys. Chem.* **1988,** *92,* 33; Ortiz, J. V. *ibid.* **1987,** *109,* 5072.

(105) Hopkinson, A. C., Lien, M. H. *J. Org. Chem.* **1981,** *46,* 998; *Tetrahedron* **1981,** *37,* 1105.

(106) Magnusson, E. *Tetrahedron* **1985,** *41,* 2945.

(107a) Hopkinson, A. C.; Lien, M. H. *J. Mol. Struct.* **1983,** *104,* 303.

(107b) Hopkinson, A. C.; Lien, M. H. *J. Organomet. Chem.* **1981,** *206,* 287.

(108a) von Schleyer, P. R.; Clark, T. *J. Chem. Soc., Chem. Commun.* **1986,** 1371.

(108b) von Schleyer, P. R.; Reed, A. E. *J. Am. Chem. Soc.* **1988,** *110,* 4453.

(108c) Rajca, A.; Wang, P.; Streitwieser, A.; von Schleyer, P. R. *Inorg. Chem.* **1989,** *28,* 3064.

(108d) Luke, B. T.; Pople, J. A.; Jespersen, M. B. K.; Apeloig, Y.; Chandrasekhar, J.; von Schleyer, P. R. *J. Am. Chem. Soc.* **1986,** *108,* 260.

(109) Clark, T.; von Schleyer, P. R. *J. Organomet. Chem.* **1980,** *191,* 347.

(110) Reiter, B.; Hassler, K. *J. Organomet. Chem.* **1994,** *467,* 21.

(111) Gilman, H.; Shiina, K.; Aoki, D.; Gaj, B. J.; Wittenberg, D.; Brennan, T. *J. Organomet. Chem.* **1968,** *13,* 323.

(112) Ilsley, W. H.; Sadurski, E. A.; Schaaf, T. F.; Albright, M. J.; Anderson, T. J.; Glick, M. D.; Oliver, J. P. *J. Organomet. Chem.* **1980,** *190,* 257.

(113) Steward, O. W.; Heider, G. L.; Johnson, J. S. *J. Organomet. Chem.* **1979,** *168,* 33.

(114) Hengge, E.; Mitter, F. K. *Monatsh. Chem.* **1986,** *117,* 721.

Not for Synthesis Only: The Reactions of Organic Halides with Metal Surfaces

JOHN S. THAYER

Department of Chemistry
University of Cincinnati
Cincinnati, Ohio

I

SOME BASIC CONSIDERATIONS

A. Introduction

The reaction of alkyl halides with metals has played a crucial role in the development of organometallic chemistry. Sir Edward Frankland's synthesis of dialkylzinc compounds (*1,2*) and subsequent use of those compounds as reagents established the foundations of organometal synthesis (*3*). Grignard's investigations into the formation of ethereal solutions of organomagnesium compounds from alkyl halides and magnesium metal so greatly expanded and diversified synthetic uses for alkylmetals (*3–5*) that the Grignard reagent has probably become the single most widely used reagent in organic chemistry. Industrial development of the Direct

59

Process Reaction and its application to the preparation of silicone precursors (6–9), plus a corresponding industrial preparative method for tetraalkyllead compounds (10), spurred further investigations into the use of direct reactions between solid metals and liquid alkyl halides to synthesize alkylmetal derivatives, generating an extensive literature on this subject (3,4,11).

Studies on mechanistic aspects of such reactions have been relatively sparse. This article will explore recent work on both mechanistic and practical aspects of the direct reaction between metals and halides.

B. Selected Physical Properties

One characteristic property that all organic halides possess is a carbon–halogen bond dipole moment. The halogen atom is always the negative pole of this bond, and determines the extent of the polarity, with some modification by the organic group. Values for these bond dipole moments (12) and the molecular dipole moments for the two simplest alkyl halide series (13,14) appear in Table I. Interestingly, the carbon–halogen polarity *increases* from fluorine to chlorine! Organic fluorides have been little studied in respect to their reactivity towards metals and are not considered in this article.

Another characteristic property of organic halides is the bond dissociation energy. Table II lists bond energy values for carbon–halogen linkages

TABLE I

SOME SELECTED DIPOLE MOMENTS

Compound	Dipole moment (D)	Reference
Bond dipole moments		
C–F	1.41	(12)
C–Cl	1.46	(12)
C–Br	1.38	(12)
C–I	1.19	(12)
Molecular dipole moments		
CH_3F	1.81	(13)
CH_3Cl	1.86	(13)
CH_3Br	1.78	(13)
CH_3I	1.59	(13)
C_2H_5F	1.96	(14)
C_2H_5Cl	2.00–2.05	(14)
C_2H_5Br	1.99–2.02	(14)
C_2H_5I	1.87–1.95	(14)

and for bonds between carbon and elements of the neighboring groups (15–19). It is worth noting that the carbon–halogen bond energies tend to be slightly higher than those for the neighboring elements of the same period. However, their markedly higher bond polarity renders them much more reactive. Organic halides are used as starting materials in many organic syntheses.

For metals, the nature of the active metal surface determines its reactivity, as do both surface cleanness and metal particle size (4,9–11). Finely divided metals, with their correspondingly larger surface areas, show markedly greater reactivity towards alkyl halides than do corresponding bulk metals (20). Sono-chemical treatment of metal surfaces removes impurities and renders such surfaces correspondingly more reactive (21). Various other factors about metal surfaces that affect their adsorption of organic molecules are discussed in a book by Albert and Yates (22).

C. Selected Chemical Aspects

Perhaps the best-known reaction of alkyl halides is oxidative addition to another atom. For elements (usually, nonmetals or metalloids) having

TABLE II

SELECTED ELEMENT–CARBON BOND ENERGIES

Bond	Energy (kJ/mol)	Reference
P–C	264	(15)
	276	(16)
	260	(17)
AS–C	229	(16)
Sb–C	214	(16)
S–C	291	(18)
	270	(17)
Se–C	250	(18)
Te–C	(220)	a
Cl–C	327	(15)
	335	(17)
	347	(19)
Br–C	285	(15)
	205	(17)
	294	(19)
I–C	213	(15)
	226	(17)
	243	(19)

a Estimated by author.

a nonbonding pair of electrons in the valence shell, the reaction yields an onium cation:

$$R-X + EL_n \rightarrow REL_n^+ + X^- \tag{1}$$

This reaction also occurs for those derivatives of representative and transition metals in which the metal atom can be oxidized, and was recently used to prepare the first methylnickel(IV) compounds (23).

The halogen atoms in organic halides contain electron pairs in their valence shells and behave as Lewis bases. In this respect, organic halides may be considered as analogous to alkyl derivatives of the corresponding elements of the two preceding periodic groups. However, although hundreds of complexes between triorganophosphines, -arsines, or -stibines and transition metals have been reported, as have a respectable number of corresponding complexes for diorganosulfides, -selenides, and -tellurides, only a very few analogous complexes between alkyl halides and transition metals have been isolated as well as characterized (24,25). One such complex is shown in Fig. 1. This complex contains an iridium atom in the oxidation state of +3, which state is very resistant to further oxidation; such resistance may account for the complex's reported stability (24). Metal–alkyl halide complexes may form as intermediates in Friedel-Crafts reactions (26,27):

$$R-X + AlX_3 \rightarrow [R-X-AlX_3] \rightarrow R^+ AlX_4^- \tag{2}$$

The lone pairs on the halogen atoms occur in p-orbitals, and readily interact with acceptor orbitals on metals. By the aforementioned analogy to Groups VA and VIA elements, one might expect to encounter electron back-donation from the metal onto the halogen atom (24,25). In methyl iodide (1), the LUMO is a σ^* orbital localized primarily on iodine and is

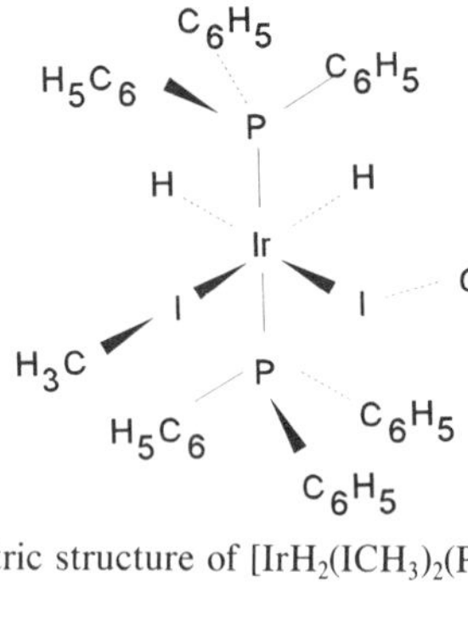

FIG. 1. Geometric structure of $[IrH_2(ICH_3)_2(P\{C_6H_5\}_3)_2]^+$ cation (24).

a more likely candidate for electron acceptance than the iodine d-orbitals (24). Comparative studies indicated that halocarbon binding was enhanced by the electrophilic character of the metal atom, suggesting that back-donation was a minor contributor. Any back-bonding, of course, weakens the carbon–halogen linkage and favors dissociation (25).

II

HALIDE ADSORPTION ON METAL SURFACES

A. Introduction

Numerous workers have investigated adsorption of alkyl halides onto metal surfaces (28), primarily in order to investigate the ease of carbon–hydrogen bond cleavage. Such studies are usually carried out under conditions of low temperature (<150 K) and high vacuum, using single crystals as substrates (28). Adsorption is molecular, and initial attachment occurs through the halogen atom (28) (Fig. 2). An adsorbed halide might simply be desorbed (released unchanged) or undergo dissociation of the carbon–halogen bond. The tendency to dissociate varies in the following order (28): $I \gg Br > Cl$. As a result, alkyl iodides have been used in the majority of reported studies. While in principle any metal might be used as substrate, elements of Groups 10 and 11 (Ni, Pd, Pt, Cu, Ag, Au) have been the metals most frequently investigated, possibly because of their importance in catalysis.

B. Carbon–Halogen Bond Scission

When methyl halide monolayers were generated on a Ni(100) surface at 100 K, CH_3Cl (2) desorbed unchanged, while CH_3Br (3) was partly

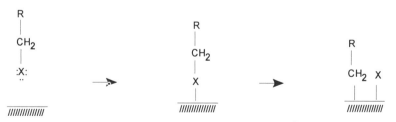

FIG. 2. Bonding of alkyl halides to metal surfaces.

dissociated, and (1) was completely dissociated (29). Molecule (2) decomposed only on Fe or W but not on Ag, Cu, Pt, Pd or Au, while (3) dissociated on Ni (but not on Ag or Pt) and (1) dissociated on all metals tested (28). Carbon–halogen bonds can be broken by thermolysis or photolysis; use of the latter technique has been reviewed (30). A thermal activation energy of approximately 14 kJ/mol for the C–I bond on Ni(100) surfaces has been reported (31). When (1) adsorbed onto a Pd(111) surface, cleavage of the carbon–iodine linkage occurred between 175 and 200 K; on a corresponding Pd(100) surface, scission occurred at 95 K (32). Methyl groups desorbed as methane; no C_2 products formed (32). Pretreatment of the surface with carbon monoxide enhanced the thermal stability of adsorbed methyl groups (32). Photodissociation studies of (3) adsorbed upon a Cu(111) surface (33) or a Cu(111) pretreated with bromine (34) gave results compatible with a mechanism involving photoexcitation of an electron from the metal d-band into the carbon–bromine σ^*-orbital. Comparative photochemical studies on Ag(111) surfaces showed that initial photolysis rates for the methyl halides increased going from Cl to I (35–37). Adsorbed halogen atoms remain strongly attached to the metal surface and require much higher temperatures for desorption.

C. Fragmentation of Adsorbed Alkyl Groups

When the carbon–halogen bond undergoes scission, the resulting alkyl radical adsorbs onto the metal surface (28). Both (3) (33,38) and (1) (39–42) generated surface-bound methyl groups. Ethyl chloride (4) (35) and ethyl iodide (5) (43, 44) formed adsorbed ethyl groups, while both 1- and 2-propyl moieties formed on Ni(100) surfaces from the corresponding propyl iodides (45). At low surface coverage, (5) on Pt(111) formed ethyl groups that were flat against the surface; increasing coverage converted them to an upright position (44). When diiodomethane was adsorbed onto the Al(111) surface at 108 K, both iodine atoms were removed; no evidence for adsorbed ICH_2 groups was reported (46).

Adsorbed alkyl groups usually undergo elimination of an α-hydrogen to give adsorbed hydrogen and methylane (cf. Fig. 3a). Ethyl and longer alkyl groups can also undergo elimination of a β-hydrogen atom to form an alkenyl group (Fig. 3b). Comparative studies indicate that beta-hydrogen elimination occurs at least 10^6 times faster than alpha-hydrogen elimination (47). Desorption usually yielded a mixture of hydrocarbons whose composition varied with temperature.

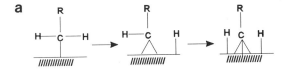

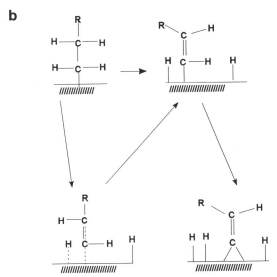

FIG. 3. Hydrogen elimination from alkyl groups adsorbed on metal surfaces: (a) α-hydrogen; (b) β-hydrogen.

III

MECHANISTIC INVESTIGATIONS OF APPLIED REACTIONS

A. *Introduction*

Detailed investigations into the mechanism of reaction between alkyl halides and metals have concentrated primarily on systems of great practical interest: the Grignard reaction, the Direct Process Reaction, and a few others. Whereas research reported in the preceding section involved solely pure gaseous halides and pure metals as single crystals, mechanistic investigations have involved a much wider variety of substrates and reaction conditions. When a solvent is involved, solvation of the metal product provides considerable driving force for the reaction. If the metal substrate is not uniform in composition (i.e., it is an alloy or intermetallic species), electron distribution will also be nonuniform, creating a charge polarity. This polarity will affect interaction with the polar carbon–halogen bond, as shown in Fig. 4; the alkyl group bonds to the atom with the greater electron density. If one metal of a two-metal alloy is an alkali metal, the polarity becomes large enough to make the system quasi-ionic, making the reaction proceed more rapidly and more completely. Interest in semiconductors involving such materials as gallium antimonide and cadmium telluride has also led to studies on the interaction of these materials with alkyl halides.

B. Grignard Reaction Mechanism Studies

In their early monograph on the Grignard reaction, Kharasch and Reinmuth noted the effects that varying the form of magnesium metal would have on reagent preparation [(4), pp. 6–15]; they also noted that the addition of copper to magnesium lowered the yield of reagent. In a detailed mechanistic study, Rogers et al. reported that reaction rates for ethereal organic halides reacting with magnesium depended both on the concentration of halide and on the magnesium surface area (48). They also found that, for alkyl iodides and most alkyl bromides, the rate of reaction was limited by transport to the magnesium surface; this was not true for the chlorides (48). Initiation of the magnesium–halide reaction proceeded through formation of corrosion pits, which would expand and overlap, so that much of the reaction occurred on a shiny surface (49). While reaction rates were not much affected by metal lattice plane or grain boundaries, initiation of reaction occurred more readily at strained dislocations (49). Reactivity increased when small quantities of lithium were alloyed with magnesium and decreased in the presence of more electronegative metals (49). Investigations using EPR spectroscopy indicated that the carbon–halogen bond energy influenced the mechanism of Mg–C bond formation and that radicals and/or ion-radicals were the most likely intermediates (50).

FIG. 4.   Effect of metal polarity (M_+ is the more electropositive metal).

High-vacuum studies on the adsorption of methyl bromide on Mg(0001) surfaces at 123 K indicated that reaction occurred readily but that no stable magnesium–carbon bonds formed (51); the methyl group desorbed as a hydrocarbon mixture, with ethane as the primary component. An activation energy of 25–33 kJ/mol was estimated (51). The presence of coadsorbed dimethyl ether did not affect the reaction. When magnesium metal and methyl halides were codeposited in an argon matrix, they were found by infrared spectroscopy to give cleavage of the carbon–halogen bond, but nothing more could be said about the nature of the product (52). Similar matrix isolation studies using methyl bromide and clusters of magnesium atoms gave evidence for the formation of $CH_3(Mg)_xBr$ species (53).

There has been some controversy over the nature of intermediates in the Grignard reaction (54–56). Available evidence indicates that organic radicals form through cleavage of the carbon–halogen linkage; everyone agrees on this. Controversy centers on the question of whether the alkyl radical remains adsorbed on the metal surface ("A model") or whether it diffuses into solution ("D model"). Evidence appears to be more consistent with the "D model" (56), but the question cannot be considered as settled.

C. The Direct Process Reaction

As previously mentioned, the Direct Process Reaction involves the reaction of (2) with silicon at elevated temperatures to form methylchlorosilanes (6–9):

$$Si + 2CH_3Cl \xrightarrow[\Delta]{Cu} (CH_3)_2SiCl_2 \qquad (3)$$

This equation is a major oversimplification; a mixture of many silicon-containing products actually forms. Their proportions may be controlled by adjusting reaction conditions. In fact, research has concentrated on determining those conditions that favor formation of one specific product. Pure silicon is not used; instead, copper must be added as a catalyst. The addition of trace metals such as zinc, aluminum, and/or tin tends to increase reaction rates and to favor the formation of dimethyldichlorosilane over other products (57). Two recent patents pertaining to the preparation of the silicon–copper catalyst have been issued ($58,59$).

A critical review of proposed mechanisms for the direct process has recently appeared (60). The authors proposed that an atom of silicon bonds to copper and behaves as a silylene (i.e., has electron density exceeding that of an ordinary silicon atom). This silyene subsequently reacts with methyl chloride to undergo oxidative addition, forming adsorbed methylsilicon(II) chloride (Fig. 5a), which then reacts further to yield various products. Other reported reactions of silicon are also consistent with this mechanism (61). By contrast (but consistent with other single-crystal work previously cited), adsorption of (1) onto a Si(100) surface resulted in the finding of the methyl group and the iodine atom to separate silicon atoms (62) [Fig. 5b]. A study of Si(100) crystals and various forms of copper catalysts indicated that Cu_3Si (6) formed in most cases, but that the reactivity and selectivity for dimethyldichlorosilane varied quite markedly (63); the most reactive surfaces had a randomly oriented (6) phase.

D. Other Applied Syntheses

The reaction of (4) with a sodium–lead alloy has long been used for the industrial preparation of tetraethyllead (7) (10,64). Traditionally the reaction has been written:

$$4NaPb + 4C_2H_5Cl \xrightarrow[\Delta]{Cu} (C_2H_5)_4Pb + 4NaCl + 3Pb \qquad (4)$$

The sodium–lead alloy has to be specifically prepared (65), and quite a variety of reaction conditions have been reported (10). The surface needs to be oxygen free. Partial replacement of sodium by potassium enhances yields, as does treatment with iodine. The system's temperature and pressure of (4) also are crucial factors in determining reaction rate. The following mechanism was proposed (10):

$$C_2H_5Cl_{(gas)} \rightarrow C_2H_5Cl_{(ads)} \qquad (5a)$$

$$C_2H_5Cl_{(ads)} + NaPb \rightarrow NaPb\cdot C_2H_5Cl \qquad (5b)$$

$$2NaPb\cdot C_2H_5Cl \rightarrow (C_2H_5)_2Pb + 2NaCl + Pb \qquad (5c)$$

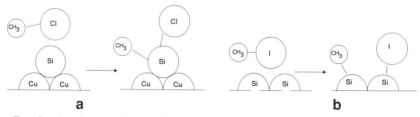

FIG. 5. Attachment of methyl halides to silicon atoms: (a) copper silicide; (b) silicon single crystal.

The species "$NaPb\cdot C_2H_5Cl$" is probably "$Na\cdot C_2H_5PbCl$." Under the conditions of the reaction, adsorbed diethylplumbylene reacts to form (7) or triethyllead species. Reaction occurs slowly or not at all if the sodium content of the alloy becomes too high:

$$NaPb > Na_9Pb_4 > Na_5Pb_2 > Na_4Pb \qquad (6)$$

Halides containing longer-chain alkyl groups reacted more slowly. Unalloyed lead metal reacted with alkyl halides, especially if it was oxide free and finely divided (10). Binary alloys of lead with other active metals have also been tried as substrates, with varying success (10).

The use of a sodium–tin alloy for the preparation of tetramethyltin preceded use of the corresponding sodium–lead alloy (64,66), but was subsequently displaced by other methods. Currently, the Direct Reaction is used to prepare dialkyltin dihalides, and generally requires an oxygen-containing solvent plus another metal as "accelerator" (66):

$$Sn + 2C_4H_9I \xrightarrow[n\text{-}C_4H_9OH]{Mg} (C_4H_9)_2SnI_2 \qquad (7)$$

The Direct Reaction between aluminum and (2) is used to prepare trimethylaluminum (67):

$$2Al + 3CH_3Cl \rightarrow (CH_3)_2AlCl + CH_3AlCl_2 \qquad (8a)$$

$$(CH_3)_2AlCl + CH_3AlCl_2 + 3Na \rightarrow (CH_3)_3Al + Al + 3NaCl \qquad (8b)$$

The growing use of trialkylgallium and -indium compounds in the synthesis of semiconductors has led to the issuance of patented methods for their preparation from metal–magnesium alloys and alkyl halides (68–71). Other such preparations are likely to appear as the need for alkylmetal compounds increases.

IV

NONSYNTHETIC APPLICATIONS

A. Introduction

Investigations into nonsynthetic aspects of the Direct Reaction have usually focused on their possible uses to extract metals from solids. The substrate may be a pure metal, a metal alloy, or a compound between a metal and a semimetal. In these applications, the yield may or may not be important. It is not even necessary that the metal–carbon bond be sufficiently stable to isolate the alkylmetals formed; even if the initially formed species exist for only a second or less, that may still be sufficient for the purposes of the process. The crucial aspect is that the metal atom reacts with the alkyl halide, and in so reacting becomes detached from the solid substrate. Metal removal through alkylmetal formation may be deliberate (as in etching or extraction) or unintended (as in corrosion). Such processes have been investigated under both nonaqueous and aqueous conditions.

B. Nonaqueous Systems

Growing interest in the use of III/V compounds, e.g., GaAs (8), as semiconducting materials has encouraged extensive investigations into processes for etching these materials. Organic halides have been used, primarily as a halogen source, in photochemical etching of indium phosphide (9) (72,73) and (8) (74,75). Rates of etching were measured for (8) and AlGaAs, using (1) [CH_3I] vapor in a carrier gas at various temperatures (76). Similar studies using CF_3Cl on these materials showed that etching rates decreased as the mole fraction of Al increased (74). Treatment of the (111) faces of (9) with unsubstituted benzyl bromides gave substitution on exposed phosphorus atoms of the P-rich (111)B face, but not on indium atoms of the In-rich (111)A face (77). To date, all reports on reactions between III/V materials and alkyl halides indicate that only the Group V atom actually forms a bond to a carbon atom of the alkyl moiety.

Corrosion studies have been rare. (8), copper, or iron were corroded by carbon tetrachloride when exposed to Co-60 radiation (78). Alkyl halides enhanced the corrosive effect of benzoic acid on iron (79). (1) was found to promote stress-corrosion cracking in zirconium alloys used in nuclear reactors (80).

C. Aqueous Systems

1. General Considerations

Addition of water to a metal–organic halide system alters the nature of the reaction medium considerably, and can cause various side reactions:

a. *Reaction with metal*—Water can react with many metals, especially if these are present as finely divided, highly reactive powders. If the water is acidic, such reactivity increases. Dissolution of metals in an acidic medium proceeds considerably faster than the corresponding dissolution by alkyl halides.

b. *Hydrolysis of the alkyl halide*—This side reaction is especially important for secondary and tertiary alkyl halides, which react with water more rapidly than their primary counterparts. In addition to destroying the halide reactant, hydrolysis generates strongly acidic hydrogen halides, resulting in the complications noted in part (a).

c. *Reaction with the metal product*—Water can react with the very polar metal–halide and metal–alkyl bonds. Both bonds are broken, generating metal–hydroxide linkages. For many metals, most notably iron, these hydroxy products show very low solubility in water, even acidic water, and will coat the surface of the metal, hampering and complicating further reaction.

The solubility of most alkyl halides in water is very low, unless some solubilizing group happens to be present (e.g., –OH, –CO$_2$H). Hence, most reaction systems have involved three separate phases (water, halide, metal substrate), with all the accompanying complications.

Interest concerning the Direct Reaction in aqueous systems has arisen from association with two burgeoning areas of research: environmental formation of alkyl (mostly methyl) derivatives of metals, and corrosion of metal surfaces exposed to water. These will be considered separately.

2. *Environmental Methylmetal Formation*

Interest and concern about environmental occurrence of methylmetal derivatives grew out of the extensive poisoning by methylmercuric derivatives in Minamata Bay (Japan) and elsewhere (*81,82*). Methylation can occur either through biological intermediation ("biological methylation") or by abiotic methyl exchange. The great majority of substrates in both types of methylation are metal ions or metal compounds. However, metallic tin and lead are known to react with (**1**) to form methylmetal compounds (*83,84*), and other metals or metalloids probably behave similarly. Methyl halides, along with other organohalogen compounds, form through biogenic processes and occur in nature in substantial quantities (*85–87*). These compounds may well play a role in the environmental cycles of metals (*88*).

3. *Metal Leaching and Corrosion*

In 1977, Mor and Beccaria investigated the roles of alkyl bromides and iodides on the corrosion of copper in seawater (*89*). They found that alkyl bromides enhanced the corrosion of copper, whereas alkyl iodides inhibited it, perhaps by lowering the pH of the system (*89*). A wide variety of binary metal species released metals into aqueous solution when treated with alkyl halides (90–92). Many of these are shown in Table III. Treatment of (8) with aqueous alkyl halides gave dissolution of arsenic as methylarsonic or cacodylic acids, but no dissolution of gallium (93).

All reported work on binary systems containing a metal and a second element indicate that the more electronegative element, which presumably has a higher electron density, forms a bond to a carbon atom in the alkyl group, while the halogen atom bonds to the other atom or is reduced to a halide ion. If the second element is a chalcogen, volatile dialkylchalconides and/or water-soluble trialkylchalconium ions form as products:

$$FeSe + 2RX \rightarrow Fe^{2+} + 2X^- + R_2Se \qquad (9a)$$

$$R_2Se + RX \rightarrow R_3Se^+ + X^- \qquad (9b)$$

For arsenic, the water-soluble methylarsonic and cacodylic acids are apparently the major products, and presumably similar products are formed also for phosphorus and antimony (93).

Various metals, when treated with aqueous (1) or other alkyl iodides under flow conditions, showed marked dissolution into water (91,94–96). Dissolution also occurred under static conditions (92,95,97). Initially, these studies used metals more active than hydrogen (Fe, Cr, Ni, Zn, Sn, Pb, Mn); however, Cu was also found to undergo similar dissolution. Preliminary investigations indicated that metallic silver readily formed silver bromide in the presence of alkyl bromides and that palladium dis-

TABLE III

METAL SPECIES REACTED WITH ALKYL HALIDES[a]

Compounds		Ores
AsS_2	Cu_5Si	Arsenopyrite (FeAsS)
FeS	FeSi	Chalcopyrite ($CuFeS_2$)
	GaAs	Enargite (Cu_3AsS_4)
HgS	GaSb	Galena (PbS)
MnS	InAs	Niccolite (NiAs)
MoS_2	InSb	Pyrrite (FeS_2)
PbS	Ni_2P	Pyrrhotite (FeS)
SnS	NiAs	Stibnite (Sb_2S_3)
FeSe		
FeTe		
PbSe		
PbTe		

[a] Taken from Thayer et al. (90).

solved to some extent when treated with certain aqueous alkyl bromides (98). We proposed the following mechanism for reaction between a metal and (1) (90):

$$M_{(s)} + CH_3I_{(aq)} \rightarrow CH_3I{\cdot}M_{(s)} \tag{10a}$$

$$CH_3I{\cdot}M_{(s)} \rightarrow CH_3I{\cdot}M_{(aq)} \tag{10b}$$

$$CH_3I{\cdot}M_{(aq)} + H_2O \rightarrow M^{2+} + 2I^- + CH_4 \tag{10c}$$

When copper or iron strips were placed in water after long exposure to benzyl bromide vapors, they showed much greater dissolution than unexposed strips (Table IV) (97). Replacement of water by deuterium oxide as the reaction medium yielded CH_3D as the primary gaseous products (97). While this would arise from the hydrolysis of a metal–methyl bond:

$$CH_3M + D_2O \rightarrow CH_3D + MOD \tag{11}$$

there was in this case insufficient evidence to indicate whether the methylmetal species was still present on the substrate surface or had already diffused into solution.

The initial step (eq. 10a) in the proposed mechanism requires adsorption/ binding of the alkyl halide to the metal surface—a process already previously discussed. Anything that might enhance this adsorption should affect the rate and extent of reaction. We have found that ω-bromo-1-alkenes enhanced copper dissolution from alloys (Table V). Comparative studies on brass foil confirmed these observations and also showed that the dissolution of zinc was affected only slightly (Table VI). The compound

TABLE IV
METAL–BENZYL BROMIDE VAPOR REACTION[a]

Elapsed time (hours)	Copper		Nickel		Molybdenum	
	Control	Reaction	Control	Reaction	Control	Reaction
2	0.10	1.42	0.11	0.50	1.44	1.74
24	0.40	2.42	0.20	1.13	4.00	6.00
48	0.44	2.44	0.22	1.50	6.00	15.20

[a] Strips of Cu, Mo, or Ni were placed in a sealed dessicator with an open container of benzyl bromide, and exposed to the vapor for eight days. Corresponding "control" strips were kept in a dessicator for the same period. All strips were then withdrawn and placed into 50-mL portions of distilled water. Aliquots were withdrawn to test for dissolved metal levels. All dissolved metal concentrations have units of mg/L. Molybdenum concentrations were determined as molybdate ion.

TABLE V

DISSOLUTION OF COPPER FROM BRONZE POWDER[a]

Alkyl bromide	[Cu], mg/L
None (control)	0.06
$CH_3CH_2(CH_2)_2Br$	0.4
$CH_2=CH(CH_2)_2Br$	2.4
$CH_3CH_2(CH_2)_3Br$	0.2
$CH_2=CH(CH_2)_3Br$	0.37
$CH_3CH_2(CH_2)_4Br$	0.01
$CH_2=CH(CH_2)_4Br$	0.03

[a] 46.97 ± 0.20 mg bronze powder (90% Cu; 10% Sn; particle size 45–53 μm) was treated with 25 mL water and 0.20 mL RBr for five days.

TABLE VI

UNIT DISSOLUTION RATES OF ZINC AND COPPER FROM BRASS FOIL IN AERATED WATER[a]

Metal	Alkyl halide	Rate (μg/S·m^2)
Zinc	None	7.00
	$CH_3CH_2(CH_2)_2Br$	7.20
	$CH_2=CH(CH_2)_2Br$	11.00
	$CH_3CH_2(CH_2)_3Br$	4.70
	$CH_2=CH(CH_2)_3Br$	7.20
Copper	None	1.00
	$CH_3CH_2(CH_2)_2Br$	2.10
	$CH_2=CH(CH_2)_2Br$	10.00
	$CH_3CH_2(CH_2)_3Br$	1.40
	$CH_2=CH(CH_2)_3Br$	6.80

[a] Data taken from Thayer (94).

7-bromoheptanenitrile gave similar results (95). We attributed this effect to the ability of copper to form a π-complex with the carbon–carbon double bond, thereby favoring binding of the reactant halide to copper and facilitating subsequent reaction (95).

V

CONCLUSIONS

Every example of the Direct Reaction, whether used for synthesis or not, involves removal of metal atoms from a solid matrix through formation of metal–carbon bonds. Most investigations have centered on isolation and/or stabilization of the alkylmetal compounds for subsequent synthetic usage. However, there is a growing number of applications that utilize the oxidizing nature of the Direct Reaction and do not require isolation of the alkylmetal product. Steady expansion of interest in the properties of semiconducting materials suggests that the use of alkyl halides as etching chemicals will likewise expand. The selective alkylation of the phosphorus atoms of (9) (77) raises the possibility that alkyl halides with a fluorescent- or otherwise-labeled alkyl group might be used to mark atoms selectively on a solid surface.

Apart from the Grignard reaction and the industrial preparation of alkyl-leads, solutions of alkyl halides as reactants have received relatively little attention. A 6:1 solution of (1)/acetic acid could be used to reclaim rhodium from used catalyst waste (99). The possibility of using aqueous alkyl halides for leaching has been proposed elsewhere (94). Other binary or ternary mixtures containing alkyl halides may well find important applications not yet suspected.

While the Direct Reaction causes the oxidization of a metal and removes it from a solid, it also results in the destruction of carbon–halogen bonds. Since many anthropogenic organic chlorides or bromides are major pollutants, reaction with metals provides one potential route for their detoxification. At least one patent has been issued for this application (100).

The Direct Reaction has been known for almost 150 years. Over that time its scope and usage have expanded enormously. While research on this reaction has been quite extensive, investigations have tended to focus on certain subareas, with little or no interaction among them. When this changes, understanding of details about a very diversified reaction should expand accordingly, and new combinations of metals and organic halides should suggest themselves for investigation. Thus the Direct Reaction, one of the oldest reactions in Chemistry, remains one of the most multifaceted and dynamic, showing every sign of continuing and developing for another 150 years.

REFERENCES

(1) Frankland, E. *J. Chem. Soc.* **1849,** *2,* 263.
(2) Frankland, E. *J. Chem. Soc.* **1849,** *2,* 294.
(3) Thayer, J. S. *Adv. Organomet. Chem.* **1975,** *13,* 1.
(4) Kharasch, M. S.; Reinmuth, O. "Grignard Reactions of Nonmetallic Substances"; Prentice-Hall: New York, 1954.
(5) Thayer, J. S. *J. Chem. Educ.* **1969,** *46,* 764.
(6) Rochow, E. G. "An Introduction to the Chemistry of the Silicones"; Wiley: New York, 1946.

(7) Rochow, E. G. *Adv. Organomet. Chem.* **1970,** *9,* 1.

(8) Liebhafsky, H. A. "Silicones Under the Monogram"; Wiley: New York, 1978.

(9) Rochow, E. G. "Silicon and Silicones"; Springer-Verlag: Berlin, 1987.

(10) Shapiro, H.; Frey, F. W. "The Organic Compounds of Lead"; Wiley: New York, 1968; pp. 34–44.

(11) Rochow, E. G. *J. Chem. Educ.* **1966,** *43,* 58.

(12) Wagniere, G. H. In "The Chemistry of the Carbon-Halogen Bond"; Patai, S., Ed.; Wiley: London, 1973; p. 21.

(13) Smith, R. P.; Ree, T.; Magee, J. L.; Eyring, H. *J. Am. Chem. Soc.* **1951,** *73,* 3932.

(14) Smith, R. P.; Mortensen, E. M. *J. Am. Chem. Soc.* **1956,** *78,* 2263.

(15) Huheey, J. A.; Keiter, E. A.; Keiter, R. L. "Inorganic Chemistry"; 4th ed.; Harper Collins: New York, 1993; p. A-30.

(16) Elschenbroich, C.; Salzer, A. "Organometallics: A Concise Introduction"; 2nd ed.; VCH Publishers: Weinheim, 1992; p. 11.

(17) Douglas, B.; McDaniel, D.; Alexander, J. "Concepts and Models of Inorganic Chemistry"; 3d ed.; Wiley: New York, 1994; p. 89.

(18) Batt, L. In "The Chemistry of Organic Selenium and Tellurium Compounds"; Patai, S.; Rappoport, Z., Eds., Wiley: Chichester, 1986; Vol. 1, p. 159.

(19) Chen, E. C. M.; Albyn, K.; Dussack, L.; Wentworth, W. E. *J. Phys. Chem.* **1988,** *93,* 6827.

(20) Rieke, R. D. *Science* **1989,** *246,* 1260.

(21) Suslick, K. S. *Adv. Organomet. Chem.* **1986,** *25,* 73.

(22) Albert, M. R.; Yates, J. T. "The Surface Scientist's Guide to Organometallic Chemistry"; American Chemical Society: Washington, DC, 1987.

(23) Klein, H.; Bickelhaupt, A.; Jung, T.; Cordier, G. *Organometallics* **1994,** *13,* 2557.

(24) Burk, M. J.; Segmuller, B.; Crabtree, R. H. *Organometallics* **1987,** *6,* 2241.

(25) Burk, M. J.; Crabtree, R. H.; Holt, E. M. *J. Organomet. Chem.* **1988,** *341,* 495.

(26) Olah, G. A. "Friedel-Crafts Chemistry"; Wiley: New York, 1973.

(27) Roberts, R. M.; Khalaf, A. A. "Friedel-Crafts Alkylation Chemistry"; Dekker: New York, 1984.

(28) Zaera, F. *Acct. Chem. Res.* **1992,** *25,* 260.

(29) Zhou, X. L.; Whike, J. M. *Surf. Sci.* **1988,** *194,* 438.

(30) Zhou, X. L.; Zhu, X. Y.; White, J. M. *Acc. Chem. Res.* **1990,** *23,* 327.

(31) Tjandra, S.; Zaera, F. *J. Vac. Sci. Technol., A* **1992,** *10,* 404.

(32) Chen, J. J.; Winograd, N. *Surf. Sci.* **1994,** *314,* 288.

(33) Lamont, C. L. A.; Conrad, H.; Bradshaw, A. M. *Surf. Sci.* **1993,** *280,* 79.

(34) Lamont, C. L. A.; Conrad, H.; Bradshaw, A. M. *Surf. Sci.* **1993,** *287,* 169.

(35) Zhou, X. L.; White, J. M. *Surf. Sci.* **1991,** *241,* 244.

(36) Zhou, X. L.; White, J. M. *Surf. Sci.* **1991,** *241,* 259.

(37) Zhou, X. L.; White, J. M. *Surf. Sci.* **1991,** *241,* 270.

(38) Radhakrishnan, G.; Stenzel, W.; Hemmen, R.; Conrad, H.; Bradshaw, A. M. *J. Chem. Phys.* **1991,** *95,* 3930.

(39) Solymosi, F.; Revesz, K. *J. Am. Chem. Soc.* **1991,** 9145.

(40) Chiang, C.; Wentzlaff, T. H.; Bent, B. E. *J. Phys. Chem.* **1992,** *96,* 1836.

(41) Lin, J.; Bent, B. E. *J. Am. Chem. Soc.* **1992,** *115,* 2849.

(42) Chen, J. J.; Winograd, N. *Surf. Sci.* **1994,** *314,* 188.

(43) Jenks, C. J.; Lin, J. L.; Chiang, C. M.; Kang, L.; Leang, P. S.; Wentzlaff, T. H.; Bent, B. E. *Stud. Surf. Sci. Catal.* **1991,** *67,* 301.

(44) Hofmann, H.; Griffiths, P. R.; Zaera, F. *Surf. Sci.* **1992,** *262,* 141.

(45) Tjandral, S.; Zaera, F. *Langmuir* **1994,** *10,* 2640.

(46) Hara, M.; Domen, K. *J. Chem. Soc., Chem. Commun.* **1990,** 1717.

(47) Jenks, C. J.; Chiang, M.; Bent, B. E. *J. Am. Chem. Soc.* **1991**, *113*, 6308.

(48) Rogers, H. R.; Hill, C. L.; Fujiwara, Y.; Rogers, R. J.; Mitchell, H. L.; Whitesides, G. M. *J. Am. Chem. Soc.* **1980**, *102*, 217.

(49) Hill, C. L.; Vander Sande, J. B.; Whitesides, G. M. *J. Org. Chem.* **1980**, *45*, 1020.

(50) Sergeev, G. B.; Zagorsky, V. V.; Badaev, F. Z. *J. Organomet. Chem.* **1983**, *243*, 123.

(51) Nuzzo, R. G.; Dubois, L. H. *J. Am. Chem. Soc.* **1986**, *108*, 2881.

(52) Ault, B. S. *J. Am. Chem. Soc.* **1980**, *102*, 3480.

(53) Imizu, Y.; Klabunde, K. J. *Inorg. Chem.* **1984**, *23*, 3602.

(54) Walborsky, H. M. *Acct. Chem. Res.* **1990**, *23*, 286.

(55) Garst, J. F. *Acct. Chem. Res.* **1991**, *24*, 95.

(56) Walling, C. *Acct. Chem. Res.* **1991**, *24*, 255.

(57) Gasper-Galvin, L. D.; Sevenich, D. M.; Friedrich, H. B.; Rethwisch, D. G. *J. Catal.* **1991**, *128*, 468.

(58) Mui, J. Y. P. *Chem. Abstr.* **1994**, *129*, 30932u.

(59) Streckl, W.; Straussberger, H.; Pachaly, B. *Chem. Abstr.* **1994**, *120*, 5469r.

(60) Lewis, K. M.; McLeod, D.; Kanner, B.; Falconer, J. L.; Frank, T. In "Studies in Organic Chemistry. 49. Catalyzed Direct Reactions of Silicon"; Lewis, K. M.; Rethwisch, D. S., Eds.; Elsevier: Amsterdam, 1993, pp. 333–440.

(61) Ono, Y.; Okano, M.; Watanabe, N.; Suzuki, E. In "Silicon for the Chemical Industry II"; Oye, H. A.; Rong, H. M.; Nygaard, L.; Schuessler, G.; Tuset, J. K., Eds.; Tapir Forlag: Trondheim, Norway, 1994; pp. 185–196.

(62) Gutleben, H.; Lucas, S. R.; Cheng, C. C.; Choyke, W. J.; Yates, J. T. *Surf. Sci.* **1991**, *257*, 146.

(63) Floquet, N.; Yilmaz, S.; Falconer, J. L. *J. Catal.* **1994**, *148*, 348.

(64) Nickerson, S. P. *J. Chem. Educ.* **1954**, *31*, 560.

(65) Shapiro, H. *Adv. Chem. Ser.* **1959**, *23*, 290–298.

(66) Omae, I. *J. Org. Chem., Libr.* **1989**, *21*, 35–48.

(67) Eisch, J. J. "Comprehensive Organomettallic Chemistry"; Wilkinson, G., Ed.; Pergamon: Oxford, 1982; Vol. 1, p. 560.

(68) Imori, T.; Kondo, K.; Ninomya, T.; Nakamura, K. *Chem. Abstr.* **1991**, *115*, 159431f.

(69) Imori, T.; Kondo, K.; Ninomya, T.; Nakamura, K. *Chem. Abstr.* **1991**, *115*, 159432g.

(70) Ninomiya, T.; Kondo, K.; Imori, T.; Nakamura, K. *Chem. Abstr.* **1991**, *115*, 280275m.

(71) Smit, C. J.; Van Hunnik, E. W. J.; Van Eijden, G. J. M. *Chem. Abstr.* **1993**, *119*, 8125m.

(72) Haigh, J.; Aylett, M. R. *Chem Abstr.* **1987**, *106*, 128994b.

(73) Durose, K.; Summersgill, J. P. L.; Aylett, M. R.; Haigh, J. *Chem. Abstr.* **1990**, *112*, 26318r.

(74) Seabaugh, A. *J. Vac. Sci. Technol., B* **1988**, *6*, 77.

(75) Buhaenko, D. S.; Francis, S. M.; Goulding, P. A.; Pemble, M. E. *Stud. Surf. Sci. Catal.* **1988**, *48*, 229.

(76) Krueger, C. W.; Wang, C. A.; Flytzani-Stephanopoulos, M. *Appl. Phys. Lett.* **1992**, *60*, 1459.

(77) Spool, A. M.; Daube, K. A.; Mallouk, T. E.; Belmont, J. A.; Wrighton, M. S. *J. Am. Chem. Soc.* **1986**, *108*, 3155.

(78) Proesch, U.; Sander, R. *Isotopenpraxis* **1990**, *26*, 469; *Chem. Abstr.* **1991**, *114*, 52708d.

(79) Yamamoto, T.; Imaizumi, K.; Kurata, Y. *Inorg. Chim. Acta* **1984**, *85*, 175.

(80) Cox, B.; Haddad, R. *J. Nucl. Mater.* **1986**, *137*, 115.

(81) Sigel, H.; Sigel, A. (Eds.). "Metal Ions in Biological Systems"; Dekker: New York, 1993; Vol. 29.

(82) Thayer, J. S. "Environmental Chemistry of the Hydrido and Organo Derivatives of the Heavy Elements"; VCH Publishers: New York, 1995.

(83) Anagnostopoulos, A.; Hadjispyrou, S. *Toxicol. Environ. Chem.* **1993**, *39*, 207.

(84) Craig, P. J.; Rapsomanikis, S. *Environ. Sci. Technol.* **1985,** *19,* 726.

(85) Harper, D. B. *Met. Ions Biol. Syst.* **1993,** *29,* 345.

(86) Harper, D. B. *Biochem. Soc. Trans.* **1994,** *22,* 1007.

(87) Gribble, G. W. *J. Chem. Educ.* **1994,** *71,* 907.

(88) Brinckman, F. E.; Olson, G. J., Thayer, J. S. In "Marine and Estuarine Geochemistry"; Sigleo, A. C.; Hattori, A., Eds.; Lewis: Chelsea, MI, 1985; pp. 227–228.

(89) Mor, E. D.; Beccaria, A. M. *Br. Corros. J.* **1977,** *12,* 243.

(90) Thayer, J. S.; Olson, G. J.; Brinckman, F. E. *Environ. Sci. Technol.* **1984,** *18,* 726.

(91) Thayer, J. S.; Olson, G. J.; Brinckman, F. E. *Appl. Organomet. Chem.* **1987,** *1,* 73.

(92) Thayer, J. S.; Brinckman, F. E. *Abstr. 4th Chem. Congr. North Am.* **1991,** Inor. 80.

(93) Craig, P. J.; Laurie, S. H.; McDonagh, R. *Appl. Organomet. Chem.* **1994,** *8, 183.*

(94) Thayer, J. S. *ChemTech.* **1990,** *20,* 188.

(95) Thayer, J. S. *Appl. Organomet. Chem.* **1993,** *7,* 525.

(96) Thayer, J. S.; Brinckman, F. E. *Appl. Organomet. Chem.* (submitted for publication).

(97) Thayer, J. S. *Abstr. 206th Natl. Am. Chem. Soc. Meet.,* **1993,** Inor. 428.

(98) Thayer, J. S. Unpublished observations.

(99) Gulliver, D. J. *Chem. Abstr.* **1989,** *111,* 43152d.

(100) Mandl, J. *Chem. Abstr.* **1990,** *112,* 11530e.

The Bonding of Metal–Alkynyl Complexes

JOSEPH MANNA, KEVIN D. JOHN, and
MICHAEL D. HOPKINS

Department of Chemistry, University of Pittsburgh
Pittsburgh, Pennsylvania

I

INTRODUCTION

The chemistry of metal–alkynyl compounds is developing at a rapid pace. At the time of the only extensive (though not comprehensive) review of this area, by Nast in 1982 (*1*), the published work on the syntheses, reactions, structures, and properties of complexes, clusters, and polymers containing terminal and bridging alkynyl ligands over the previous 30 years consisted of approximately 200 papers. Since then, nearly 500 papers (200 of which have appeared in this decade) have been published on terminal metal–alkynyl complexes alone. The increasing interest in these compounds has made the need for a new survey of their chemistry an urgent one, although the size of the field, and the diversity of interests within it, has made a comprehensive review an imposing (monograph-scale) task.

The scope of research in metal–alkynyl chemistry is broad; in addition to spanning the general areas of traditional interest to organometallic chemists, there is an emerging interest in metal–alkynyl complexes regarding their potential materials-science applications. Special attention has been devoted to several subtopics within the field. The $C \equiv C$ bonds of alkynyl ligands, like those of organic alkynes, constitute a natural focal point for reactivity studies, and have been shown to participate in a rich set of chemical transformations. Among the ligand-localized reactions of metal–alkynyl complexes, those that lead to the formation of clusters in which the alkynyl ligand bridges multiple metal centers (2), and in which the alkynyl ligand is transformed into vinylidene or other ligands (3), have been extensively studied. The linear geometry of the $-C \equiv C-$ unit and its potential to act as a "molecular wire" that might enable electronic communication between its appended substituents has made the synthesis and study of polymers derived from metal–alkynyl building blocks (e.g., 1) an active area of research (4a–e). These materials hold the promise to

$$\left[\begin{array}{c} PBu^n_3 \\ | \\ Pt \equiv\!\!\!= R \equiv\!\!\!= \\ | \\ PBu^n_3 \end{array} \right]_n$$

1

possess interesting electronic and structural properties; the encouraging discoveries of metal–alkynyl complexes and polymers that possess large nonlinear-optical responses (5), that are electrical conductors (6a–d), and that form liquid-crystal phases (7a–f) suggests that this will be a productive area of research for some time to come.

These diverse topics of study have more in common than their simple reliance on alkynyl ligands. A question of central importance to much of this field, and one for which a comprehensive answer has not yet been achieved, is: What is the nature of the bonding within the $M-C \equiv C-R$ fragment? Or, stated more narrowly: Are metal–alkynyl bonds characterized by appreciable (or any) M–CCR π-interactions, and, if so, for what metals, electron counts, ancillary ligands, and alkynyl R-groups? From the standpoint of the alkynyl-localized reactions of these compounds, understanding the electron distribution within the $C \equiv C$ bond and the degree to which it is attenuated by the attached ML_n fragment are essential for rationalizing the relative electro- and nucleophilicity of the two carbon centers. Such questions are equally important to understanding the properties of the advanced materials that are derived from metal–alkynyl building blocks, because significant π-electron delocalization within the $M-C \equiv C-R$ (or $M-C \equiv C-M$) fragment is a prerequisite for the presence of large optical nonlinearities and electrical conductivities.

The question of potential M—CCR π-bonding within the metal–alkynyl linkage arises from the simple frontier-orbital relationship between alkynyl ligands and the cyanide ligand, which is a known π-acceptor in some electronic environments (8). In fact, it was this simple analogy that motivated Nast to conduct his pioneering studies on metal–alkynyl complexes (1). A principal conclusion drawn from his spectroscopic and magnetic studies of a wide variety of high-symmetry metal–alkynyl complexes and ions from across the periodic table is that the cyanide and alkynyl ligands occupy similar positions in the spectrochemical series, lying toward the strong-field limit. The conventional interpretation of these data is that alkynyl ligands, like cyanide, are good σ-donors and modest π-acceptors (i.e., poorer than CO), as compared to other ligands within the series. Despite these early studies, the effort to refine this simple model and to understand the nature of metal–alkynyl bonding in fine detail has not kept pace with studies directed toward developing the synthetic, reaction, and materials-science chemistry of these complexes. Only a few of the many classes of metal–alkynyl complexes have been the subjects of theoretical calculations or detailed electronic- or photoelectron-spectroscopic studies, for example.

Until high-quality theoretical calculations and sensitive experimental probes of electronic structure are extended to the full range of metal–alkynyl complexes, any attempt to provide a general description of the nature of metal–alkynyl bonding is premature. In the absence of such a description, chemists have relied on the results of more readily accessible experimental techniques, especially X-ray crystallography and infrared spectroscopy, for insight into questions of electronic structure. [NMR spectroscopy is also an obvious probe of metal–alkynyl interactions; its applications in this regard have recently been reviewed (9).] In principle, the presence or absence of M–CCR π-interactions in metal–alkynyl complexes should be manifested in M–C or C$\equiv$C bond distances and vibrational frequencies, with the latter potentially being highly sensitive probes by analogy to the textbook history of correlating the $\nu(\text{C}\equiv\text{N})$ and $\nu(\text{C}\equiv\text{O})$ frequencies of metal–cyanide and metal–carbonyl complexes with the extent of M $\rightarrow$ CN and M $\rightarrow$ CO π-backbonding. Of course, there are other electronic factors, in addition to M–CCR π-bonding, that influence the bond distances and vibrational frequencies within the M—C$\equiv$C—R fragment, so in order for these data to be properly interpreted it is essential that the limiting cases of metal–alkynyl bonding be clearly defined and that benchmark metrical and spectroscopic values corresponding to these cases be established. Despite the facts that hundreds of papers report vibrational spectroscopic and X-ray crystallographic data for metal–alkynyl complexes, and that these data have been widely invoked to explain various aspects of metal–alkynyl bonding, this task has not, for the most part, been satisfactorily accomplished.

The purpose of this review is to attempt to discern, from consideration of the full range of data from these techniques, which metrical and vibrational-spectroscopic parameters may be meritoriously applied to the discussion of the electronic structures of metal–alkynyl complexes, and what conclusions may be drawn from them. Because this review aims to extract the intrinsic parameters that characterize $M—C\equiv C—R$ bonding, the discussion is limited in scope to compounds possessing terminal (η^1) alkynyl ligands bound to a single metal center (**2**); complexes, clusters, oligomers,

$$L_nM \!\!=\!\!\!=\!\!\!= \!\!R$$

2

and polymers in which the bonding picture is complicated (albeit in interesting ways) because the alkynyl ligand bridges two or more metal centers, either in a σ-only or a side-on fashion (**3–5**), have, for the most part, been

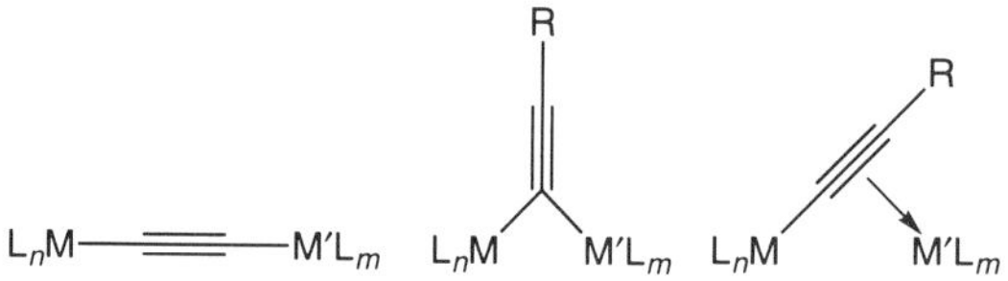

3   **4**   **5**

excluded. In addition to the primary literature and review articles cited earlier on these clusters and polymers, a recent extensive review by Beck *et al.* on multimetallic compounds containing bridging hydrocarbyl ligands is noteworthy in this regard (*10*). The coverage of the literature extends into 1994 and is complete through 1993; where appropriate, there is overlap with Nast's review (*1*).

II

THEORETICAL AND EXPERIMENTAL BACKGROUND

We begin with a brief overview of the findings from the limited number of theoretical calculations, photoelectron and electronic-spectroscopic studies, and other physical measurements on terminal metal–alkynyl complexes in order to provide a context for discussing the results of X-ray crystallographic and vibrational-spectroscopic studies. The simple analogy between cyanide and alkynyl ligands that seems to provide a reasonable framework for interpreting the results of Nast's work (1) has not always been upheld by the physical studies to be described below, both because the specific complexes in question vary among these studies, and because the various techniques employed in them are not equally sensitive (or successful) probes of metal–ligand interactions. As will become apparent, a single perspective on the nature of metal–alkynyl bonding cannot be distilled from this body of work.

A. Molecular-Orbital Calculations and Photoelectron Spectroscopy

We describe the results from photoelectron-spectroscopic and molecular-orbital calculations together because these studies are frequently conducted in tandem. The primary conclusion from this work is that, in most of the complexes studied, the alkynyl ligand is good σ-donor and π-donor, but is a poor π-acceptor.

The first and most influential molecular-orbital calculation on metal–alkynyl complexes is that of Kostić and Fenske, who applied the Fenske-Hall method to the complexes $FeCp(C{\equiv}CH)(PH_3)_2$ and $FeCp(C{\equiv}CH)(CO)_2$ (11). They concluded that the M–CCH bonds in these complexes are nearly pure σ in character. The large energy gap (ca. 15 eV) between the occupied metal orbitals and $\pi^*(C{\equiv}CH)$ levels severely limits the π-accepting quality of the latter, with the total electron population for the pair of π^* orbitals being 0.22 e$^-$ for $FeCp(C{\equiv}CH)(PH_3)_2$ and 0.14 e$^-$ for $FeCp(C{\equiv}CH)(CO)_2$. The filled $\pi(C{\equiv}CH)$ orbitals, in contrast, mix extensively with the higher-lying occupied metal orbitals; these filled–filled interactions result in the destabilization of the metal-based orbitals. The HOMOs of both complexes possess substantial coefficients at the alkynyl β-carbon; this was noted to be consistent with the alkynyl-localized reactivity of these complexes.

The conclusions of Kostić and Fenske for $FeCp(C{\equiv}CH)(CO)_2$ received firm experimental support a decade later from the work of Lichtenberger and co-workers, who carried out photoelectron-spectroscopic measurements and Fenske-Hall and extended-Hückel calculations on the complexes $FeCp(C{\equiv}CR)(CO)_2$ (R = H, Ph, t-Bu, $C{\equiv}CH$) and $FeCp^*(C{\equiv}CBu^t)(CO)_2$ ($12a,b$). Comparisons among the ionization energies of these compounds and to those of the methyl, chloro, and cyano derivatives of $FeCpX(CO)_2$, as well as the nature of the vibronic structure observed in the spectra of $FeCp(C{\equiv}CH)(CO)_2$ and $FeCp(C{\equiv}CBu^t)(CO)_2$, led to several important conclusions. First, whereas cyanide stabilizes

π-symmetry metal levels via Fe $\rightarrow$ CN π-backbonding, the alkynyl ligands destabilize them as a result of RCC $\rightarrow$ Fe π-donation. The extent of alkynyl π-donation depends strongly on the nature of the alkynyl R-group, being greater for the t-Bu, Ph, and C$\equiv$CH substituents than for the parent ethynyl ligand (R = H). The σ-donor abilities of alkynyl ligands are also a function of R-group; among these FeCpX(CO)$_2$ compounds, σ-donation by the X ligands increases according to Cl < C$\equiv$CH, C$\equiv$C—C$\equiv$CH, C$\equiv$N < C$\equiv$CPh, C$\equiv$CBu^t < Me. Overall, the interactions between the ethynyl ligand and the FeCp(CO)$_2$ fragment are very similar to those of the chloride ligand with this metal center. The butadiynyl ligand (C$\equiv$C–C$\equiv$CH) is described as being a better π-donor than halide ligands.

Photoelectron-spectroscopic studies by Oskam *et al.* on the series of *trans*-substituted complexes M(C$\equiv$CR)$_2$(PEt$_3$)$_2$ (M = Pd, Pt; R = H, Me), M(C$\equiv$N)$_2$(PEt$_3$)$_2$ (M = Pd, Pt), and Pd(C$\equiv$N)Cl(PEt$_3$)$_2$, together with Hartree-Fock-Slater molecular-orbital calculations on the model PH$_3$ derivatives, yielded a description of metal–alkynyl bonding that is essentially congruent with that for the FeCp(C$\equiv$CR)(CO)$_2$ class of compounds: M $\rightarrow$ CCR π-backbonding is unimportant—the electron population of π^*(CCR) is no greater than 0.05 e$^-$—and net destabilizing interactions between the occupied π-symmetry metal and π(CCR) orbitals are strong (*13*). However, in contrast to Lichtenberger's work (*12a,b*), which this study preceded, Oskam and co-workers concluded that the cyanide ligand is also a π-donor in these compounds, rather than a π-acceptor; thus, their work supports a cyano/alkynyl analogy, although not the traditional version that classifies these ligands as π-acceptors.

An exception to this picture of alkynyl ligands as π-donors is provided by Thompson and co-workers, who reported *ab initio* calculations on the d^0 complexes TiCp$_2$(C$\equiv$CR)$_2$ (R = H, Me) (*14*). They noted that there is little evidence in their computational results for RCC $\rightarrow$ M π-donation in these compounds, but that there is significant mixing between the alkynyl and Cp π levels. Unfortunately, attempts to measure photoelectron spectra of these compounds were unsuccessful.

B. *Electronic Spectroscopy*

Electronic-absorption band maxima and extinction coefficients have been reported for a large number of metal–alkynyl complexes. In some papers, shifts in the energies of these bands within a class of L$_n$M—CCR complexes as a function of M, L, or R have been cited as indicating (or as being consistent with) the presence of M $\rightarrow$ CCR π-backbonding. Although qualitative assignments of electronic-absorption bands as arising from generic d–d, charge-transfer, or $\pi \rightarrow \pi^*$ transitions are not uncommon, with very few exceptions the experiments necessary to assign these bands definitively to transitions between specific orbitals or electronic and spin states have not been carried out (i.e., single-crystal polarized-light absorption measurements, cryogenic measurements for the observation of vibronic structure, and/or photophysical measurements). Casual comparisons of electronic-absorption band energies or intensities among a series of compounds in the absence of such experiments are not without the risk of misinterpretation because a variety of important effects not associated with the presence or absence of M–CCR π-bonding may be operative, such as mixings between electronic states of identical symmetry but different orbital character. We limit our survey to those papers in which electronic transitions are assigned with some certainty, and that comment specifically on the nature of metal–alkynyl bonding.

Work in our laboratory directed toward developing a class of transition-metal complexes that are structurally and electronically analogous to conjugated organic compounds and polymers led to the synthesis and spectroscopic study of alkynyl-substituted, quadruply metal–metal bonded compounds of the type $M_2(C\equiv CR)_4(PMe_3)_4$ (**6**; M = Mo, W; R =

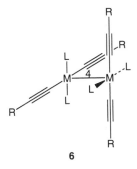

6

H, alkyl, Ph, SiMe$_3$) (*15a–e*). These dimetallapolyyne compounds are particularly suitable for the study of metal–alkynyl π-interactions by electronic spectroscopy because the formally metal–metal localized δ HOMO and δ* LUMO are of π-symmetry with respect to the alkynyl ligands, and hence the $^1(\delta \rightarrow \delta^*)$ absorption band should manifest any M–CCR π-interactions (Scheme 1). The $^1(\delta \rightarrow \delta^*)$ transition energies of these alkynyl derivatives are substantially smaller (M = Mo, $\bar{\nu}_{max}$ = 13,500–15,500 cm^{-1}), and their band intensities higher (ε_{max} = 4500–9500 M^{-1} cm^{-1}), than those of Mo$_2$Cl$_4$(PMe$_3$)$_4$ and Mo$_2$Me$_4$(PMe$_3$)$_4$ ($\bar{\nu}_{max} \cong$ 17,000 cm^{-1}, ε_{max} = 2000–3000 M^{-1} cm^{-1}). The red-shift and intensity of the $^1(\delta \rightarrow \delta^*)$ band are dependent on the alkynyl R-group, increasing with

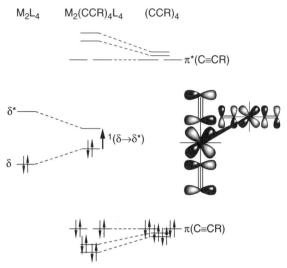

SCHEME 1   Frontier orbital interactions in $M_2(CCR)_4(PMe_3)_4$ complexes.

the R-group π-conjugation according to R = H < alkyl < $SiMe_3$ < Ph (15b,d). These shifts parallel those of tetraalkynylethene compounds [$(RC\equiv C)_2C=C(C\equiv CR)_2$], which are the organic analogues of the dimetallapolyynes (15b). The $^1(\delta \rightarrow \delta^*)$ bands of the alkynyl complexes exhibit considerable vibronic structure, displaying progressions in the $\nu(M\overset{4}{-}M)$, $\nu(M-C)$, and $\nu(C\equiv C)$ modes. While these results (and complementary NMR (15c) and resonance Raman (15d) spectroscopic data) clearly indicate that significant M_2-CCR π-interactions are present in these complexes, the question of which alkynyl orbital(s) the metal δ and δ^* orbitals mix with [π(CCR) and/or π^*(CCR)] cannot be addressed at this point. Nonetheless, these findings provide an interesting counterpoint to those from the photoelectron-spectroscopic studies of $FeCp(C\equiv CR)(CO)_2$ (12a,b) in that the alkynyl ligands are decidedly not halidelike in their interactions with the frontier metal orbitals in these compounds.

We have also prepared and studied alkynyl-substituted methylidyne complexes of the type trans-$W(C\equiv CR)(\equiv CH)(dmpe)_2$ (7) (16a–c). Deducing the nature of the frontier orbital M–CCR π-interactions from the

$$HC\equiv W-C\equiv CR$$

7

lowest-energy electronic transition (assigned as $[d_{xy} \rightarrow (d_{xz}, d_{yz})]$) of these metallabutadiynes is straightforward because only the (d_{xz}, d_{yz}) levels are of π-symmetry with respect to the alkynyl ligand; the d_{xy} level is nonbonding. The transition energies of WX($\equiv$CH)(dmpe)$_2$ complexes lie in the order X = C$\equiv$CPh < C$\equiv$CSiMe$_3$ < C$\equiv$CH < n-Bu (17) < halide ($16a$–c). The fact that the chloride and alkynyl derivatives exhibit opposing shifts of the $[d_{xy} \rightarrow (d_{xz}, d_{yz})]$ band relative to the σ-only n-butyl derivative indicates that the alkynyl ligand is not acting as a π-donor toward the frontier orbitals to any significant extent in these compounds, but rather mixes strongly with the (d_{xz}, d_{yz}) level via its π^*(CCR) orbitals. A recent extension of this study to include the cyanide derivative W(C$\equiv$N)($\equiv$CH)(dmpe)$_2$ reveals that the energy and intensity of its $[d_{xy} \rightarrow (d_{xz}, d_{yz})]$ band are closely similar to those observed for the alkynyl derivatives ($16c$). The W–CCR bond lengths of the alkynyl compounds also appear to manifest a metal-alkynyl π-bonding interaction (Section III.C.1), as does the ^{1}H-NMR spectrum of W(C$\equiv$CH)($\equiv$CH)(dmpe)$_2$, which exhibits spin–spin coupling over five bonds between the ethynyl and methylidynyl hydrogen atoms.

The electronic spectra of complexes of the type trans-M(C$\equiv$CR)$_2$L$_2$ (M = Ni, Pd, Pt) have been studied by several groups, and while experiments aimed at providing rigorous assignments for these spectra have not been undertaken, discussion of these studies is warranted by the considerable interest in the electronic properties of polymers derived from this class of compounds ($4a,e$). The absorption spectra of an extensive series of trans-M(C$\equiv$CR)$_2$L$_2$ (M = Ni, Pd, Pt; R = H, Me, CH$_2$F, CH=CH$_2$, Ph, C$\equiv$CH, C$\equiv$CMe, C$\equiv$CPh; L = PMe$_3$, PEt$_3$, SbEt$_3$) complexes were reported by Masai et al. (18). The spectra are dominated by three intense ($\varepsilon \cong 10^4$) bands in the 200–350-nm region; no d–d bands were observed. The energy of the lowest-lying band is strongly dependent on the nature of the alkynyl R-group, red-shifting with increasing R-group conjugation (from R = H to C$\equiv$CPh) by up to 0.5 eV, depending on M and L, and is linearly related to the first ionization potential of the free alkyne (HC$\equiv$CR). Based on these data and a molecular-orbital calculation, Masai et al. concluded that the lowest-energy band arises from a transition between the π(CCR) and π^*(CCR) orbitals, with the transition possessing charge-transfer character as a result of mixing between the π^*(CCR) and metal $(n + 1)p$-orbitals. The R-group dependence of the transition energy was also interpreted as indicating that the RC$\equiv$C—M—C$\equiv$CR fragment is π-conjugated. Subsequent studies of the absorption spectra of polymers of the general class [M(C$\equiv$C—R—C$\equiv$C)(PR$_3'$)$_2$]$_n$ (M = Ni, Pd, Pt) by the Osaka group have yielded conclusions consistent with those for the monomeric compounds ($7d,19$).

The observation of luminescence from compounds containing the $[\text{Pt}(C\equiv CR)L_2]^+$ fragment is increasingly common (20–23), and vibronically structured emission spectra have been reported in several instances. DeGraff, Lukehart, Demas, and their coworkers discovered that the complexes $\text{Pt}(C\equiv CR)_2(PEt_3)_2$ (R = H, Ph) are intensely phosphorescent in glassy matrices at 77 K, and that the vibronic structure in the emission spectra consists primarily of a progression in the $\nu(C\equiv C)$ mode; the emission band of $\text{Pt}(C\equiv CPh)_2(PEt_3)_2$ exhibits additional progressions arising from phenyl-localized vibrational modes (21). The authors infer from these data that the emissive state is MLCT in character, with the excitation involving promotion of a metal electron to a $\pi^*(CCR)$ orbital. Although the authors did not reconcile their results with the photoelectron-spectroscopic data and molecular-orbital calculations of Oskam et al. on these compounds (13), the frontier-orbital scheme implied by this study is qualitatively consistent with Oskam's; the emission data shed no light on the question of π-bonding in the Pt–CCR linkage, however. Che et al. discovered that $\text{Pt}(C\equiv CPh)_2(\text{dppm-}P)_2$ and $[\text{Pt}(C\equiv CPh)_2(\mu\text{-dppm})_2\text{Au}][PF_6]$ exhibit vibronically structured luminescence at 77 K similar to that observed for $\text{Pt}(C\equiv CR)_2(PEt_3)_2$ (22). These data and extended-Hückel calculations on the model dmpm derivatives also suggest assignment of the emissive state as MLCT; the calculations indicate that $\pi(CCR) \rightarrow$ Pt π-donation is important, and that Pt $\rightarrow \pi^*(CCR)$ π-backbonding is not. Luminescence spectra of $[\text{Pt}\{(C\equiv C)_m\}_2(PBu_3^n)_2]_n$ polymers, reported by Lewis, Marder, Friend et al., also exhibit vibronic structure similar to that for $\text{Pt}(C\equiv CR)_2(PEt_3)_2$ (6b,23), although these workers attributed the emission to an alkynyl-localized π–π^* excited state. Based on these data and complementary electronic-absorption spectroscopic studies, they concluded that the metal–alkynyl backbones of these polymers are π-conjugated.

C. Other Physical Studies

The nature of metal–alkynyl bonding has been studied by a variety of other physical methods, although, as for the techniques described earlier, a single electronic model for this bond has not emerged. Experimental studies by Brindza, Bercaw, and co-workers of the relative Ru–X bond-dissociation energies within an extensive series of compounds of the type $\text{RuCp}^*X(PMe_3)_2$ (24) revealed that ruthenium–carbon bond strengths decrease according to Ru–CN > Ru–CCPh > Ru–C(sp^3). With few exceptions, the relative Ru–X energies are a linear function of the X–H bond strengths of the protonated free ligand. The fact that the relative Ru–CCPh bond strength fits this correlation would seem to imply that it is not detectably strengthened by a Ru–C π-bonding component. In contrast, the Ru–CN bond is anomalously strong; this was interpreted as arising from substantial Ru $\to$ CN π-backbonding. A related experimental study by Bercaw *et al.* on ScCp$_2^*$X complexes (*25*) showed that metal–carbon bond strengths decrease according to Sc–CCR > Sc–C(aryl) > Sc–C(alkyl); a similar conclusion has been reached through theoretical calculations (*26*).

From the results of ^{57}Fe Mössbauer spectroscopy, Birchall and Myers concluded that the alkynyl ligand is a strong-field ligand, being a σ-donor and strong π-acceptor in ions of the type [Fe(C$\equiv$CR)$_6$]$^{4-}$ (*27*). In contrast, a study by Leigh *et al.* using this technique led to the conclusion that the alkynyl ligand is primarily a σ-donor in *trans*-Fe(C$\equiv$CPh)X(L–L)$_2$ (X = Cl, Br; L–L = dmpe, depe) complexes (*28*).

Wrackmeyer, Sebald, and co-workers have conducted detailed multinuclear NMR-spectroscopic studies of metal–alkynyl complexes with the aim of probing the nature of the M–CCR bond; this work has been recently reviewed (*9*), so we mention only selected results. Germane to the discussion in the preceding sections is the conclusion that for *cis*- and *trans*-M(C$\equiv$CR)$_2$(PR$_3'$)$_2$ (M = Ni, Pd, Pt) complexes the ^{13}C chemical shifts and $^1J_{\equiv CH}$ and $^1J_{C\equiv C}$ coupling constants (and, for the platinum complexes, the ^{195}Pt chemical shifts and the $^1J_{PtC}$ coupling constants) are consistent with the presence of M $\to$ CCR π-backbonding, with the extent of backbonding decreasing according to Ni > Pd > Pt (*29*). A ^{19}F-NMR study by Parshall of the *m*- and *p*-fluorophenyl derivatives of *trans*-Pt(C$\equiv$CPh)(C$_6$H$_4$F)(PEt$_3$)$_2$ also indicated that the phenylethynyl ligand is a strong π-acceptor in these complexes (*30*). We have noted that the compounds 1,1,2',2'-W$_2$(C$\equiv$CMe)$_2$Cl$_2$(PMe$_3$)$_4$ (*15c*) and *trans*-W(C$\equiv$CH)($\equiv$CH)(dmpe)$_2$ (*16b*) exhibit long range ($\geq$5 bonds) spin–spin couplings in their ^{1}H-NMR spectra, as do their conjugated organic analogs, which is consistent with π-conjugation between their alkynyl and metal-centered multiple bonds. ^{1}H-NMR spectroscopic studies of paramagnetic complexes by Köhler *et al.* on compounds of the type VCp$_2'$(C$\equiv$CR) (Cp' = Cp or substituted Cp; R = aryl) (*31a–c*) and by Balch *et al.* on Fe(TPP)(C$\equiv$CPh) (*31d*) indicated that the unpaired metal-centered electron of these complexes is delocalized onto the alkynyl ligand, suggesting the presence of some type of M–CCR π-interaction. A contrasting picture is provided by Al-Najjar *et al.*, who reported that ^{13}C-NMR data for [Au(C$\equiv$CPh)X]$^-$ (X = Cl, Br, I, C$\equiv$CPh) ions are consistent with only weak Au $\to$ CCPh π-backbonding (*32*).

Electrochemical (primarily cyclic voltammetric) data have been reported for many metal–alkynyl complexes (*33*). In some studies, the redox potentials have been stated to be either consistent or inconsistent with

$M \rightarrow CCR$ π-backbonding, depending on the complex, based on comparisons to electrochemical data for related halide, cyanide, or other compounds. We will not elaborate on these findings here, and note that such comparisons need not straightforwardly provide information about the nature of metal–alkynyl π-bonding because the redox-active orbitals probed by these measurements are not necessarily of π-symmetry with respect to the alkynyl ligand(s); in few of these studies are the orbitals involved in the electron-transfer process specifically identified. Solvation effects on electrochemical potentials also need to be accounted for if these data are to be treated as providing direct insight into orbital energies.

III

X-RAY CRYSTALLOGRAPHIC STUDIES

A. *Overview*

The number of X-ray crystal structures of terminal metal–alkynyl complexes has increased ten-fold since the last review 13 years ago (*1*). Metrical data for the $M—C \equiv C—R$ fragments of terminal metal–alkynyl complexes are set out in Table I (*34–151*). Bridging alkynyl complexes of the type $L_nM—C \equiv C—ML_n$ (and higher $L_nM—C_x—ML_n$ analogs) are not included in this survey; these complexes have been the subject of recent reviews (*4b,10*).

In addition to their obvious role in establishing connectivity and stereochemistry, the data from structural studies of metal–alkynyl complexes have been used to evaluate the M–C bond order, the extent of $M \rightarrow CCR$ π-backbonding and π-conjugation within the $M—C \equiv C—R$ fragment, and the *trans* influences of alkynyl ligands. By analogy to the valence-bond canonical structures for metal–carbonyl complexes that represent the two limits of $M \rightarrow CO$ π-backbonding (**8**), the corresponding structures for metal–alkynyl complexes (**9**) suggest that the study of M–C, $C \equiv C$, and

$$\ddot{M}-C\equiv\overset{+}{O}: \longleftrightarrow M=C=\overset{+}{\ddot{O}}: \qquad \ddot{M}-C\equiv C-R \longleftrightarrow M=C=C=\overset{-}{R}$$

$$\textbf{8} \qquad\qquad\qquad \textbf{9}$$

C–R bond lengths is of particular importance for gaining insight into the nature of the metal–alkynyl bond. As will be described more fully later on, some of the structural perturbations of these linkages are subtle, and their observation thus requires high-quality X-ray diffraction data.

TABLE I

METRICAL DATA FOR TERMINAL METAL–ALKYNYL COMPLEXES

Compound	$d(\text{M-C})$ (Å)	$d(\text{C}{\equiv}\text{C})$ (Å)	$d(\text{C-R})$ (Å)	$\angle(\text{M-C}{\equiv}\text{C})$ (°)	$\angle(\text{C}{\equiv}\text{C-R})$ (°)	Ref.
$\text{Ti}\{\eta^5\text{-C}_5\text{H}_4\text{SiMe}_3\}_2\}(\text{C}{\equiv}\text{CSiMe}_3)_2$	a	a	a	$177.0(a)$	$176.5(a)$	(34)
$\text{ZrCp}_2(\text{C}{\equiv}\text{CMe})_2$	$2.249(3)$	$1.206(4)^b$	$1.462(5)$	$177.0(3)$	$179.4(4)$	(35)
$\text{ZrCp}_2\{\text{C}{\equiv}\text{CPh}\}\{\eta^2\text{-N(Ph)NHPh}\}$	$2.258(6)$	$1.208(8)^b$	$1.454(9)$	$179.0(7)$	$179.0(7)$	(36)
$\text{ZrCp}(\text{C}{\equiv}\text{CBu}^t)(\text{CH}{=}\text{CHBu}^t)\text{Cl}(\text{dmpe})$	$2.273(9)$	$1.20(1)$	a	$173.8(8)$	a	(37)
$\text{VCp}_2(\text{C}{\equiv}\text{CBu}^t)$	$2.075(5)$	$1.191(7)^b$	$1.474(7)$	$177.0(4)$	$177.8(6)$	(38)
$\text{V}(\eta^5\text{-C}_5\text{Me}_4\text{Et})_2(\text{C}{\equiv}\text{CC}_6\text{H}_2\text{-1,3,5-Me}_3)$	$2.032(13)$	$1.234(17)$	$1.447(14)$	$176.6(10)$	$175.5(12)$	$(31c,39)$
$trans\text{-Mo}(\text{C}{\equiv}\text{CPh})_2(\text{dppe})_2$	$2.093(8)$	$1.237(12)$	$1.422(12)$	$175.6(7)$	$177.8(9)$	$(33f,40)$
$\text{Mo}(\text{C}{\equiv}\text{CBu}^t)(\text{H})_3(\text{dppe})_2$	$2.175(10)$	$1.209(14)$	$1.458(15)$	$176.9(8)$	$177.1(11)$	(41)
$\text{Mo}(\eta^7\text{-C}_7\text{H}_7)(\text{C}{\equiv}\text{CPh})(\text{dppe})$	$2.138(5)$	$1.205(6)^b$	$1.434(7)$	$178.5(4)$	$177.9(5)$	$(33g)$
$[\text{Mo}(\eta^7\text{-C}_7\text{H}_7)(\text{C}{\equiv}\text{CPh})(\text{dppe})][\text{BF}_4]$	$2.067(9)$	$1.196(11)$	$1.445(12)$	$174.6(8)$	$175.0(10)$	$(33g,42)$
$\text{Mo}_2(\text{C}{\equiv}\text{CH})_4(\text{PMe}_3)_4$	$2.153(4)$	$1.183(6)^b$	$0.950(39)$	$171.7(3)$	$172.1(24)$	$(15b)$
	$2.161(4)$	$1.174(6)^b$	$0.909(35)$	$171.9(4)$	$166.7(24)$	
$\text{Mo}_2(\text{C}{\equiv}\text{CMe})_4(\text{PMe}_3)_4$	$2.147(4)$	$1.203(5)^b$	$1.471(6)$	$172.1(3)$	$177.6(4)$	$(15e)$
	$2.150(5)$	$1.888(7)^b$	$1.469(9)$	$172.3(3)$	$178.4(5)$	
	$2.140(4)$	$1.195(5)^b$	$1.473(6)$	$172.2(3)$	$178.8(6)$	
	$2.141(4)$	$1.195(6)^b$	$1.471(6)$	$175.8(3)$	$176.9(4)$	
	$2.149(4)$	$1.199(7)^b$	$1.473(7)$	$174.5(4)$	$178.1(4)$	
	$2.137(4)$	$1.185(6)^b$	$1.465(8)$	$167.5(4)$	$178.5(6)$	
	$2.149(4)$	$1.198(6)^b$	$1.476(6)$	$171.8(4)$	$176.8(4)$	
	$2.149(4)$	$1.193(5)^b$	$1.474(5)$	$173.0(4)$	$178.1(5)$	
$\text{Mo}_2(\text{C}{\equiv}\text{CPr}^i)_4(\text{PMe}_3)_4$	$2.26(4)-$	$1.05(6)-$	a	$153(4)-$	$166(5)-$	$(15a)$
	$2.49(4)$	$1.18(7)$		$164(4)$	$175(5)$	
$\text{W}(\text{C}{\equiv}\text{CCO}_2\text{Me})_2(\text{H})_2(\text{dppe})_2$	$2.038(21)$	$1.22(3)$	$1.40(4)$	$172.8(16)$	$168.5(23)$	$(33f,l)$
	$2.041(25)$	$1.24(4)$	$1.41(6)$	$174.5(20)$	$175.9(24)$	
$\text{W}(\text{C}{\equiv}\text{CPh})_2(\text{NPh})(\text{PhC}{\equiv}\text{CPh})(\text{PMe}_3)_2$	$2.148(15)$	$1.165(19)$	a	a	a	(43)
	$2.152(13)$	$1.165(17)$	a	a	a	
$\text{WCp}(\text{C}{\equiv}\text{C-}c\text{-Pr})(\text{CO})_2(\text{PMe}_3)$	$2.134(11)$	$1.205(15)$	$1.417(18)$	$177.6(10)$	$176.8(13)$	(44)

(continued)

TABLE I (continued)

METRICAL DATA FOR TERMINAL METAL–ALKYNYL COMPLEXES

Compound	$d(\text{M–C})$ (Å)	$d(\text{C}\equiv\text{C})$ (Å)	$d(\text{C–R})$ (Å)	$\angle(\text{M–C}\equiv\text{C})$ (°)	$\angle(\text{C}\equiv\text{C–R})$ (°)	Ref.
$trans$-W(C≡CH)(CH)(dmpe)$_2$	2.263(5)	1.192(8)^b	0.822(68)	176.8(5)	179.5(56)	(16c)
$trans$-W(C≡CSiMe$_3$)(CH)(dmpe)$_2$	2.246(6)	1.228(9)^b	1.814(7)	178.2(5)	179.8(7)	(16b)
$trans$-W(C≡CPh)(CH)(dmpe)$_2$	2.250(4)	1.222(5)^b	1.443(5)	171.6(3)	178.5(4)	(16c)
$trans$-W(C≡CC$_6$H$_4$-4-CCPr^n)(CH)(dmpe)$_2$	2.245(8)	1.210(11)	1.437(11)	178.2(7)	178.0(9)	(16c)
W$_2$(C≡CMe)$_4$(PMe$_3$)$_4$	2.102(10)	1.203(15)	1.490(16)	174.3(10)	179.2(13)	(15e)
	2.121(12)	1.186(19)	1.473(22)	172.1(8)	177.1(10)	
	2.129(11)	1.217(16)	1.475(17)	172.1(9)	176.3(15)	
	2.129(11)	1.189(17)	1.474(19)	176.2(9)	178.5(10)	
	2.120(12)	1.183(17)	1.496(20)	174.9(10)	176.1(13)	
	2.124(10)	1.169(17)	1.495(21)	168.0(12)	175.8(13)	
	2.128(9)	1.176(13)	1.496(15)	173.3(11)	177.3(14)	
1,1,2′,2′,-W$_2$(C≡CMe)$_2$Cl$_2$(PMe$_3$)$_4$	2.139(11)	1.184(16)	1.497(17)	170.5(11)	178.3(11)	(15c)
	2.122(13)	1.213(19)	1.459(23)	173.1(11)	177.4(16)	
	2.141(12)	1.212(17)	1.441(20)	173.8(10)	174.1(15)	
fac-Mn(C≡CBu^t)(dppe)(CO)$_3$	1.996(6)	1.214(8)^b	1.522(10)	179.0(5)	178.0(8)	(45)
mer-Mn(C≡CPh){P(OPh)$_3$}$_2$(CO)$_3$	2.002(6)	1.198(9)^b	1.450(8)	169.2(6)	177.7(7)	(46)
Mn(C≡CPPh$_3$)Br(CO)$_4$	1.981(14)	1.216(14)	1.679(13)	176.3(12)	164.0(12)	(47)
MnCp(C≡CPMe$_2$Ph)(CO)$_2$	1.895(5)	1.221(7)	1.683(6)	179.6(4)	166.6(5)	(48)
fac-Re(C≡CPh)(CO)$_3$(py)$_2$	2.108(9)	1.23(1)	1.42(1)	175(1)	178(1)	(49)
ReCp(C≡CMe)(NO)(PPh$_3$)	2.066(7)	1.192(11)	1.484(12)	175.8(7)	176.8(9)	(50)
Re$_2$(C≡CPh)(CO)$_7$(μ-H)(μ-dmpm)	2.126(7)	1.200(9)^b	1.432(9)	178.7(6)	178.9(6)	(51)
$trans$-Fe(C≡CMe)$_2$(dmpe)$_2$	1.968(6)	1.152(9)^b	1.465(11)	a	a	(52a)
$trans$-Fe(C≡CPh)$_2$(dmpe)$_2$	1.925(6)	1.209(9)^b	1.438(9)	a	a	(52a,b)
$trans$-Fe(C≡CPh)$_2$(depe)$_2$	1.918(3)	1.222(4)^b	1.435(4)	a	a	(52a)
$trans$-Fe(C≡CC$_6$H$_4$-4-CCH)$_2$(dmpe)$_2$	1.933(3)	1.193(4)^b	1.438(4)	a	a	(52a)
$trans$-Fe(C≡CPh)Cl(dmpe)$_2$	1.897(3)	1.192(3)^b	1.442(4)	178.4(2)	177.1(3)	(53)
$trans$-Fe(C≡CPh)Cl(dmpe)$_2$	1.880(5)	1.216(8)^b	1.421(8)	177.8(5)	177.5(6)	(54)
[$trans$-Fe(C≡CMe)(C=CHMe)(dmpe)$_2$][BPh$_4$]	1.963(7)	1.174(11)	1.467(14)	176.0(7)	177.7(9)	(55)

[trans-Fe(C≡CPr^i)(C=CHPr^i)(dmpe)$_2$][BPh$_4$]	a	a	a	a	a	(55)
trans,cis-Fe{C≡CCH(OMe)$_2$}I{P(OMe)$_3$}$_2$(CO)$_2$	1.948(4)	1.201(6)^b	1.464(6)	170.7(5)	178.8(6)	(56)
Fe(C≡CPh){P(CH$_2$CH$_2$PPh$_2$)$_3$}	1.92(1)	1.21(1)	1.45(1)	173.8(9)	174(1)	(33a)
[Fe(C≡CPh){P(CH$_2$CH$_2$PPh$_2$)$_3$}][BF$_4$]	1.88(2)	1.20(3)	1.47(3)	175(2)	177(2)	(33a)
FeCp(C≡CPh)(CO)$_2$	1.920(6)	1.201(9)^b	1.444(9)	174.4(4)	176.8(5)	(57)
FeCp(C≡CPh)(dppm)	1.900(7)	1.206(10)^b	1.438(10)	177.4(7)	175.9(8)	(58)
	1.909(7)	1.201(10)^b	1.430(10)	179.3(6)	177.4(8)	
FeCp*(C≡CH)(CO)$_2$	1.921(3)	1.173(4)^b	0.865(33)	178.70(26)	177.7(24)	(59)
FeCp*(C≡CPh)(CO)$_2$	1.924(7)	1.214(9)^b	1.431(9)	175.89(60)	175.46(70)	(59)
[Ru(C≡CPh)(PMe$_2$Ph)$_4$][PF$_6$]	2.051(a)	1.203(a)	a	a	a	(60)
trans-Ru(C≡CPh)$_2$(dppe)$_2$	2.061(5)	1.207(7)^b	1.434(7)	178.1(5)	174.4(6)	(61)
	2.064(5)	1.194(7)^b	1.449(8)	174.3(5)	168.3(6)	
trans-Ru(C≡CH)Cl(dppm)$_2$	1.906(9)	1.162(9)^b	0.899(0)	177.0(6)	180(0)	(62)
trans-Ru(C≡CH)$_2$(CO)$_2$(PEt$_3$)$_2$	2.078(1)	1.199(2)^b	a	a	a	(63a)
trans-Ru(C≡CSiMe$_3$)$_2$(CO)$_2$(PEt$_3$)$_2$	2.062(2)	1.221(2)^b	1.812(2)	178.1(1)	176.1(1)	(63b)
trans-Ru(C≡CPh)$_2$(CO)$_2$(PEt$_3$)$_2$	2.074(3)	1.200(4)^b	1.438(4)	a	a	(63a)
trans-Ru(C≡CCCPh)$_2$(CO)$_2$(PEt$_3$)$_2$	2.078(2)	1.194(2)^b	1.386(3)	177.9(1)	176.9(2)	(63b)
trans-Ru(C≡CCCSiMe$_3$)$_2$(CO)$_2$(PEt$_3$)$_2$	2.057(2)	1.226(2)^b	1.370(2)	176.5(2)	178.9(2)	(63b)
[cis,trans-Ru(C≡C-n-C$_6$H$_{13}$)(CO)(py)$_2$(PPh$_3$)$_2$][ClO$_4$]	2.01(3)	1.13(5)	1.54(5)	175(3)	171(5)	(64)
[trans,mer-Ru(C≡CPh)(PPh$_3$)$_2$(CNBu^t)$_3$][PF$_6$]	2.03(3)	1.17(4)	1.46(4)	175(3)	178(4)	(65)
mer-Ru(C≡CPh){η^3-PhC≡CC=CHPh}{PPh((CH$_2$)$_3$PCy$_2$)$_2$}	2.037(3)	1.205(5)^b	a	178.1(3)	a	(66a,b)
Ru(C≡CCO$_2$Me){OC(OMe)C=CHCH=CHCO$_2$Me}(CO)(PPh$_3$)$_2$	2.064(4)	1.203(5)^b	1.444(6)	179.6(3)	168.2(5)	(67)
Ru(C≡CPh)(CO){C$_6$H$_3$-3-Me-6-C(O)p-Tol}(PMe$_2$Ph)$_2$	2.120(5)	1.192(7)^b	1.432(7)	176.2(5)	178.9(6)	(68)
RuCp(C≡CPh)(PPh$_3$)$_2$	2.016(3)	1.215(4)^b	1.456(4)	178.0(2)	171.9(3)	(69)
RuCp(C≡CPh)(PPh$_3$)$_2$	2.017(5)	1.214(7)^b	1.462(8)	177.4(4)	170.6(5)	(70)
			1.418(8)^c		170.6(5)^c	
RuCp{C≡CC(OC(O)CF$_3$)=CMe$_2$}(PPh$_3$)$_2$	2.02(1)	1.20(1)	1.42(2)	175.1(11)	175.2(14)	(71)

(continued)

TABLE I (*continued*)

METRICAL DATA FOR TERMINAL METAL–ALKYNYL COMPLEXES

Compound	d(M–C) (Å)	d(C≡C) (Å)	d(C–R) (Å)	∠(M–C≡C) (°)	∠(C≡C–R) (°)	Ref.
RuCp{C≡CC(O)Me}(PPh$_3$)$_2$	1.996(3)	1.212(5)^b	1.427(5)	176.1(3)	169.8(3)	(71)
RuCp(C≡CPh)(dppe)	2.009(3)	1.204(5)^b	1.444(5)	178.1(3)	176.3(4)	(72)
(R, R)-RuCp(C≡CPh){PPh$_2$CH(Me)CH(Me)PPh$_2$}	2.038(7)	1.172(9)^b	a	179.4(7)	172(1)	(70)
Ru(η^6-C$_6$Me$_6$){C≡CC≡CCPh$_2$(OSiMe$_3$)}(PMe$_3$)Cl	1.93(3)	1.26(4)	1.40(5)	174(3)	175(4)	(73)
						(74)
Ru$_2$Cp$_2^*$(C≡C-p-Tol)$_2$(μ-SPr^i)$_2$	2.04(3)	1.16(4)	a	170(2)	a	(75)
	2.00(2)	1.14(4)	a	173(2)	a	
Ru$_2$(C≡CPh)$_2$\{μ-NC$_5$H$_4$-2-N(C$_6$F$_5$)\}$_4$	1.955(12)	a	a	a	a	(33c)
	1.951(11)	a	a	a	a	
Ru$_2$(C≡CPh)(μ-NC$_5$H$_4$-2-NPh)$_4$	2.08(3)	1.14(3)	1.51(3)	172(3)	173(3)	(33p)
Ru$_2$(C≡CPh)$_2$(μ-NPhCH$_2$NPh)$_4$	a	a	a	a	a	(33b)
Os(C≡CCO$_2$Me){OC(OMe)CH=CH}(CO)(PPr$^i_3$)$_2$	1.977(4)	1.212(6)^b	1.422(9)	176.9(4)	165.8(6)	(76)
			1.517(24)^c		163.4(10)^c	
Os$_3$(C≡CPh)(CO)$_6$(μ-C≡CPh)(μ-PPh$_2$)$_2$(NH$_2$Et)$_2$	2.075(12)	a	a	178.3(6)	178.5(8)	(77)
	a	1.243(28)	1.494(31)	175.0(2)	177.3(22)	(78)
trans,*mer*-Co(C≡CPh)$_2$(H)(PMe$_3$)$_3$	a	a	1.548(33)	173.6(21)	173.1(25)	
Co(C≡CH)(PMe$_3$)$_4$	1.890(6)	1.185(10)^b	a	180.0(1)	a	(79)
[*trans*-Co{C≡CCO$_2$Et}(H)(depe)$_2$][BPh$_4$]	1.911(10)	1.192(15)	1.393(17)	176.2(9)	169.9(11)	(80)
trans-Co(C≡CSiMe$_3$){N$_4$}	1.948(5)	1.127(7)^b	1.851(6)	175.3(5)	173.0(5)	(81)
[Co(C≡CSiMe$_3$)(H){P(CH$_2$CH$_2$PPh$_2$)$_3$}][BPh$_4$]	a	a	a	a	178$^{(a)}$	(82)
[*mer*-Co(C≡CBu^t)(CH=CHBu^t)(NCMe)(PMe$_3$)$_3$][BPh$_4$]	1.869(4)	1.209(6)^b	1.476(6)	177.4(3)	178.6(4)	(83)
CoCp(C≡CCN)$_2$(PPh$_3$)	1.857(12)	1.192(14)	1.358(15)	174(1)	178(1)	(84)
Rh(C≡CSiMe$_3$)(PMe$_3$)$_3$	1.877(7)	1.198(10)^b	1.363(13)	173(1)	175(1)	(85)
Rh(C≡CPh)(PMe$_3$)$_4$	2.02(1)	1.19(2)	1.79(2)	179.8(6)	176.5(5)	(85)
{Rh(PMe$_3$)$_4$}$_2$(μ-C≡CC$_6$H$_4$-4-C≡C)	2.002(3)	1.212(4)^b	1.432(4)	178.0(1)	176.1(2)	(85)
Rh(C≡CPh){η^2-C(CN)$_2$=C(CN)$_2$}(NCMe)(PPh$_3$)$_2$	2.014(4)	1.209(6)^b	1.438(6)	177.3(4)	177.7(5)	(5b,86a,b)
	1.939(18)	1.179(28)	1.508(26)	174.6(17)	168.7(21)	(87)

Compound						Ref.
[Rh(C≡CH)(H){N(CH₂CH₂PPh₂)₃}][BPh₄]	1.96(1)	1.20(2)	^{a}	177(1)		(88a,b)
[cis-Rh(C≡CCH₂CH₂OH)(H)(PMe₃)₄]Cl	2.035(3)	1.196(4)^b	^{a}	178.5(5)		(89)
trans,mer-Rh(C≡CPh)₂(H)(PMe₃)₃	2.019(4)	1.204(6)^b	1.439(6)	173.9(2)		(90)
	2.031(4)	1.213(6)^b	1.440(6)	178.4(2)		
{trans-Rh(C≡CSiMe₃)₂(H)(dmpe)}₂(μ-dmpe)	2.023(6)	1.200(9)^b	1.810(7)	178.1(6)		(90)
	2.031(6)	1.203(9)^b	1.797(7)	176.9(5)		
trans-Rh(C≡CSiMe₃)₂(H)(py)(PPr$^i_3$)₂	2.032(6)	1.207(9)^b	1.797(7)	177.0(5)		(91)
	2.020(6)	1.219(9)^b	1.801(7)	176.8(6)		
Rh(C≡CCMe=CH₂)(H)Cl(py)(PPr$^i_3$)₂	1.958(4)	1.206(5)^b	1.432(6)	177.1(4)		(92)
Rh₂(C≡CH)(μ-NC₅H₄-2-NPh)₄	2.021(10)	^{a}	^{a}	175.6(9)		(33i)
Ir(C≡CPh)(cod)(PCy₃)	1.998(6)	1.200(10)^b	1.447(11)	176.5(8)		(93)
Ir(C≡CPh)(CO)(μ-dppm)₂CuCl	2.04(1)	1.19(1)	1.46(1)	172.8(6)		(94)
Ir{(C≡CPh){C=CC=C-1,2,3,4-(CO₂Me)₄} (CO)(PPh₃)₂	2.056(9)	1.198(13)	^{a}	^{a}		(95)
Ir{C=C(B₁₀C₂H₁₁){CH=CH(B₁₀C₂H₁₁)}Cl(CO)(PPh₃)₂	1.99(2)	1.21(2)	1.40(^{a})	175(2)	175(^{a})	(96)
trans-Ni(C≡CPh)₂(PEt₃)₂	1.87(1)	1.18(2)	1.46(2)	176.5(15)	176.5(15)	(97)
trans-Ni(C≡CPh)₂(PEt₃)₂	1.879(11)	1.218(14)	1.464(15)	177.27(110)	176.92(133)	(98)
trans-Ni(C≡CC₆H₄-2-CCH)(PBu$^n_3$)₂	1.859(17)	1.218(24)	1.469(24)	175.9(14)	178.9(16)	(99)
trans-Pd(C≡CSiMe₃)Br(PPh₃)₂	1.974(6)	1.16(1)	1.852(8)	175.0(6)	170.6(7)	(100)
{trans-PdCl(PEt₃)₂}₂{μ-C≡CC(NPh)C(NPh)}	1.939(8)	1.200(12)	1.450(12)	176.8(9)	173.8(11)	(101)
trans-Pd(C≡CPh){C(CO₂Me)=CH(CO₂Me)}(PEt₃)₂	2.04(2)	1.20(2)	1.47(2)	^{a}	^{a}	(102)
	2.03(2)	1.21(3)	1.41(3)	^{a}	^{a}	
trans-Pd(C≡CC₆H₄-2-CCH)(NCS)(PEt₃)₂	1.952(7)	1.20(1)	1.46(1)	176(1)	174(1)	(103)
cis-Pd(C≡CSiMe₃)Br(NMe₂CH₂C₆H₄-2-PPh₂)	1.945(4)	1.193(6)^b	1.834(4)	171.7(4)	172.3(4)	(104)
cis-Pd(C≡CPPh₃)(PPh₃)Cl₂	1.91(1)	1.23(2)	1.71(1)	175.5(9)	175.5(8)	(105)
trans-Pt(C≡CMe)₂(PEt₃)₂	2.062(8)	1.039(11)	1.541(13)	174.7(7)	173.5(9)	(106)
trans-Pt(C≡CPh)₂(PEt₃)₂	1.98(1)	1.21(1)	1.43(1)	177.8(9)	176(1)	(107)
trans-Pt(C≡CC≡CPh)₂(PMe₂Ph)₂	2.009(5)	1.175(8)^b	1.406(8)	177.8(5)	177.16(9)	(5b,e)
trans-Pt(C≡CCMe=CH₂)(PPh₃)₂	2.024(6)	1.18(1)	1.44(1)	177.4(5)	176.5(7)	(108)
trans-Pt(C≡CC₆H₄-4-OMe)(C≡CC₆H₄-4-NO₂) (PMe₂Ph)₂	^{a}	^{a}	^{a}	^{a}	^{a}	(5f)

(continued)

TABLE I (*continued*)

METRICAL DATA FOR TERMINAL METAL–ALKYNYL COMPLEXES

Compound	d(M–C) (Å)	d(C≡C) (Å)	d(C–R) (Å)	$\angle$(M–C≡C) (°)	$\angle$(C≡C–R) (°)	Ref.
trans-Pt(C≡CPh)Cl(PEt$_2$Ph)$_2$	1.98(2)	1.18(3)	1.44(3)	163(2)	174(2)	(*109*)
trans-Pt(C≡CCl)Cl(PPh$_3$)$_2$	1.933(12)	1.154(19)	1.690(15)	178(1)	^{a}	(*110*)
trans-Pt(C≡CCMe=CH$_2$)Cl(PPh$_3$)$_2$	2.14(3)	1.09(4)	1.45(2)	151(3)	170(2)	(*111*)
trans-Pt(C≡CCMe=CH$_2$)Cl(PPh$_3$)$_2$	1.84(2)	1.07(8)	1.59(8)	169(5)	178(2)	(*112*)
	1.88(6)	1.22(3)	1.52(3)	175(1)	173(6)	
trans-Pt{C≡CC(OH)Me$_2$}Cl(PPh$_3$)$_2$	1.95(2)	1.18(2)	1.50(3)	175(1)	176(2)	(*113*)
trans-Pt{C≡CC(OH)MeEt}Cl(PPh$_3$)$_2$	1.988(9)	1.15(2)	1.52(2)	174(1)	178(1)	(*113*)
trans-Pt(C≡C-*c*-C$_6$H$_{10}$-1-OEt)Cl(PPh$_3$)$_2$	1.935(15)	1.22(2)	1.49(3)	172(1)	178(2)	(*114*)
trans-Pt{C≡CCMe$_2$(OOH)}Br(PPh$_3$)$_2$	1.990(9)	1.15(1)	1.49(1)	175.7(8)	173(1)	(*115*)
trans-Pt{C≡CCMeEt(OH)}(H)(PPh$_3$)$_2$	1.90(1)	1.25(2)	1.58(4)	170(1)	169(2)	(*116*)
{*trans*-Pt(NCS)(PEt$_3$)$_2$}$_2$(μ-C≡CC$_6$H$_4$-4-C≡C)	1.921(12)	1.22(2)	1.45(2)	173(2)	178(2)	(*117*)
trans-Pt(C≡CBu^t)(CMe=CH$_2$)(PPh$_3$)$_2$	2.04(1)	1.21(2)	1.46(2)	^{a}	178(1)	(*118*)
trans-Pt(C≡CCMe=CH$_2$){C(CMe=CH$_2$)=CH$_2$} (PPh$_3$)$_2$	1.99(1)	1.20(2)	1.46(2)	179(1)	178(2)	(*119*)
trans-Pt{C≡CPh}(CPh=CH$_2$)(PPh$_3$)$_2$	1.99(2)	1.18(3)	1.48(2)	171(2)	176(2)	(*120*)
trans-Pt(C≡CCMe){C(CMe=C(CN)$_2$)=C-*c*-CH=CHC(=C(CN)$_2$)CH=CH}(PMe$_3$)$_2$	2.009(15)	1.15(3)	1.54(3)	178.5(15)	175.9(20)	(*121*)
[*trans*-Pt(C≡CBu^t)(PEt$_3$)$_2${C(1-py)=CHBu^t}][SbF$_6$]	2.00(1)	1.20(2)	^{a}	176(1)	178(2)	(*122*)
[*trans*-Pt(C≡CPh)(NHNC$_6$H$_4$-4-F)(PPh$_3$)$_2$][BF$_4$]	1.95(3)	1.21(5)	^{a}	178(^{a})	172(^{a})	(*123*)
cis-Pt(C≡CH)$_2$(PEt$_3$)$_2$	1.987(15)	1.205(25)	^{a}	178.7(17)	^{a}	(*124*)
	1.949(16)	1.241(30)	^{a}	178.1(16)	^{a}	
cis-Pt(C≡CPh)$_2$(PPh$_3$)$_2$	2.10(3)	1.17(5)	1.48(4)	162(4)	169(4)	(*125*)
	2.04(3)	1.14(6)	1.48(6)	172(4)	171(5)	
	2.07(4)	1.21(5)	1.47(5)	165(4)	173(4)	
	2.02(4)	1.14(6)	1.43(5)	176(4)	173(5)	
cis-Pt{C≡CC(OH)Me$_2$}(PPh$_3$)$_2$	2.02(2)	1.17(2)	1.51(2)	177(2)	178(2)	(*113*)
	1.99(2)	1.19(2)	1.51(2)	178(1)	177(2)	
cis-Pt(C≡CCO$_2$Me)(CO$_2$Et)(PPh$_3$)$_2$	1.991(8)	1.180(11)	1.434(13)	176.3(7)	174.3(9)	(*126*)

Compound						Ref.
cis-Pt(C≡CCN)(CN)(PPh₃)₂	1.96(3)	1.24(5)	1.31(5)	172.2(29)	173.7(37)	(127)
cis-Pt(C≡CBuᵗ)(CMe=CH₂)(PPh₃)₂	1.97(2)	1.22(2)	1.46(2)	176(2)	175(2)	(128)
cis-Pt{C≡C(C₅H₄FeCp)}₂(PPh₃)₂	2.010(11) 1.991(10)	1.159(16) 1.139(16)	1.485(17) 1.497(17)	176.5(10) 174.5(10)	a a	(129)
trans-Pt(C≡C-c-C₆H₁₀-1-OH)₂(H)₂(PPh₃)₂	2.004[a]	1.15[a]	1.55[a]	a	a	(130)
cis,trans-Pt(C≡CPh)₂(PEt₃)₂{N=N-N-1,4-(C₆H₄-4-NO₂)}	1.99(1)	1.20(2)	1.43(3)	172.5(9)	173(1)	(131)
Pt(C≡C-p-Tol)(μ-C≡C-p-Tol)(μ-dppm)₂W(CO)₃	2.00(1)	1.19(2)	1.47(2)	177.4(7)	175.0(2)	(132a,b)
[Pt(C≡CMe)(μ-C≡CMe)(μ-dppm)₂Rh(CO)][PF₆]	2.000(8)	1.19(1)	1.45(1)	176.5(9)	a	(133)
trans-Pt(C≡CBuᵗ)(μ-C≡CBuᵗ)(μ-dppm)AuCl)₂	1.96(2)	1.13(3)	1.53(5)	175(2)	177(3)	(134)
[Pt(C≡CBuᵗ)(μ-C≡CBuᵗ)(μ-dppm)₂PtCl][ClO₄]	2.007(7) 1.971(6)	1.211(10)[b] 1.194(9)[b]	1.491(11) a	174.1(6) 177.3(4)	175.5(8) a	(20a)
Pt₂(C≡CBuᵗ)₂(μ-dppm)₂(μ-Au)	2.01(3) 2.02(3)	1.14(3) 1.10(3)	1.59(5) 1.54(3)	172(2) 177(2)	176(3) 175(3)	(135)
[{Pt(C≡CBuᵗ)}₂(μ-C≡CBuᵗ)(μ-dppm)₂][BPh₄]	1.911(14) 1.995(13)	1.231(23) 1.195(20)	1.570(28) 1.457(23)	175.4(13) 175.5(13)	176.5(15) 176.7(17)	(136)
[{Pt(C≡CPh)}₂(μ-C≡CPh)(μ-dppm)₂][ClO₄]	1.961(7)	1.18(1)	a	176.5(7)	178.8(6)	(20a)
[Pt(C≡CPh)(μ-N₂C₃Br₃)(μ-dppm)₂Rh(CO)][PF₆]	1.971(12)	1.177(17)	a	176.3(8)	a	(137)
[Pt(C≡C-p-Tol)(μ-dppm)₂(μ-H)Ir{C≡C(C₆H₃Me)CH₂CH=CHCH₂}][BF₄]	1.985(14)	a	a	a	179.0(11)	(138)
[trans-Pt(C≡CPh)(PEt₃)₂{μ-η²:η⁴-CPh=CC(Ph)CH(Ph)}Pt(PEt₃)₂][BF₄]	2.07(2) 1.96(2)	1.11(2)	1.45(2)	177(1)	177(1)	(139)
[trans-Pt(C≡CPh)(PEt₃)₂{μ-η²:η⁴-CPh=CC(p-Tol)CH(p-Tol)}Pt(PEt₃)₂][BF₄]	a	1.20(3)	1.45(4)	174(3)	170(3)	(140)
[Pt(C≡CBuᵗ)(PPh₃){Au(PPh₃)}₆][Au(C≡CBuᵗ)₂]	a	a	a	a	a	(141)
[trans-Pt(C≡CBuᵗ)(μ-dppm)₂Au][PF₆]	1.958(11) 1.954(11)	1.23(2) 1.23(2)	1.44(1) a	174.0(9) 177(1)	176(1) a	(22)
trans-Pt(C≡CPh)₂(μ-dppm)₂AgI	2.00(3) 2.00(4)	1.22(4) 1.20(5)	a a	a a	a a	(142)

(continued)

TABLE I (continued)

METRICAL DATA FOR TERMINAL METAL–ALKYNYL COMPLEXES

Compound	d(M–C) (Å)	d(C≡C) (Å)	d(C–R) (Å)	∠(M–C≡C) (°)	∠(C≡C–R) (°)	Ref.
trans-Pt(C≡CPh)(PPh₃)₂{μ-C≡CPh}{cis-Pt(C₆F₅)₂(CO)}	1.917(12)	1.206(18)	1.478(16)	176.7(12)	177.2(13)	(143)
Pt₂(C≡CPh)(μ-C=CHPh)(PEt₃)₃Cl	2.045(16)	1.17(2)	1.45(2)	173.3(10)	178.3(19)	(20c)
Pt₂(C≡CPh)(μ-C=CHPh)(PEt₃)₃I	2.012(10)	1.171(14)	1.475(14)	171.3(10)	178.4(10)	(20c)
[Pt₂(C≡CPh)(μ-C=CHPh)(PEt₃)₄][BF₄]	2.17(2)	1.14⁽ᵃ⁾	ᵃ	163⁽ᵃ⁾	172⁽ᵃ⁾	(144)
Pt₃(C≡CPh)(PCy₃)₂{μ-C≡CPh}(μ-SiMe₂)	2.01(1)	1.20(2)	1.43(2)	174(1)	174(1)	(145)
Au(C≡CPh)(PPh₃)	1.97(2)	1.18(2)	1.46(2)	175.7(16)	176.5(18)	(146)
	2.02(2)	1.16(2)	1.47(2)	170.8(19)	174.0(20)	
Au(C≡CPh)(NH₂Prⁱ)	1.935(19)	1.210(28)	1.479(28)	174.2(18)	178.6(21)	(147)
Au(C≡CC₆F₅)(PPh₃)	1.993(14)	1.197(16)	1.442(20)	175.4(10)	178.4(12)	(148)
Au(C≡CCH₂OMe){P(p-Tol)₃}	2.024(2)	1.169(3)ᵇ	1.474(4)	170.3(2)	178.2(3)	(149)
Au(C≡CBuᵗ)(CNC₆H₃-2-Me-4-CCH)	1.955(10)	1.206(15)	1.463(16)	173.6(10)	173.9(12)	(150)
{Au(C≡CPh)}₂{μ-PPrⁱ₂-4,4'-C₆H₄-C₆H₄-PPrⁱ₂}	1.997(9)	1.179(11)	1.451(11)	172.7(9)	177.2(9)	(151)

ᵃ Not reported.
ᵇ Data included in Figure 2.
ᶜ Second datum due to disordered atom.

Examination of the data in Table I reveals that only one-third of the approximately 150 metal–alkynyl structures therein have been determined to a level of precision sufficient for meaningful discussions of the finer details of electronic structure; by our (admittedly arbitrary) definition, these are structures for which the estimated standard deviation (esd) for the $C\equiv C$ bond distance is ≤ 0.010 Å. The dearth of precise structures is a reflection of the experimental difficulty, using X-ray diffraction methods, of locating light atoms that lie near a heavy-metal center. This point is well illustrated by the results of the two independent crystal-structure determinations for the complex $trans$-Fe($C\equiv CPh$)Cl(dmpe)$_2$; despite consistencies among the space group and unit cell dimensions, and the low R-factor to which each structure is refined, the Fe–C, $C\equiv C$ and CC–C(Ph) bond distances reported in the two studies differ by 0.017, 0.024, and 0.021 Å, respectively (53,54). Not surprisingly, precise bond distances and bond angles are particularly uncommon for M—C$\equiv$C—R fragments of alkynyl complexes of third-row transition metals. We will confine our discussion to those metrical data with small esd's.

The following discussion centers on the electronic-structural implications of bond lengths within the M—C$\equiv$C—R fragments of metal–alkynyl complexes. We will not dwell on the bond angles within these subunits, and instead note at the outset that neither the M—C$\equiv$C nor the C$\equiv$C—R bond angle typically differs significantly from the expected 180°: The median for both angles is 177° among those structures for which the esd's $\leq 1.0°$. Thus, these parameters appear to have little diagnostic value regarding the nature of metal–alkynyl bonding for complexes with terminal alkynyl ligands, other than to denote the M—C$\equiv$C—R unit as being alkyne-like. Small deviations from linearity for bond angles involving the C$\equiv$C fragment are not surprising in view of the fact that bending force constants of sp-hybridized carbon atoms are small compared to those for sp^2- and sp^3-hybridized carbon centers (152). Bending along the M—C$\equiv$C and C$\equiv$C–R coordinates has typically been interpreted as being the result of intermolecular interactions in the crystal, although specific crystal-packing details have rarely been presented. In principle, the structures—the M—C$\equiv$C—H bond angles, especially—of parent ethynyl complexes may additionally be susceptible to perturbation by an intermolecular interaction between the terminal hydrogen atom of the ethynyl ligand and the C$\equiv$C bond of an adjacent molecule. Such an arrangement is found for acetylene, which packs within the crystal in a T shape in which the C$\equiv$C axis of one molecule is roughly perpendicular to that of an adjacent one; the intermolecular H$\cdots$C distance is 2.74–3.39 Å, depending on temperature (153). At present, the only reported observation of a similar intermolecular arrangement of the CCH moieties of an ethynyl complex is for $Mo_2(C\equiv CH)_4(PMe_3)_4$, for which the $H\cdots C\equiv C$ distance is 3.01 Å at 218 K (Figure 1) (*15b*).

B. C≡C Bond Distances

The $C\equiv C$ bond distance has been invoked as a benchmark parameter for establishing the presence or absence of $M \rightarrow CCR$ π-backbonding in alkynyl complexes; bond distances both longer and shorter than "normal" (*vide infra*) have been taken as evidence for the π-acceptor character of alkynyl ligands. However, in metal–carbonyl complexes, only small differences (ca. 0.03 Å) in $C\equiv O$ bond distances are observed among complexes for which the $C\equiv O$ stretching frequencies (and hence, presumably, the magnitude of $M \rightarrow CO$ π-backbonding) vary significantly (*154*). Since carbon monoxide is unquestionably a stronger π-acceptor than are alkynyl ligands, the structural perturbation of the $C\equiv C$ bond by $M–C$ π-bonding would not be expected to be large.

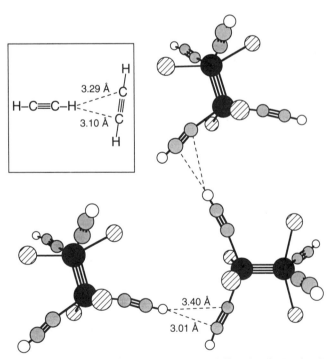

FIG. 1. Intermolecular interactions between ethynyl ligands of $Mo_2(C\equiv CH)_4(PMe_3)_4$ (*15b*). *Inset:* Intermolecular interactions in crystalline acetylene at $T = 131$ K (*153*).

Displayed in Fig. 2 is the distribution of $C\equiv C$ bond lengths for metal–alkynyl complexes in which the alkynyl R-group is hydrogen or is attached through a carbon or silicon atom; only $C\equiv C$ bond lengths for which the esd is 0.010 Å or less are included. The distribution of $C\equiv C$ distances is quite narrow, with two-thirds of them lying in the range 1.190–1.214 Å; the mean $C\equiv C$ distance for substituted (L_nM—$C\equiv C$—C or L_nM—$C\equiv C$—Si) compounds is 1.201(16) Å, where the standard deviation is that pertaining to the distribution, not the average of the crystallographic esd's (155). The $C\equiv C$ bond lengths of trimethylsilylethynyl complexes $[L_nM$—$C\equiv C$—$SiMe_3$; $d(C\equiv C)_{mean}$ = 1.210(13) Å, ignoring the outlying point at 1.127 Å (81)] may be slightly longer than those of L_nM—$C\equiv C$—C(R) compounds, but there are too few data at present, and the distances are too widely distributed, to justify treating them separately from the latter complexes; in any case, removing them from the distribution does not change the mean $C\equiv C$ bond length. Even fewer data are available for complexes of the parent ethynyl ligand $[L_nM$—$C\equiv C$—H;

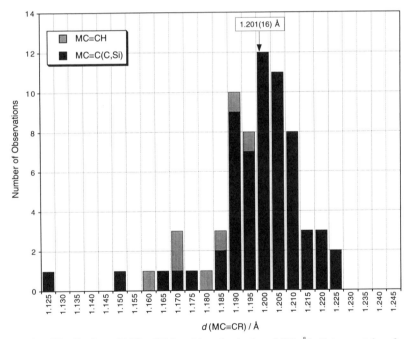

FIG. 2. Distribution of $MC\equiv CR$ bond lengths (esd $\leq$ 0.010 Å). Data are taken from Table I. In structures where two $C\equiv C$ bonds are identical under crystallographically imposed symmetry, only a single bond length is reported. The distance labels correspond to the left edges of the bins.

$d(C\equiv C)_{mean}$ = 1.181(13) Å], although reference to Fig. 2 suggests fairly strongly that the $C\equiv C$ bonds of these compounds are shorter than those of substituted alkynyl complexes.

The logical way to interpret these bond lengths is to compare them to those of organic alkynes. In many papers, the reference point for what constitutes a "normal" $C\equiv C$ bond distance is that of acetylene, which, at 1.2033(2) Å [in the gas phase (*156*); other distances have been reported in crystallographic studies (*153*)], is nearly identical to the mean $C\equiv C$ bond length for substituted metal–alkynyl complexes [1.201(16) Å]. The bond length of the terminal alkyne ($HC\equiv CR$) corresponding to the alkynyl ligand in question ($L_nM-C\equiv CR$) is also a commonly used reference. However, a more meaningful interpretation of these data is afforded by establishing $C\equiv C$ bond-length ranges for organic alkynes, and comparing these with the distribution of $C\equiv C$ bond lengths in metal–alkynyl complexes. Figure 3 displays the distributions of $C\equiv C$ bond lengths for organic alkynes of the type $HC\equiv CR$ and $RC\equiv CR'$ (where R is an organic substituent attached to the $C\equiv C$ core through a carbon atom, and R' is attached through either a carbon or a silicon atom) whose crystal structures are recorded in the Cambridge Structural Database. These histograms include data only from crystallographically ordered, well-refined structures (R ≤ 10%) for which the average esd of all carbon–carbon distances within the molecule is 0.010 Å or less. As reported in an earlier compilation of the bond lengths of organic molecules (*157*), the mean $C\equiv C$ bond distance for organic terminal alkynes [1.171(14) Å] is shorter than that for disubstituted alkynes [1.192(12) Å]; both distributions are narrow. Viewed in this context, it is apparent that the $C\equiv C$ bonds of metal–alkynyl compounds are, in fact, slightly elongated (ca. 0.01 Å) relative to organic alkynes, and that the observation of shorter $C\equiv C$ bonds for the parent metal–ethynyl compounds relative to substituted metal–alkynyl compounds (L_nMCCR, R ≠ H) is consistent with the data for organic terminal and disubstituted alkynes. There are too few data for silyl-substituted alkynes to indicate whether their $C\equiv C$ bonds are elongated relative to those of alkynes with a $C-C\equiv C-C$ core, although, as for trimethylsilylethynyl metal complexes, there is a preliminary indication to that effect. These distributions also make clear that acetylene is an inappropriate reference compound because its gas-phase $C\equiv C$ bond length is unusually long, and that comparing the $C\equiv C$ bond lengths of terminal alkynes to those of substituted metal–alkynyl compounds is likely to significantly overestimate the elongation of the latter.

One possible explanation for the generally longer $C\equiv C$ bonds of metal–alkynyl complexes relative to organic alkynes is that they are the result of M → CCR π-backbonding [or RCC → M π-bonding, as could be

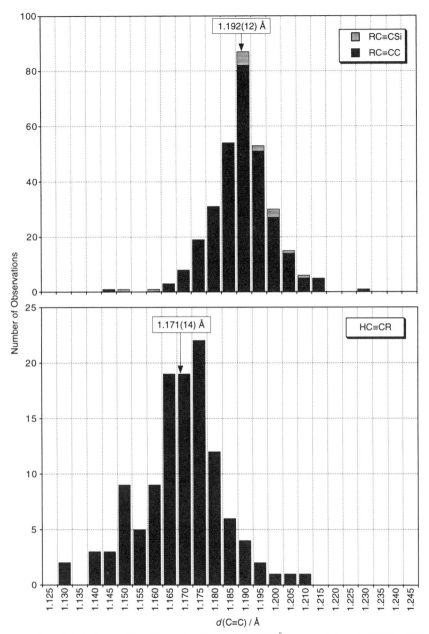

FIG. 3. Distributions of C≡C bond lengths (esd ≤ 0.010 Å) in organic terminal (RC≡CH) and disubstituted (RC≡CR, RC≡CR') alkynes. Data are taken from the Cambridge Structural Database. In structures with two or more crystallographically identical C≡C bonds, only a single bond length is included. Two outlying bond distances, at 1.12 and 1.36 Å, are not shown in the top figure. The distance labels correspond to the left edges of the bins.

103

inferred from the results of photoelectron-spectroscopic studies (12–13)].
Unfortunately, the narrow range of C≡C bond lengths, and their large
individual uncertainties, makes this hypothesis essentially impossible to
test via comparisons among L_nM–CCR complexes for which the ML_n
fragments possess different d-electron counts or π-basicities, because, at
the 3σ confidence level, nearly all C≡C bond lengths for the compounds
in Fig. 2 are indistinguishable, and some of those that are statistically
distinct from one another lack an obvious chemical reason for being
so [e.g., d(C≡C) = 1.152(9) Å for Fe(C≡CMe)$_2$(dmpe)$_2$; d(C≡C) =
1.222(4) Å for Fe(C≡CPh)$_2$(depe)$_2$] ($52a$). The sensitivity of the C≡C
bond length to subtle differences in metal–alkynyl bonding is clearly very
low, so small differences among these distances must be interpreted with
caution. On a positive note, the extreme narrowness of the C≡C bond-
length distribution suggests that bond lengths that clearly lie outside this
range could be viewed as substantive evidence for significant M–CCR
π-interactions. The only structural study to date that has provided compelling evidence for elongated C≡C bonds is that of Carty and co-workers
for complexes of the type $trans$-Ru(C≡CR)$_2$(CO)$_2$(PEt$_3$)$_2$ (R = H, Ph,
SiMe$_3$, C≡CH, C≡CSiMe$_3$) ($63a,b$). Long C≡C bonds are observed in
the structures of the complexes for which R = SiMe$_3$ [1.221(2) Å] and
C≡CSiMe$_3$ [1.226(2) Å]. Corresponding contractions of the Ru–C bonds
are also observed (Section III.C.3), consistent with multiple-bond character in these linkages.

An alternative explanation for the 0.01-Å elongation of metal-alkynyl
C≡C bonds relative to those of organic alkynes is that they are the result
of metal–carbon σ-bonds that are polarized in a $M^{\delta+}$–$C^{\delta-}$ fashion. The
lone pair of the alkynyl anion, [C≡CR]$^-$, is slightly C≡C σ-antibonding
in character; significant residual negative charge on the alkynyl ligand
could thus result in a longer-than-normal C≡C distance. An upper limit
to the bond-length elongation due to this effect is provided by the C≡C
distance of the free ethynyl anion, [C≡CH]$^-$. Ab $initio$ molecular-struc-
ture calculations on this ion yield d(C≡C) = 1.2463(10) Å (158), which
is ca. 0.04 Å longer than the C≡C bond of acetylene [1.2033(2) Å] (156),
and ca. 0.08 Å longer than the average bond length of organic terminal
alkynes [1.171(14) Å, Fig. 3]; these same calculations predict correctly
the experimentally observed vibrational frequencies of this ion (159), suggesting that the calculated bond distance is a reliable one. While the
$M^{\delta+}$–$C^{\delta-}$ bond-polarity effect undoubtedly contributes to the general
bond-length elongation of metal–alkynyl complexes, it is unlikely that it
alone accounts for the long C≡C bonds observed for complexes such as
$trans$-Ru(C≡CR)$_2$(CO)$_2$(PEt$_3$)$_2$ ($63a,b$) given that this is far from being
the most electropositive metal center among those whose distances are

represented in Fig. 2. The $C\equiv C$ bond lengths of alkali or alkaline-earth metal–alkynyl compounds would seem to be logical reference compounds to test this point. Unfortunately, nearly all either are polymeric or involve both σ- and π-type metal–carbon interactions.

C. M–C Bond Distances

As is the case for transition-metal carbonyl complexes (154), the M–C bond distance should be a more sensitive structural indicator of the nature of the M–CCR bond than is the length of the multiple bond, because bond distances change more between bond orders 1 and 2 than between bond orders 2 and 3. As for the $C\equiv C$ bond lengths described earlier, classifying a M–CCR bond length as "normal" or "short" requires a reference point; some approaches to this are described next.

1. Comparisons to Metal–Alkyl Compounds

The best and simplest means of placing the M–C bond distances of metal–alkyl complexes in a context that allows insight into whether or not they manifest M–CCR π-interactions is to compare them to M–C bond distances of complexes where such effects are expected to be insignificant, as for metal–alkynyl compounds. For these comparisons to be most meaningful, the pair of complexes under consideration should obviously be closely similar with respect to oxidation state, overall charge, geometry, and ancillary ligands. Differences in M–C bond distances between alkynyl and alkyl complexes also need to be corrected for the 0.08-Å difference between the single-bond covalent radii of sp- and sp^3-hybridized carbon atoms $[r(C_{sp}) = 0.69$ Å, $r(C_{sp^3}) = 0.77$ Å] (157); only in those cases where the bond-length difference exceeds 0.08 Å by a statistically significant amount would invoking a M–CCR π-bonding explanation appear to be justified. We emphasize these points because one widely cited study that concludes that alkynyl ligands participate in little or no M $\rightarrow$ CCR π-backbonding does so on the basis of comparisons between M–C bond distances of neutral alkynyl complexes and cationic carbonyl compounds, and the use of 0.60 Å (one-half the $C\equiv C$ bond length) as the single-bond covalent radius for sp-hybridized carbon (69); this latter error has been repeated in a subsequent analysis (70).

A complication with this type of interpretation is that it is formally valid only if the M–C bonds of the alkynyl and alkyl complexes in question are of comparable covalency, because bonds become stronger with increasing ionic character. The electronegativity of carbon is a sensitive function of its hybridization $[\chi(C_{sp}) = 2.99, \chi(C_{sp^3}) = 2.48$; Pauling scale] (160); thus, the M–C bond ionicities of analogous metal–alkyl and metal–alkynyl complexes will differ, perhaps significantly enough to be manifested structurally. There are several equations for estimating empirically the ionic character of a bond, due originally to Pauling, and, related to these, equations for correcting covalent bond lengths for ionic character (161), of which the earliest is the well-known formula of Schomaker and Stevenson [Eq. (1)] (162):

$$d(\text{A–B})_{\text{calc}} = r(\text{A}) + r(\text{B}) - 0.09|\chi(\text{A}) - \chi(\text{B})| \tag{1}$$

The successes and failures of this and related equations have been well documented (163); in no case are any claimed to be of more than qualitative significance. Although many of these relationships were developed from data for main-group compounds, their application is not limited to these species. For example, Labinger and Bercaw have advanced a classical electronegativity argument to explain the fact that M–H bonds are typically stronger than M–C(alkyl) bonds (164).

An updated version of the Schomaker-Stevenson equation, which was generated by fitting more accurate bond distances, is given in Eq. (2) (165):

$$d(\text{A–B})_{\text{calc}} = r(\text{A}) + r(\text{B}) - 0.07[\chi(\text{A}) - \chi(\text{B})]^2 \tag{2}$$

The problem with applying this or related equations to the calculation of M–C bond lengths for metal–alkynyl complexes of the type $\text{ML}_n(\text{CCR})$ is that one typically knows neither $r(\text{ML}_n)$ nor $\chi(\text{ML}_n)$, either or both of which may differ significantly from the tabulated values for bare M or M^{x+}. If, however, M–C distances for two $\text{ML}_n\text{R}'$ ($\text{R}' = $ alkyl, alkynyl) complexes are available, then $r(\text{ML}_n)$ can be factored out and the bond-length difference $\Delta d(\text{M–C}) = [d(\text{M–C}_{sp^3}) - d(\text{M–C}_{sp})]$ can be expressed as a function of $\chi(\text{ML}_n)$ [Eq. (3)]; if an experimental value of $\Delta d(\text{M–C})$ for a pair of $\text{ML}_n\text{R}'$ complexes gives a chemically reasonable value of $\chi(\text{ML}_n)$, then M–C π-bonding need not be invoked, even if $\Delta d(\text{M–C})$ exceeds 0.08 Å.

$$\Delta d(\text{M–C})_{\text{calc}} = r(\text{C}_{sp^3}) - r(\text{C}_{sp}) - 0.07[\chi(\text{ML}_n) - \chi(\text{C}_{sp^3})]^2$$
$$+ 0.07[\chi(\text{ML}_n) - \chi(\text{C}_{sp})]^2 \tag{3}$$

Figure 4 shows plots of the relationship between $\Delta d(\text{M–C})_{\text{calc}}$ and $\chi(\text{ML}_n)$, as derived from Eq. (3), using several different values each for the electronegativities of alkynyl and alkyl ligands. One line is based on the Pauling electronegativities of C_{sp} ($\chi = 2.99$) and C_{sp^3} ($\chi = 2.48$) (160), which undoubtedly overestimate the difference between the ionic character of their respective M–C bonds for most pairs of alkynyl and alkyl ligands. The other lines are based on the group electronegativities (160,

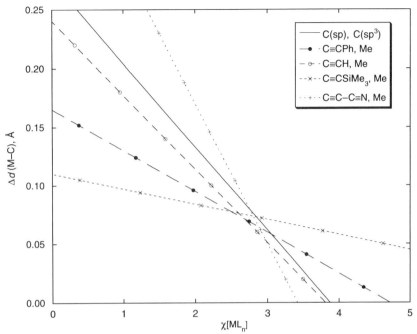

FIG. 4. Relationship between Δd(M–C) and χ(ML$_n$), as provided by Eq. (3). The solid line is that obtained using the Pauling electronegativities for sp- and sp^3-hybridized carbon atoms; other lines are obtained using group electronegativities for the indicated ligands.

166) of specific alkynyl ligands [χ(C≡CH) = 2.76, χ(C≡CPh) = 2.56, χ(C≡CSiMe$_3$) = 2.40, χ(C≡CCN) = 3.17] and that of methyl (χ = 2.31); while the group electronegativities are probably more realistic than simply using χ(C$_{sp}$) and χ(C$_{sp^3}$) they too must be calculated, introducing yet another layer of approximation to the process. Different lines could be generated using electronegativities for other alkyl groups, other bond-distance–bond-ionicity equations, or other electronegativity scales. We wish to emphasize clearly that our choice of Eq. (2) and its associated parameter sets is based neither on a strong conviction that they are generally correct nor on independent evidence that they treat M–C bond lengths appropriately; we present these correlations only as one means by which M–C bond-length data may be intrepreted, and to encourage discussion along these lines in future papers on metal–alkynyl structural chemistry.

Collected in Table II ($167–176$) are the M–C bond lengths (esd ≤ 0.010 Å) for ML$_n$(CCR) metal–alkynyl complexes for which an analogous ML$_n$R′ (R′ = alkyl) complex has also been structurally characterized.

TABLE II

METAL–CARBON BOND LENGTHS FOR ANALOGOUS METAL–ALKYL AND METAL–ALKYNYL COMPLEXES

Metal–alkyl complex	d(M–C)^a (Å)	Ref.	Metal–alkynyl complex	d(M–C)^a (Å)	Ref.	Δd^b (Å)
ZrCp$_2$Me$_2$	2.277[5]	(167)	ZrCp$_2$(C≡CMe)$_2$	2.249(3)	(35)	0.03
Mo$_2$Me$_4$(PMe$_3$)$_4$	2.245[4]	(168)	Mo$_2$(C≡CH)$_4$(PMe$_3$)$_4$	2.157[4]	(15b)	0.09
			Mo$_2$(C≡CMe)$_4$(PMe$_3$)$_4$	2.145[4]	(15e)	0.10
trans-W(Bu^n)(CH)(dmpe)$_2$	2.405(6)	(17)	trans-W(C≡CH)(CH)(dmpe)$_2$	2.263(5)	(16c)	0.14
			trans-W(C≡CPh)(CH)(dmpe)$_2$	2.250(4)	(16c)	0.16
			trans-W(C≡CSiMe$_3$)(CH)(dmpe)$_2$	2.246(6)	(16b)	0.16
			trans-W(C≡CC$_6$H$_4$-4-CCPr^n)(CH)(dmpe)$_2$	2.2458	(16c)	0.16
fac-Mn{CH$_2$SiMe(CH$_2$PPh$_2$)$_2$}(CO)$_3$	2.198(5)	(169)	fac-Mn(C≡CBu^t)(dppe)(CO)$_3$	1.996(6)	(45)	0.20
mer-Mn{P(OPh)$_2$(OC$_6$H$_4$-2-CH$_2$)}{P(OPh)$_3$}(CO)$_3$	2.207(4)	(170)	mer-Mn(C≡CPh){P(OPh)$_3$}$_2$(CO)$_3$	2.002(6)	(46)	0.21
Fe{η^5-C$_5$H$_4$PPh$_2$(WCpMe(CO)$_2$)}Me(CO)$_2$	2.06(1)	(171)	FeCp(C≡CPh)(CO)$_2$	1.920(6)	(57)	0.14
			FeCp*(C≡CH)(CO)$_2$	1.921(3)	(59)	0.14
			FeCp*(C≡CPh)(CO)$_2$	1.924(7)	(59)	0.14
(S_{Ru}, R_C)-RuCpMe(PPh$_2$CHMeCH$_2$PPh$_2$)	2.169(6)	(172)	RuCp(C≡CPh)(PPh$_3$)$_2$	2.016(3)	(69)	0.15
			RuCp(C≡CPh)(dppe)	2.009(3)	(70)	0.16
			RuCp{C≡CC(O)Me}(PPh$_3$)$_2$	1.996(3)	(72)	0.17
Co{η^5-C$_5$H$_2$Me$_3$}Me$_2$(PPh$_3$)	1.997(5)	(173)	CoCp(C≡CCN)$_2$(PPh$_3$)	1.867[10]	(84)	0.13
cis-PdMe$_2$(tmeda)	2.028[3]	(174)	Pd(C≡CSiMe$_3$)Br(NMe$_2$CH$_2$C$_6$H$_4$-2-PPh$_2$)	1.945(4)	(104)	0.08
trans-PtMeCl(PPh$_3$)$_2$	2.08(1)	(175)	trans-Pt{C≡CC(OH)MeEt}Cl(PPh$_3$)$_2$	1.988(9)	(113)	0.09
Au{CH$_2$CH$_2$C(O)Ph}(PPh$_3$)	2.077(9)	(176)	Au(C≡CCH$_2$OMe){P(p-Tol)$_3$}	2.024(2)	(149)	0.05
			{Au(C≡CPh)}$_2${μ-PPr$_2^i$-4,4′-C$_6$H$_4$-C$_6$H$_4$PPr$_2^i$}	1.997(9)	(151)	0.08

^{a} Values in square brackets are the means of crystallographic esd's.
^{b} Δd(M–C) = [d(M–C$_{sp^3}$) − d(M–C$_{sp}$)].

The paucity of available comparisons, despite the number of entries in Table I, is again testimony to the experimental difficulties that beset the determination of M–C bond lengths.

The first entry in Table II is in some respects the most intriguing. Erker and co-workers have reported a high-quality structure for $ZrCp_2$ $(C\equiv CMe)_2$ (35), the Zr–C bond lengths of which [2.249(3) Å] are 0.03 Å shorter than those for $ZrCp_2Me_2$ [2.277(5) Å] (167), where the value in parentheses is the average of the crystallographic eds's. Since $M \rightarrow CCR$ π-backbonding is absent for these d^0 systems, and theoretical calculations on the related complex $TiCp_2(CCH)_2$ indicate that the alkynyl ligand is not acting as a π-donor (14), the fact that $\Delta d(Zr-C)$ is significantly different from the pure single-bond value of 0.08 Å is surprising; that it is considerably *smaller* than 0.08 Å is striking under any circumstances. From bond-ionicity arguments (Fig. 4), the 0.03-Å difference between the lengths of the Zr–Me and Zr–CCMe bonds can be rationalized only if the group electronegativity of the $[ZrCp_2R']$ fragment is >3.5! While this value seems implausibly high, it finds some support in the observations of Labinger and Bercaw (164), who conclude from their study of metal–hydride and metal–alkyl bond strengths that the electronegativity of $[ScCp_2^*]$ is quite high, being roughly equal to that of methyl. This pair of zirconium compounds represents the only case for which $\Delta d(M-C) \ll 0.08$ Å; structural studies of other pairs of metal–alkynyl and metal–alkyl compounds from Group 3 or 4 would provide insight into the generality or singularity of this observation.

Studies in our research group on conjugated transition-metal compounds and polymers have resulted in the structural characterization of several quadruply metal–metal bonded dimetallapolyyne complexes of the type $M_2(C\equiv CR)_4(PMe_3)_4$ (M = Mo, W; R = H, alkyl, aryl, $SiMe_3$) (15a–e). The electronic, resonance-Raman, and NMR spectra of these species provide clear evidence for M–CCR π-interactions between the M_2 core and the alkynyl ligands (Scheme 1, Section II-B). Comparison of the M–C bond lengths of $Mo_2(C\equiv CH)_4(PMe_3)_4$ [2.157(4) Å] (15b) and $Mo_2(C\equiv C-Me)_4(PMe_3)_4$ [2.145(4) Å] (15e) with those of the analogous complex $Mo_2Me_4(PMe_3)_4$ [2.245(4) Å] (168) yields $\Delta d(Mo-C)$ values of 0.09 Å and 0.10 Å, respectively (Table II). This indicates that the M–CCR π-interactions in these complexes are of insufficient magnitude to be manifested structurally. From Fig. 4, the group electronegativity of the $[Mo_2R_3'(PMe_3)_4]$ fragment obtained from these values of $\Delta d(Mo-C)$ is 2.1–2.4; this compares favorably with $\chi(Mo^{II}) = 2.19$ (160).

The class of metallabutadiyne complexes of the type *trans*-$W(C\equiv CR)(\equiv CH)(dmpe)_2$ (R = H, $SiMe_3$, Ph, C_6H_4-4-CCPr^n) provides a second example from our laboratory for which there is strong spectroscopic evidence for metal–alkynyl π-conjugation (Section II-B). The W–C bond lengths of these complexes lie in the range 2.245(8)–2.263(5) Å ($16a$–c), and are considerably shorter [Δd(W–C) = 0.14–0.16 Å, Table II] than the W–C_{sp^3} bond length for $trans$-W(Bu^n)($\equiv$CH)(dmpe)$_2$ [2.405(6) Å] (17). This is consistent with the presence of a significant π-bonding component to the HCW–CCR linkage; a bond-ionicity interpretation for the 0.16-Å bond-length difference observed for the R = SiMe$_3$, Ph, and C$_6$H$_4$-4-CCPr^n derivatives unreasonably suggests χ[W($\equiv$CH)(dmpe)$_2$] $\cong$ 0 (Fig. 4), in the absence of M–CCR π-bonding (and steric effects for the n-Bu ligand). Consistent with this finding, the C$\equiv$C bond lengths of $trans$-W(C$\equiv$CPh)($\equiv$CH)(dmpe)$_2$ [1.222(5) Å] ($16c$) and W(C$\equiv$CSiMe$_3$)($\equiv$CH) (dmpe)$_2$ [1.228(9) Å] ($16b$) are among the longest known for metal–alkynyl complexes (Table I), although, as noted in Section III-B, the interpretation of these data as being indicative of M–CCR π-bonding is not unambiguous.

The largest apparent contractions of metal–alkynyl bonds relative to metal–alkyl bonds are found for complexes of manganese. Comparison of the Mn–C bond lengths of the Mn^I alkynyl complexes fac-Mn(C$\equiv$CBu^t)(dppe) (CO)$_3$ [1.996(6) Å] (45) and mer-Mn(C$\equiv$CPh){P(OPh)$_3$}$_2$(CO)$_3$ [2.002(6) Å] (46) with those of the alkyl complexes fac-Mn(CO)$_3${CH$_2$SiMe(CH$_2$PPh$_2$)$_2$}$_2$ [2.198(5) Å] (169) and mer-Mn(CO)$_3${P(OPh)$_3$}{P(OPh)$_2$(OC$_6$H$_4$-2-CH$_2$)} [2.207(4) Å] (170), respectively, gives Δd(Mn–C) $\cong$ 0.20 Å for both pairs of complexes (Table II). An unfortunate complication with this comparison is that the Mn–C bond distances of the alkyl reference compounds may be lengthened by ring strain, which would inflate Δd(Mn–C). There is no independent evidence of M $\rightarrow$ CCR π-backbonding from spectroscopic studies of these alkynyl complexes that would lead one to expect short Mn–C bonds for them.

The structures of complexes of the FeCpR'(CO)$_2$ type are worthy of mention because FeCp(C$\equiv$CR)(CO)$_2$ complexes are among the few alkynyl compounds to have been the subject of both theoretical calculations (11) and photoelectron-spectroscopic studies. ($12a,b$). The Fe–C bond lengths of the structurally characterized FeCp(C$\equiv$CR)(CO)$_2$ complexes ($57,59$) exhibit remarkable consistency (Table II), with an average Fe–CCR distance of 1.922 Å. Unfortunately, the 2.06(1)-Å Fe–C(Me) bond distance of Fe{η^5-C$_5$H$_4$(PPh$_2$WCpMe(CO)$_2$)}Me(CO)$_2$ (171), the alkyl reference compound, is not sufficiently precise to allow with any certainty the assessment of the nature of the Fe–CCR bonding in these compounds.

The series of related ruthenium complexes of the type RuCp(C$\equiv$CR)L$_2$ (R = Ph, L = PPh$_3$, ½ dppe; R = C(O)Me, L = PPh$_3$) ($69,70,72$), are more useful in this context. Comparison of the Ru–C bond distance of

RuCpMe(PPh$_2$CHMeCH$_2$PPh$_2$) [2.169(6) Å] (172) with those of the alkynyl complexes yields Δd(Ru–C) = 0.15–0.17 (Table II), which would appear to indicate that the Ru–C bond order of the alkynyl complexes is greater than 1; for Δd(Ru–C) to be the result of bond-ionicity factors alone in the phenylethynyl complexes would require χ[RuCpL$_2$] < 0.5 (Fig. 4). The 0.02-Å difference between the Ru–C distances of the two RuCp(C≡CR)(PPh$_3$)$_2$ complexes [R = Ph, 2.016(3) Å; R = C(O)Me, 1.996(3) Å] might be interpreted as implying that this π-bonding is due to Ru → CCR π-backbonding, since the CCC(O)Me ligand should be a better π-acceptor than CCPh. However, the electronegativities of the two ligands are sufficiently different (χ[C≡CPh] = 2.56, χ[C≡CC(O)Me] = 2.76) that this need not be the case; for RuCp{C≡CC(O)Me}(PPh$_3$)$_2$, a group electronegativity for the [RuCp(PPh$_3$)$_2$] fragment of ca. 1.1 would account for Δd(Ru–C) = 0.17 Å.

The cyanoethynyl ligand (C≡C–C≡N) should be among the best of all alkynyl ligands in terms of potential π-acceptor character, both because of its large group electronegativity (χ = 3.17) and the viability of a M=C=C=C=N canonical structure. High-quality structural data are available for one metal complex of this ligand: The mean Co–C bond distance of CoCp(C≡C—C≡N)$_2$(PPh$_3$) is 1.867(14) Å (84), which can be compared to that of Co{η^5-C$_5$H$_2$Me$_3$}Me$_2$(PPh$_3$) [1.997(5) Å] (173) to yield Δd(Co–C) = 0.13 Å (Table II). This might be taken as evidence for the presence of a significant Co–C π-bonding interaction. However, the large electronegativity of CCCN also implies that the difference between the M–C bond ionicities of complexes of this ligand and analogous metal–alkyl complexes will be greater than for other pairs of metal–alkynyl and metal–alkyl complexes. Reference to Fig. 4 indicates that, for this ligand, a M–C bond-length contraction of 0.15 Å corresponds to χ[CoCpR′(PPh$_3$)] = 2.2; this is in reasonable agreement with the value for Co^{II} (χ = 1.88) (160). Unfortunately, the C≡C, CC–CN, and C≡N distances of this compound are not of sufficiently high precision to provide insight into whether the large M–C bond contraction seen for this compound is the result of π-backbonding.

Among the alkynyl complexes of Groups 9 and 10, there are no examples at present for which there is structural evidence for M–C π-bonding; the compounds Pd(C≡CSiMe$_3$)Br(NMe$_2$CH$_2$C$_6$H$_4$-2-PPh$_2$) ($104,177$), trans-Pt{C≡CC(OH)MeEt}Cl(PPh$_3$)$_2$ (113), Au(C≡CCH$_2$OMe){P(p-Tol)$_3$} (149), and {Au(C≡CPh)}$_2$(μ-PPr$_2^i$-4,4′-C$_6$H$_4$-C$_6$H$_4$-PPr$_2^i$) (151) possess M–C bond lengths that differ from those of corresponding alkyl complexes by Δd(M–C) ≅ 0.08 Å (Table II). None of these alkynyl complexes possess unusually elongated C≡C bonds, consistent with the absence of structurally observable M–C π-bonding.

Unless one completely discounts the bond-ionicity arguments set out earlier, it appears that there are few instances where comparisons between metal–alkynyl and metal–alkyl M–C bond lengths provide unambiguous evidence for M–CCR π-bonding interactions, and thus interpreting structural data from such a standpoint is probably well grounded only if one has independent spectroscopic (or other) evidence for these interactions in the compounds under consideration.

2. *Comparisons to Metal–Cyano Complexes*

As pointed out in Section II, the electronic structures of analogous metal–cyano and metal–alkynyl complexes are viewed as being either quite similar or distinctly different, depending on the specific complexes under consideration and the experimental method used to probe their bonding. The possibility that differences between the natures of the M–CN and M–CCR bonds might be manifested in the molecular structures of these complexes makes comparisons of their M–C bond lengths of interest. The nominal single-bond covalent radius of both ligands is 0.69 Å, based on the C–C bond lengths of HC≡C—C≡CH [1.384(2) Å] and N≡C—C≡N [1.391(1) Å] (*178*), suggesting that, all other factors being equal, the M–C bond lengths of pairs of these complexes should be nearly identical. A first-order prediction of the relative M–C bond lengths of these complexes, however, would be d(M–CN) $<$ d(M–CCR), for two reasons: First, the electronegativities of the two ligands differ significantly (χ[CN] = 3.32, χ[CCR] = 2.4–2.8) (*160,166*), which, based on the arguments outlined in the preceding section, should lead to a contraction of the M–CN bond length in complexes for which χ[ML$_n$] $<$ χ[CCR] $<$ χ[CN]; and second, the available evidence points to CN being a stronger π-acceptor than alkynyl ligands, which should shorten the M–CN bonds of complexes in which there are metal $d\pi$-electrons of appropriate energy for backbonding. (Conversely, alkynyl ligands should be better π-donors, on electronegativity grounds, which might favor shorter M–CCR bonds for high-valent complexes.)

The M–C bond lengths for pairs of analogous metal–alkynyl and metal–cyano complexes are set out in Table III (*179–181*). Interestingly, the first-order prediction for these bond lengths is not borne out; the lengths of the M–C bonds for the pairs of complexes VCp$_2$R′, *trans*-PtR$_2'$(PR$_3''$)$_2$ (*182*), and AuR′(PR$_3''$) (R′ = CN, CCR) are statistically indistinguishable from one another. The only complexes for which significant differences in M–C bond lengths are observed are those of the *trans*-WR′(≡CH)(dmpe)$_2$ class (*16a–c*). Interestingly, the W–CN bond is significantly *longer* than the W–CCR bonds of the alkynyl complexes. Spec-

TABLE III

METAL–CARBON BOND LENGTHS FOR ANALOGOUS METAL–CYANO AND METAL–ALKYNYL COMPLEXES

Metal–cyano complex	d(M–C) (Å)	Ref.	Metal–alkynyl complex	d(M–C) (Å)	Ref.	Δd^a (Å)
VCp$_2^*$(C≡N)	2.088(7)	(179)	VCp$_2^*$(C≡CBu^t)	2.075(5)	(38)	0.01
$trans$-W(C≡N)(CH)(dmpe)$_2$	2.292(4)	(16c)	$trans$-W(C≡CH)(CH)(dmpe)$_2$	2.263(5)	(16c)	0.03
			$trans$-W(C≡CPh)(CH)(dmpe)$_2$	2.250(4)	(16c)	0.04
			$trans$-W(C≡CSiMe$_3$)(CH)(dmpe)$_2$	2.246(6)	(16b)	0.05
			$trans$-W(C≡CC$_6$H$_4$-4-CCPr^n)(CH)(dmpe)$_2$	2.245(8)	(16c)	0.05
$trans$-Pt(C≡N)$_2$(PPh$_3$)$_2$	1.991(4)	(180)	$trans$-Pt(C≡CCCPh)$_2$(PMe$_2$Ph)$_2$	2.009(5)	(5b,e)	−0.02
			$trans$-Pt(C≡CCMe═CH$_2$)$_2$(PPh$_3$)$_2$	2.024(6)	(108)	−0.03
Au(C≡N)(PPh$_3$)	2.003(7)	(181)	Au(C≡CCH$_2$OMe){P(p-Tol)$_3$}	2.024(2)	(149)	−0.02
			{Au(C≡CPh)}$_2$(μ-PPr$_2^i$-4,4'-C$_6$H$_4$-C$_6$H$_4$PPr$_2^i$)	1.997(9)	(151)	0.01

^{a} $\Delta d = [d$(M–CN) − d(M–CCR)].

troscopic and theoretical studies aimed at explaining this observation are in progress in our laboratory.

3. *Influence of Alkynyl R-Group on M–C Distance*

Metal complexes with alkynyl R-groups that are strongly electron donating or withdrawing might possess M–C bond lengths that differ from those for complexes whose alkynyl ligands lack these characteristics. One compound where such an effect might be operative, CoCp $(C\equiv C-C\equiv N)_2(PPh_3)$ *(84)*, was discussed in Section III-C-1. Two other studies are of note in this regard. Carty and co-workers have reported high-quality structural data for complexes of the type *trans*-Ru $(C\equiv CR)_2(CO)_2(PEt_3)_2$ (R = H, Ph, SiMe$_3$, C$\equiv$CH, C$\equiv$CSiMe$_3$) *(63a,b)*. As noted in Section III-B, the C$\equiv$C bond lengths of *trans*-Ru(C$\equiv$CSiMe$_3$)$_2$ $(CO)_2(PEt_3)_2$ [1.221(2) Å] and *trans*-Ru(C$\equiv$CCCSiMe$_3$)$_2$(CO)$_2$(PEt$_3$)$_2$ [1.226(2) Å] are unusually long, suggesting greater Ru $\rightarrow$ CCR π-backbonding for these species than for the other derivatives. Consistent with this, the Ru–C bond lengths of the former compounds [R = SiMe$_3$, 2.062(2) Å; R = C$\equiv$CSiMe$_3$, 2.057(2) Å] are significantly shorter than those of the ethynyl [2.078(1) Å], phenylethynyl [2.074(3) Å], and butadiynyl [2.078(2) Å] complexes. This ordering of Ru–C bond lengths would not be expected on electronegativity grounds (Section III-C-1). Unfortunately, a Ru–C$_{sp^3}$ distance for an alkyl derivative of the type *trans*-RuR$_2'$(CO)$_2$(PR$_3''$)$_2$ is not available for comparison.

Alkynyl-phosphorane ligands (C$\equiv$CPR$_3$), such as those of the complexes Mn(C$\equiv$CPPh$_3$)Br(CO)$_4$ *(47)*, MnCp(C$\equiv$CPMe$_2$Ph)(CO)$_2$ *(48)*, and *cis*-Pd(C$\equiv$CPPh$_3$)(PPh$_3$)Cl$_2$ *(105)*, potentially support the limiting canonical structures M—C$\equiv$C—PR$_3$ and M=C=C=PR$_3$. Moreover, they are unique among alkynyl ligands from a formal charge standpoint in that they are neutral rather than anionic. While this obviates comparisons of their M–C bond lengths to those of metal–alkyl, –cyano, and even other alkynyl complexes, it does present the only case where straightforward comparisons between alkynyl and carbon monoxide complexes can be made. For the pair of compounds MnCp(C$\equiv$CPMe$_2$Ph)(CO)$_2$ and MnCp(CO)$_3$, the Mn–CCR bond length [1.895(5) Å] *(48)* is observed to be 0.1 Å longer than those of the Mn–CO bonds of the latter compound [1.793(3) Å] *(183)*. In contrast, the Mn–CO bond lengths of MnCp(C$\equiv$CPMe$_2$Ph)(CO)$_2$ [1.743(7) Å] are 0.05 Å shorter than those of MnCp(CO)$_3$. These data clearly indicate that the Mn–CCPR$_3$ bond order is lower than the Mn–CO bond order, and that the electron density available at the [MnCp(CO)$_2$] fragment for Mn $\rightarrow$ CO π-backbonding is greater for the alkynyl complex than for MnCp(CO)$_3$, suggesting that the alkynyl ligand is a poorer π-acceptor and/or a better σ-donor than CO. Additional insight into the multiplicity of the Mn–C bond order for $MnCp(C{\equiv}CPMe_2Ph)(CO)_2$ can be gleaned from the bond lengths of the MnC–C–P linkages within the ligand. Reference to the Cambridge Structural Database indicates that organic compounds of type $R_2'C{=}C{=}PR_3$ (three structures reported) possess mean $C{=}C$ and $P{=}C$ bond lengths of 1.268(40) Å and 1.687(14) Å, respectively. By comparison, $d(C{-}CP) = $ 1.221(7) Å and $d(CC{-}P) = $ 1.683(6) Å for $MnCp(C{\equiv}CPMe_2Ph)(CO)_2$ (48), suggesting that the $Mn{=}C{=}C{=}PR_3$ canonical structure is a significant contributor to the bonding in this complex.

4. *Comparisons among Metal–Alkynyl Redox Congeners*

A potentially important probe of the bonding in alkynyl complexes, and, in particular, the extent of $M \rightarrow CCR$ π-backbonding, is to determine how the $M{-}C{\equiv}C{-}R$ fragment is structurally perturbed by the oxidation or reduction of the complex. The $M{-}C{\equiv}O$ bond lengths of metal–carbonyl complexes are extremely sensitive to oxidation state, with M–C bonds lengthening and $C{\equiv}O$ bonds contracting as the oxidation state of the metal increases (and, concomitantly, $M \rightarrow CO$ π-backbonding decreases) (154).

To date, there have been only two studies of this type for metal–alkynyl complexes. Whiteley and co-workers have reported the structures of the 18- and 17-electron complexes $[Mo(\eta^7{-}C_7H_7)(C{\equiv}CPh)(dppe)]^{n+}$ ($n = 0,1$) (33g,42). Unlike the observation for metal–carbonyl complexes, the Mo–CCPh bond distance of the cation [2.067(9) Å] is 0.07 Å shorter than that of the neutral complex [2.138(5) Å]. This contraction of the M–C bond contrasts with the behavior of the Mo–P bonds of these redox partners, which lengthen by 0.06 Å upon oxidation. The $C{\equiv}C$ bond distances of the neutral and cationic complexes are statistically indistinguishable, being 1.205(6) and 1.196(11) Å, respectively, as are the distances from the metal center to the centroid of the cycloheptatrienyl ligand (ca. 1.62 Å). The elongation of the Mo–P bonds upon oxidation was interpreted as reflecting lessened $Mo \rightarrow P$ π-backbonding in the oxidized compound, whereas the contraction of the Mo–C bond upon oxidation was said to be consistent with the smaller radius of Mo^I relative to that of Mo^0 and the larger electrostatic contribution to the $Mo^I{-}C$ bond as compared to the $Mo^0{-}C$ bond. Thus, $Mo \rightarrow CCR$ π-backbonding appears to be of little or no importance to the bonding in these compounds. Infrared-spectroscopic data (Section IV-B) and electrochemical data were noted to support the conclusion that the Mo–CCR bond is predominantly of σ-character.

Bianchini and co-workers have prepared and structurally characterized redox congeners of the type $[\text{Fe}(C\equiv CPh)\{P(CH_2CH_2PPh_2)_3\}]^{n+}$ ($n = 0$, 1) ($33a$). As is observed for the Mo–P bonds of the $[\text{Mo}(\eta^7\text{-}C_7H_7)(C\equiv CPh)(dppe)]^{n+}$ ($n = 0$, 1) complexes from earlier, the Fe–P bonds of the cation are longer than those of the neutral complex [Δd (Fe–P) $= 0.07$–0.12 Å]. Unfortunately, the large esd's for the lengths of the Fe–C and $C\equiv C$ bonds of these complexes [$n = 0$, d(Fe–C) $= 1.92(1)$ Å, $d(C\equiv C) = 1.21(1)$ Å; $n = 1$, d(Fe–C) $= 1.88(2)$ Å, $d(C\equiv C) = 1.20(3)$ Å] prohibit any quantitative differentiation of the natures of their Fe–CCR bonding. Infrared-spectroscopic data (Section IV-B) and electrochemical data for these compounds were interpreted as indicating that Fe $\rightarrow$ CCR π-backbonding is insignificant.

D. C–R Bond Distances

The R-groups of alkynyl ligands that might be expected to exhibit variable C–R bond lengths as a function of the magnitude of M–CCR π-bonding include PR_3' and CN, since, as noted in Section III-C-3, canonical structures of the type $L_nM{=}C{=}C{=}R$ can realistically be invoked for their complexes. The variable M–C and $C\equiv C$ bond lengths of trimethylsilylethynyl complexes ($L_nM\text{—}C\equiv C\text{—}SiMe_3$) suggest that C–Si bond lengths might also manifest a π-bonding component. In contrast, one can infer from these considerations that the C–R bond lengths of complexes of alkyl- and aryl-substituted alkynyl ligands should deviate little from the expectation values of their organic counterparts, and the data bear this out: The MCC–C(Ph) bond distances of phenylethynyl complexes, for which more precise data (22 distances with esd ≤ 0.010 Å) are available than for any other type of metal–alkynyl complex, span a range of only 0.041 Å [$1.421(8)$–$1.462(8)$ Å; $d(C\text{–}C)_{mean} = 1.439(9)$ Å], which is of marginal statistical significance and little different from the RCC–C(Ph) bond-length distribution for organic alkynes [$d(C\text{–}C)_{mean} = 1.435(9)$ Å for 55 distances with esd ≤ 0.010 Å].

The limited C–R bond-length data for R $=$ CN and R $=$ PR_3' alkynyl complexes have been noted earlier. For trimethylsilylethynyl complexes, the MCC–Si bond lengths span a range of 0.055 Å [$1.797(7)$–$1.852(8)$ Å, Table I; $d(C\text{–}Si)_{mean} = 1.819(22)$ Å], which, despite encompassing only nine distances (esd ≤ 0.010 Å), is broader than that for the MCC–C(Ph) distances of phenylethynyl complexes. Whether this is an indication that greater M–CCR π-bonding obtains for trimethylsilylethynyl complexes than for those in which the alkynyl R-group is an alkyl or aryl substituent is uncertain at this juncture, both because there is insufficient detailed information about the electronic structures of these complexes to make any correlation to the d(C–Si) distance, and because, among organic alkynes, the distribution of CC–Si bond lengths for silyl-substituted alkynes is also broader than that for alkyl- or aryl-substituted alkynes [1.809–1.864 Å, d(C–Si)$_{mean}$ = 1.834(14) Å for 14 distances with esd ≤ 0.010 Å]. The ambiguity concerning the interpretation of this bond length is well illustrated by the complex $trans$-Ru(C≡CSiMe$_3$)$_2$(CO)$_2$(PEt$_3$)$_2$ ($63b$). The short Ru–CC and long C≡C bonds of this compound have been noted earlier as being consistent with the presence of significant Ru → CCR π-backbonding, yet its CC–Si bond length [1.812(2) Å] is not significantly shorter than that of the mean distance for all trimethylsilylethynyl metal complexes. Despite these uncertainties, the CC–Si bond distance bears close examination in future studies, particularly in conjunction with ^{29}Si-NMR chemical shifts and MCC–Si nuclear spin–spin coupling constants, which have been reported to correlate with the extent of C–Si π-bonding (184), because of the preliminary indication that it is shorter for metal-alkynyl complexes than in organic alkynes.

E. The Structural Trans Influences of Alkynyl Ligands

An indirect measure of the nature of the metal–alkynyl bond is the $trans$ influence exerted by the alkynyl ligand. There are several experimental parameters for assessing this influence, including structural (primarily bond length) and vibrational- and NMR-spectroscopic data for the ligand in question and for those oriented $trans$ and cis to it (185). Well over a decade has passed since the $trans$ influence of the alkynyl ligand has been a subject of detailed study, and several earlier papers provide leading references to this work and summarize the main conclusions (113, $116,119,186a,b$). We will not review these older results, but instead provide an update on the structural $trans$ influence of the alkynyl ligand in view of the wealth of recent crystallographic results. We note that the implications of the following observations for understanding the nature of metal–alkynyl bonding are not straightforward to set out because there is not a simple way experimentally to separate and individually quantify the factors that contribute to the structural $trans$ influence of a ligand, such as its electronegativity and σ- and π-bonding character.

The original assessments of the $trans$ influences of alkynyl ligands were made for square-planar Pt^{II} complexes. Set out in Table IV (187–201) are crystallographically determined Pt–Cl bond lengths (esd ≤ 0.006 Å) for a variety of compounds of the type $trans$-Pt(X)Cl(PPh$_{3-n}$R$_n$)$_2$ (n = 0 or 1), and the difference Δd(Pt–Cl) = [d(XPt–Cl) − d(ClPt–Cl)] using

TABLE IV

STRUCTURAL DATA FOR *trans*-Pt(X)Cl(PR$_3$)$_2$ COMPLEXES

PtCl(PR$_3$)$_2$	X	d(Pt–Cl) (Å)	Δd(Pt–Cl)^a (Å)	Ref.
PtCl(PPh$_3$)$_2$	C(O)Pr^n	2.450(4)	0.147	(187)
PtCl(PPh$_3$)$_2$	C(O)CH$_2$CH=CH$_2$	2.439(5)	0.136	(188)
PtCl(PPh$_3$)$_2$	Me	2.431(3)	0.128	(175)
PtCl(PPh$_3$)$_2$	C(O)-n-C$_6$H$_{13}$	2.431(3)	0.128	(189)
PtCl(PPh$_2$Me)$_2$	C(O)Me	2.430(2)	0.127	(190)
PtCl(PPh$_3$)$_2$	C(O)CH$_2$CH$_2$Ph	2.427(6)	0.124	(191)
PtCl(PPh$_3$)$_2$	CH$_2$CH=CH$_2$	2.425(2)	0.122	(192)
PtCl(PPh$_3$)$_2$	C(O)C(O)Ph	2.421(3)	0.118	(193)
PtCl(PPh$_3$)$_2$	C$_6$H$_4$-2-NMe$_2$	2.416(3)	0.113	(194)
PtCl(PPh$_2$Me)$_2$	Me	2.412(2)	0.109	(195)
PtCl(PPh$_3$)$_2$	C(CMe=CH$_2$)=CH$_2$	2.410(3)	0.107	(119)
PtCl(PPh$_3$)$_2$	Ph	2.408(5)	0.105	(196)
PtCl(PPh$_3$)$_2$	C$_6$Cl$_4$-2-Au(PPh$_3$)	2.406(2)	0.103	(197)
PtCl(PPh$_2$Me)$_2$	C(O)CF$_3$	2.390(1)	0.087	(190)
PtCl(PPh$_3$)$_2$	CH$_2$CN	2.390(3)	0.087	(198)
PtCl{P(p-Tol)$_3$}$_2$	H	2.384(3)	0.081	(199)
PtCl(PPh$_3$)$_2$	C≡CCl	2.373(3)	0.070	(110)
PtCl(PPh$_2$Me)$_2$	C$_2$F$_5$	2.361(3)	0.058	(195)
		2.365(4)	0.062	
PtCl(PPh$_3$)$_2$	C≡CCMe=CH$_2$	2.363(6)	0.060	(112)
PtCl(PPh$_3$)$_2$	C≡CCMe$_2$OH	2.356(5)	0.053	(113)
PtCl(PPh$_3$)$_2$	C≡C-c-C$_6$H$_{10}$-1-OEt	2.345(5)	0.042	(114)
PtCl(PPh$_3$)$_2$	C≡CCMeEt(OH)	2.336(4)	0.033	(113)
PtCl{μ-dppf-W(CO)$_5$}$_2$	Cl	2.304(5)	0.001	(200)
PtCl{PPh$_2$CH$_2$C(CF$_3$)$_2$OH}$_2$	Cl	2.303(1)	0	(201)

^{a} Δd(Pt–Cl) = [d(XPt–Cl) – d(ClPt–Cl)].

d(Pt–Cl) for *trans*-PtCl$_2${PPh$_2$CH$_2$C(CF$_3$)$_2$(OH)}$_2$ as a reference. For acyl, alkyl, and aryl derivatives of *trans*-Pt(X)Cl(PPh$_{3-n}$R$_n$)$_2$, the data reveal a near continuum of Pt–Cl bond lengths over the range Δd(Pt–Cl) = 0.10–0.15 Å. The incorporation of strong electron withdrawing groups into these ligands [i.e., X = C(O)CF$_3$, CH$_2$CN, C$_2$F$_5$] lessens their *trans* influence, as indicated by the significantly shorter Pt–Cl bonds of their complexes (Δd = 0.06–0.09 Å). The Pt–Cl bond lengths of alkynyl complexes are distinctly shorter than those *trans* to other carbon-ligated groups [Δd(Pt–Cl) = 0.03–0.07 Å], indicating they have smaller *trans* influences. The differences among the Pt–Cl bond lengths of the alkynyl derivatives are of marginal statistical significance, so it is not possible to differentiate among their *trans* influences on the basis of these data. Overall, the *trans* influences of the X ligands can be ordered as follows:

$$C(O)R \geq C_{sp^3} \geq C_{sp^2} > H \gg CCR \gg Cl$$

This order is consistent with that derived in earlier studies ($113,116,$ $119,186$).

An important issue regarding the generality of these findings is the extent to which the *trans* influence of a ligand is dependent on the nature of the ligand *trans* to it, and on the nature of the metal. There are few other appropriate classes of structurally characterized metal–alkynyl complexes with which to address this question. Comparisons among the Au–P bond lengths (esd ≤ 0.006 Å) of complexes of the type $AuX(PPh_3)$ (Table V) ($202-205$) reveal a similar ordering to that provided by the *trans*-$Pt(X)Cl(PR_3)_2$ compounds. Moreover, the difference $\Delta d(Au-P) = [d(XAu-P) - d(ClAu-P)]$ [as determined relative to the Au–P bond length of $AuCl(PPh_3)$] for alkynyl ligands is 0.04–0.05 Å, which is similar to the values of $\Delta d(Pt-Cl)$ for these ligands noted earlier (0.03–0.07 Å). However, the difference between the *trans* influences of alkyl and alkynyl ligands, as judged by $\Delta d(Au-P)$, is much smaller than that manifested by the Pt–Cl bond lengths of *trans*-$Pt(X)Cl(PR_3)_2$ complexes of these ligands. The $AuX(PPh_3)$ compounds also provide comparisons of the cyanide and alkynyl ligands; the similar Au–P bond lengths for these compounds suggest that their *trans* influences are essentially identical (Table V).

The only relevant data for other than late-metal alkynyl complexes are provided *trans*-$FeCl_2(dmpe)_2$ [$d(Fe-Cl) = 2.352(1)$ Å] (206), *trans*-$Fe(C{\equiv}CPh)Cl(dmpe)_2$, for which two independent crystal structures have

TABLE V

GOLD–PHOSPHORUS BOND DISTANCES FOR $AuX(PPh_3)$ COMPLEXES

X	$d(Au-P)$ (Å)	$\Delta d(Au-P)^a$ (Å)	Ref.
$CH_2CH_2C(O)Ph$	2.292(3)	0.057	(176)
$CH_2C(O)Ph$	2.283(1)	0.049	(176)
C_6H_3-2,6-$(OMe)_2$	2.284(1)	0.048	(202)
$C{\equiv}CPh$	2.282(4)	0.047	(146)
	2.276(5)	0.041	
$C{\equiv}N$	2.278(2)	0.043	(181)
$C{\equiv}CCH_2OMe$	$2.274(1)^b$	0.039	(149)
$C{\equiv}CC_6F_5$	2.274(3)	0.039	(148)
Br	2.252(6)	0.017	(203)
I	2.249(2)	0.014	(204)
Cl	2.235(3)	0	(205)

^{a} $\Delta d(Au-P) = [d(XAu-P) - d(ClAu-P)]$.
^{b} Distance for $Au(C{\equiv}CCH_2OMe)\{P(p\text{-Tol})_3\}$.

been reported $[d(\text{Fe–Cl}) = 2.389(1)$ Å (53), $2.386(2)$ Å (54); $d(\text{Fe–C}) = 1.897(3)$ Å (53), $1.880(5)$ Å $(54)]$ and $trans\text{-Fe}(\text{C}\equiv\text{CPh})_2(\text{dmpe})_2$ $[d(\text{Fe–C}) = 1.925(6)$ Å$]$ $(52a,b)$. The facts that the Fe–Cl bond of the mono(phenylethynyl) derivative is ca. 0.035 Å longer than those of the dichloro complex, and that its Fe–C bond is ca. 0.03 Å shorter than those of the bis(phenylethynyl) compound, are consistent with the findings for $trans\text{-}\text{Pt}(\text{X})\text{Cl}(\text{PR}_3)_2$ and $\text{AuX}(\text{PPh}_3)$ in indicating the $trans$ influence of alkynyl ligands to be significantly greater than that of chloride.

IV

VIBRATIONAL SPECTROSCOPIC STUDIES

A. *Overview*

The considerations set out in Section III that justify inspecting the M–C, C≡C, and C–R bond distances of metal–alkynyl complexes for insight into the nature of the M–CCR bond also apply to the respective stretching frequencies of these linkages. Compared to the study of bond distances, there are several potential advantages to using vibrational frequencies in this regard, including their greater precision and sensitivity to electronic effects and the relative ease with which the experiments are performed. Consider, for example, the C≡C bond distances and stretching frequencies of metal–alkynyl complexes. The 150 or so X-ray crystallographic studies of metal–alkynyl compounds have yielded roughly 50 structures with precise C≡C bond distances; these distances span only a 0.10-Å range, with two-thirds of them lying in a 0.025-Å range (Fig. 2). In contrast, C≡C stretching frequencies are known for the majority of all metal–alkynyl complexes; more than 350 such frequencies have been reported for phenylethynyl complexes alone, and they cover a range of nearly 180 cm^{-1} (*vide infra*).

Of course, in order to use vibrational spectroscopy to study metal–alkynyl bonding the vibrational modes of interest must be assigned. This is fairly easy for $\nu(\text{C}\equiv\text{C})$ because of its distinctive frequency range, but, with few exceptions, is difficult for $\nu(\text{M–C})$ and $\nu(\text{C–R})$ because bands attributable to the ancillary ligands and alkynyl R-groups congest these frequency regions. Unfortunately, there are insufficient M–C and C–R stretching-frequency data to warrant detailed discussions of their significance regarding the nature of metal–alkynyl bonding. While the frequency of the $\nu(\text{C–R})$ mode is not expected to be an especially sensitive reporter of the nature of M–CCR bonding, the paucity of ν(M–C) frequency data is a significant loss in this regard; a detailed study of these modes for complexes whose synthetic accessibility and symmetry allowed both isotopic labeling of the alkynyl ligand and a normal-coordinate analysis would be an important addition to the field.

B. C≡C Stretching Frequencies

The long-standing use of C≡O and C≡N stretching frequencies to gauge qualitatively the extent of M → CO and M → CN π-backbonding in metal–carbonyl and –cyano complexes has naturally led to C≡C stretching frequencies of metal–alkynyl complexes being interpreted within the framework of M → CCR π-backbonding. The alkynyl R-group complicates the interpretation of C≡C stretching frequencies, as compared to those of the C≡O and C≡N oscillators, because of the possibility that the nominal ν(C≡C) mode might consist not of just the C≡C stretch but of C–R and R-group symmetry coordinates as well (207). This is well established for terminal alkynes, for which both the C≡C and the terminal C—H stretching coordinates contribute to "ν(C≡C)." Clearly, the safest approach to interpreting C≡C stretching frequencies is to limit comparisons among them to within groups of complexes that share a common alkynyl R-group.

As was discussed for the C≡C bond distance (Section III-B), a good way to approach the interpretation of the C≡C stretching frequencies of metal–alkynyl complexes is to examine their distribution. Although frequency data for metal–alkynyl complexes are abundant, they are divided among compounds with many different alkynyl R-groups. There are too few data for complexes of most alkynyl ligands to provide distributions within which trends can be discussed in a statistically significant way; unfortunately, this includes complexes with electron-withdrawing alkynyl R-groups, for which M → CCR π-backbonding (and hence ν(C≡C) frequency shifts) should be greatest. Complexes of the phenylethynyl ligand are the major exception to this, since over 350 C≡C stretching frequencies have been reported for them; Table VI (208–336) sets out these data. The frequencies therein are those specifically assigned by the reporting authors as being attributable to ν(C≡C); we have not included frequencies for compounds for which the vibrational data are given as unassigned listings, even when these contained "obvious" candidates for ν(C≡C), or for which the frequency is referred to as being "approximate." We note the experimental conditions under which the spectra were acquired, although solvent and medium effects appear to be of minor importance in the few

TABLE VI

C≡C Stretching Frequencies for Metal–Phenylethynyl Complexes

Compound	$\nu(C{\equiv}C)^a$ (cm^{-1})	Ref.
TiCp$_2$(C≡CPh)$_2$	2070^c	(208a,b)
	2065^b	(209a,b)
Ti(η^5-C$_5$H$_4$Me)$_2$(C≡CPh)$_2$	2045^b	(209a,b)
Ti(η^5-C$_5$H$_4$SiMe$_3$)$_2$(C≡CPh)$_2$	2066^c	(209b)
Ti(η^5-C$_5$H$_4$Me)(C≡CPh)(NMe$_2$)$_2$	2080^b	(210)
ZrCp$_2$(C≡CPh)$_2$	2075^c	(35)
	2073^b	(210)
Zr(η^5-C$_5$H$_4$Me)$_2$(C≡CPh)$_2$	2070^c	(35)
	2078^b	(210)
Zr(η^5-C$_5$H$_4$Bu^t)$_2$(C≡CPh)$_2$	2078^c	(35)
ZrCp$_2$(C≡CPh)Cl	2074^c	(35)
Zr(η^5-C$_5$H$_4$Me)$_2$(C≡CPh)Cl	2070^c	(35)
HfCp$_2$(C≡CPh)$_2$	2080^c	(35)
	2083^b	(210,211)^c
Hf(η^5-C$_5$H$_4$SiMe$_3$)$_2$(C≡CPh)$_2$	2072^c	(212)
VCp$_2$(C≡CPh)	2060^c	(208a)
VCp$_2$(C≡CPh)$_2$	2060^c	(208a)
V(η^5-C$_5$H$_4$Et)$_2$(C≡CPh)	2040^d	(31b)
[N(PPh$_3$)$_2$][Cr(C≡CPh)(CO)$_5$]	2080^e	(213)
K$_3$[Cr(C≡CPh)$_3$(CO)$_3$]	2176^c	(214)
Li[CrPc(C≡CPh)$_2$]	2065, 2075	(215)
CrCp(C≡CPh)(CO)$_3$	2110^c	(216)
[N(PPh$_3$)$_2$][Mo(C≡CPh)(CO)$_5$]	2079^e	(213)
[N(PPh$_3$)$_2$][Mo(C≡CPh)(CO)$_4$(PPh$_3$)]	2058^e	(213)
K$_3$[Mo(C≡CPh)$_3$(CO)$_3$]	2190^c	(217)
trans-Mo(C≡CPh)$_2$(dppe)$_2$	2020^c	(33f)
Mo(C≡CPh)$_2$H$_2$(depe)$_2$	2040^c	(33f)
MoCp(C≡CPh)(CO)$_3$	2110^b	(218)
	2100^b	(219)
MoCp$_2$(C≡CPh)$_2$	2090	(220)
Mo(η^7-C$_7$H$_7$)(C≡CPh)(CO)$_2$	2078^e	(33n)
Mo(η^7-C$_7$H$_7$)(C≡CPh)(dppe)	2045^e	(33n)
[Mo(η^7-C$_7$H$_7$)(C≡CPh)(dppe)][BF$_4$]	2032^e	(33n)
[N(PPh$_3$)$_2$][W(C≡CPh)(CO)$_5$]	2080^e	(213)
[N(PPh$_3$)$_2$][W(C≡CPh)(CO)$_4$(PPh$_3$)]	2065^e	(213)
Li[fac-W(C≡CPh)(dppe)(CO)$_3$]	2030^k	(221)
trans-W(C≡CPh)$_2$(dppe)$_2$	2040^c	(33f)
W(C≡CPh)$_2$H$_2$(dppe)$_2$	2040^c	(33f)
trans-W(C≡CPh)(CH)(dmpe)$_2$	2060	(16c)
WCp(C≡CPh)(CO)$_3$	2105^e	(222)
	2110^b	(218)
	2110^c	(216)

TABLE VI (*continued*)

Compound	$\nu(C{\equiv}C)^a$ (cm^{-1})	Ref.
WCp(C≡CPh)(CO)$_2$(PMe$_3$)	2082^c	(223)
WCp$_2$(C≡CPh)$_2$	2080	(220)
W(η^7-C$_7$H$_7$)(C≡CPh)(CO)$_2$	2080^e	(33n)
WTp'(C≡CPh)(HCCPh)(CO)	2070^c	(224)
Ba[Mn(C≡CPh)$_4$]	2048	(225)
Mn(C≡CPh)(CO)$_5$	2120^f	(226)
Mn(C≡CPh)(CO)(dppe)$_2$	2056^e	(227a,b)
cis-Mn(C≡CPh)(PCy$_3$)(CO)$_4$	2106^e	(227a)
	2111^g	(227b)
trans,mer-Mn(C≡CPh)(PEt$_3$)$_2$(CO)$_3$	2092^e	(227b)
trans,mer-Mn(C≡CPh)(PCy$_3$)$_2$(CO)$_3$	2093^e	(227a,b)
cis-Mn(C≡CPh){P(OMe)$_3$}(CO$_2$)(dppe)	2084^e	(227b)
cis-Mn(C≡CPh){P(OEt)$_3$}(CO$_2$)(dppe)	2084^e	(227b)
cis-Mn(C≡CPh){P(OPh)$_3$}(CO$_2$)(dppe)	2093^e	(227b)
cis-Mn(C≡CPh){P(OPh)$_3$}(CO$_2$)(dppm)	2089^e	(227b)
fac-Mn(C≡CPh)(dppm)(CO)$_3$	2099^e	(227b)
fac-Mn(C≡CPh)(dppe)(CO)$_3$	2097^e	(227a,b)
fac-Mn(C≡CPh)(phen)(CO)$_3$	2089^e	(227b)
fac-Mn(C≡CPh)(bpy)(CO)$_3$	2094^e	(227b)
fac-Mn(C≡CPh)(CNBu^t)(PCy$_3$)(CO)$_3$	2093^e	(227b)
Re(C≡CPh)(CO)$_5$	2120^f	(226)
	2112^f	(228)
fac-Re(C≡CPh)(py)$_2$(CO)$_3$	2090^b	(49)
Re(C≡CPh)Cl{P(O)(OMe)$_2$}{P(OMe)$_3$}$_2$(PPh$_3$)	2015^c	(229)
Re(C≡CPh)Cl{P(O)(OMe)$_2$}{P(OMe)$_3$}(NCMe)$_2$	2040^c	(229)
ReCp(C≡CPh)(NO)(PPh$_3$)	2082^h	(50)
Re$_2$(C≡CPh)(CO)$_7$(μ-dppm)(μ-H)	2098^i	(51)
trans-Fe(C≡CPh)$_2$(dmpe)$_2$	2037	(52a,b)
trans-Fe(C≡CPh)$_2$(depe)$_2$	2046^i	(230)
	2035	(52a)
	2038^e	(61)
Fe(C≡CPh)$_2$(dbpe)$_2$	2038	(231)
trans-Fe(C≡CPh)Cl(dmpe)$_2$	2044	(52a,53b)^b
trans-Fe(C≡CPh)(H)(dmpe)$_2$	2036^j	(232)
Fe(C≡CPh)(H)(dmpe)$_2$	2035^c	(233)
[PPh$_4$]$_2$[Fe(C≡CPh)(CN)$_4$(NO)]	2097	(234)
[PPh$_4$]$_2$[fac-Fe(C≡CPh)$_3$(CN)$_2$(NO)]	2056	(234)
Fe(C≡CPh){P(CH$_2$CH$_2$PPh$_2$)$_3$}	2030	(33a)
[Fe(C≡CPh){P(CH$_2$CH$_2$PPh$_2$)$_3$}][BPh$_4$]	2035	(235)
FePc(C≡CPh)	2080^c	(215)
Li$_2$[FePc(C≡CPh)$_2$]	2080^c	(236)
	2066, 2076	(237)
Na[FePc(C≡CPh)]	2084 or 2112	(238)

(*continued*)

TABLE VI (*continued*)

Compound	$\nu(C\equiv C)^a$ (cm^{-1})	Ref.
E[FePc(C≡CPh)(py)]		
E = Li	2088	(237)
E = Na	2036 or 2082	(238)
FeCp(C≡CPh)(CO)$_2$	2105^b	(219,222)
	2104^{c,e}	(59)
	2108^e	(12b)
	2121^f	(228)
	2121^b	(218)
FeCp(C≡CPh)(CO)(PPh$_3$)	2085^b	(219)
FeCp(C≡CPh)(dppm)	2071^c	(58)
FeCp*(C≡CPh)(CO)$_2$	2094^c	(59)
	2095^e	(59)
FeCp*(C≡CPh)(dppe)	2054^e	(33o)
[FeCp*(C≡CPh)(dppe)][PF$_6$]	2022	(33d)
trans-Ru(C≡CPh)$_2$(PMe$_3$)$_4$	2070^c	(239)
	2055	(240)
trans-Ru(C≡CPh)$_2$(dppe)$_2$	2061^e	(61)
trans-Ru(C≡CPh)$_2$(depe)$_2$	2054^e	(61)
trans-Ru(C≡CPh)Cl(dppm)$_2$	2075^b	(62a)
trans-Ru(C≡CPh)$_2$(CO)$_2$(PEt$_3$)$_2$	2093^e	(63a)
cis,cis,trans-Ru(C≡CPh)$_2$(CO)$_2$(PPr$^i_3$)$_2$	2095^j	(241)
Ru(C≡CPh)$_2$(CO)(PPh$_3$)$_3$	2075^c	(242)
trans,cis-Ru(C≡CPh)$_2$(PMe$_3$)$_2$(CO)(PPr$^i_3$)	2070^j	(241)
trans-Ru(C≡CPh)$_2$(CO){P(OMe)$_3$}(PPr$^i_3$)$_2$	2090^j	(241)
Ru(C≡CPh)$_2$(CO)(PPr$^i_3$)$_2$	2060^j	(241,243)^b
Ru(C≡CPh)Cl(pzH-3,5-Me$_2$)(CO)(PPh$_3$)$_2$	2085^b	(244)
Ru(C≡CPh)(H)(CO)(PPh$_3$)$_3$	2012^c	(242)
cis,trans-Ru(C≡CPh)(H)(CO)$_2$(PPr$^i_3$)$_2$	2090^j	(241)
Ru(C≡CPh)(H)(CO){P(OMe)$_3$}(PPr$^i_3$)$_2$	2078^j	(241)
cis,trans-Ru(C≡CPh)(CH=CHPh)(CO)$_2$(PPr$^i_3$)$_2$	2098^b	(243)
Ru(C≡CPh)(CH=CHPh)(CO)(PPr$^i_3$)$_2$	2065^b	(243)
Ru(C≡CPh)(O$_2$CCF$_3$)(CO)(PPh$_3$)$_2$	2090^c	(245)
Ru(C≡CPh)(O$_2$CCH$_3$)(CO)(PPh$_3$)$_2$	2100^c	(245)
Ru(C≡CPh)(NO$_3$)(CO)(PPh$_3$)$_2$	2120	(246)
mer-Ru(C≡CPh)(η^2-H$_2$BH$_2$)(PMe$_3$)$_3$	2050^c	(247)
[*cis,trans*-Ru(C≡CPh)(CO)(py)$_2$(PPh$_3$)$_2$]X	2100^c	(64)
X = BF$_4$, ClO$_4$, PF$_6$		
[*cis,trans*-Ru(C≡CPh)(py)(CO)$_2$(PPh$_3$)$_2$][PF$_6$]	2110^c	(65)
Ru(C≡CPh)Cl{PPh(CH$_2$CH$_2$CH$_2$PCy$_2$)$_2$}	2060^c	(248)
[Ru(C≡CPh){P(CH$_2$CH$_2$PPh$_2$)$_3$}][BPh$_4$]	2062	(249)
Ru(C≡CPh)(η^3-PhC≡C—C=CHPh){PPh(CH$_2$CH$_2$CH$_2$PCy$_2$)$_2$}	2060^c	(66b,248)
Li[RuPc(C≡CPh)]	2062	(236)
RuCp(C≡CPh)(PMe$_3$)$_2$	2105	(250)
RuCp(C≡CPh)(PPh$_3$)$_2$	2068	(251a,b)
	2076^l	(252)
RuCp(C≡CPh)(PMe$_3$)(PPh$_3$)	2070	(250)

TABLE VI (continued)

Compound	$\nu(C\equiv C)^a$ (cm^{-1})	Ref.
RuCp(C≡CPh){P(OMe)$_3$}$_2$	2085	(253)
RuCp(C≡CPh)(dppe)	2083	(254)
	2082	(251b)
RuCp(C≡CPh)(dppm)	2080	(254)
	2077	(251b)
RuCp(C≡CPh)(dppf)	2112^c	(255)
RuCp*(C≡CPh)(dppe)	2071^e	(330)
Ru$_2$Cp*$_2$(C≡CPh)$_2$(μ-SPr^i)$_2$	2100^c	(75,256)
Ru(η^6-C$_6$H$_6$)(C≡CPh)(PMe$_3$)Cl	2090	(257)
trans-Os(C≡CPh)$_2$(dppm)$_2$	2069^e	(61)
cis-Os(C≡CPh)$_2$(PMe$_3$)$_4$	2050	(239)
Os(C≡CPh)$_2$(CO)(PPr$^i_3$)$_2$(pzH)	2070	(76)
Os(C≡CPh)$_2$(CO)(PPr$^i_3$)$_2$	2060^j	(241,258)
	2060	(76)
trans-Os(C≡CPh)$_2$(CO){P(OMe)$_3$}(PPr$^i_3$)$_2$	2080^j	(241)
trans-Os(C≡CPh)$_2$(CO)(PMe$_3$)(PPr$^i_3$)$_2$	2090^j	(241)
trans,cis-Os(C≡CPh)$_2$(PMe$_3$)$_2$(CO)(PPr$^i_3$)	2070^i	(241)
trans-Os(C≡CPh)$_2$(CO)$_2$(PPr$^i_3$)$_2$	2090^j	(241)
trans-Os(C≡CPh)$_2$(CO)(HC≡CPh)(PPr$^i_3$)$_2$	2090^j	(241)
Os(C≡CPh)(H)(CO)(pzH)(PPr$^i_3$)$_2$	2150	(76)
Os(C≡CPh)(H)(CO){P(OMe)$_3$}(PPr$^i_3$)$_2$	2090	(76)
Os(C≡CPh)(H)(CO){P(OMe)$_3$}(PMeBu$^t_2$)$_2$	2060^e	(241)
Os(C≡CPh)(H)(CO)$_2$(PPr$^i_3$)$_2$	2095	(76)
Os(C≡CPh)(H)(CO)(PPr$^i_3$)$_2$	2090^j	(76)
Os(C≡CPh)(H)(H$_2$)(CO)(PPr$^i_3$)$_2$	2090^j	(76)
OsCp(C≡CPh)(PPh$_3$)$_2$	2066	(251b)
OsCp(C≡CPh)(dppe)	2090	(259)
[OsCp*(C≡CPh)(CO)(PPh$_3$)][BF$_4$]	2095^e	(260)
Os(η^6-C$_6$H$_6$)(C≡CPh)I(PMe$_3$)	2095^e	(261)
Os(η^6-C$_6$H$_6$)(C≡CPh)I(PMe$_2$Bu^t)	2090^e	(261)
Os(η^6-C$_6$H$_6$)(C≡CPh)I(PMeBu$^t_2$)	2085^e	(261)
Os(η^6-C$_6$H$_6$)(C≡CPh)I(PPr$^i_3$)	2085^c	(262)
E[Q]$_3$[Co(C≡CPh)$_6$]	2045	(263)
E = Na, K; Q = PPh$_4$, AsPh$_4$		
Na$_2$[Co(C≡CPh)$_4$(PEt$_3$)$_2$]	2044	(264)
Na[PPh$_4$][Co(C≡CPh)$_4$(PPh$_3$)$_2$]	2050	(264)
Co(C≡CPh)(PMe$_3$)$_4$	2018	(265)
Co(C≡CPh)$_2$(dppe)$_2$	2068	(264)
	2064^{j,m}	
[trans-Co(C≡CPh)$_2$(PMe$_3$)$_4$]X		
X = BF$_4$	2055	(83)
X = BPh$_4$	2060	(83)
[trans-Co(C≡CPh)(H)(depe)$_2$][BPh$_4$]	2096	(80)
fac,cis-Co(C≡CPh)$_2$(H)(PMe$_3$)$_3$	2068, 2080	(78)

(continued)

TABLE VI (*continued*)

Compound	$\nu(C{\equiv}C)^a$ (cm^{-1})	Ref.
mer,cis-Co(C≡CPh)$_2$(H)(PMe$_3$)$_3$	2058, 2080	(78)
mer, trans-Co(C≡CPh)$_2$(H)(PMe$_3$)$_3$	2058, 2080	(78)
Co(C≡CPh)$_2$(H)(PMe$_3$)$_3$	2080^i	
	2055, 2077	(83)
	2080^i	
mer-Co(C≡CPh)(H)(Ph)(PMe$_3$)$_3$	2083	(78)
Co(C≡CPh){P(CH$_2$CH$_2$PPh$_2$)$_3$}	2050	(33j,82)
[Co(C≡CPh){P(CH$_2$CH$_2$PPh$_2$)$_3$}][BPh$_4$]	2070	(33j)
[Co(C≡CPh)(H){P(CH$_2$CH$_2$PPh$_2$)$_3$}][BPh$_4$]	2085	(82)
CoPc(C≡CPh)	2128^c	(215)
Li$_2$[CoPc(C≡CPh)]$_2$	2096	(215)
Li[CoPc(C≡CPh)]$_2$	2092, 2103	(215)
Co(Salen)(C≡CPh)(py)	2120^b	(266)
Co(Salen)(C≡CPh)(N$_2$C$_4$H$_6$)	2110^b	(267)
trans-Co{N$_4$}(C≡CPh)$_2$	2113^b	(268)
trans-Co{N$_4$}(C≡CPh)Cl	2126^b	(268)
trans-Co{N$_4$}(C≡CPh)I	2120^c	(81)
[*trans*-Co{N$_4$}(C≡CPh)(NC$_5$H$_4$-4-Bu^t)][PF$_6$]	2110^c	(81)
[NBu$^n_4$][*trans*-Rh(C≡CPh)Cl(PPr$^i_3$)$_2$]	2030^k	(269)
Rh(C≡CPh)(PMe$_3$)$_4$	2081	(85)
Rh(C≡CPh)(dmpe)$_2$	2077^g	(85)
trans-Rh(C≡CPh)(py)(PPr$^i_3$)$_2$	2055^g	(270)
trans-Rh(C≡CPh)(C$_2$H$_4$)(PPr$^i_3$)$_2$	2060^j	(270)
cis-Rh(C≡CPh)(PCy$_3$)(cod)	2081	(271)
Rh(C≡CPh)(PPh$_3$)$_3$	2100	(272a,b)
Rh(C≡CPh)(CO)(PCy$_3$)$_2$	2090^e	(271)
Rh(C≡CPh)(CO)(PPh$_3$)$_2$	2092	(272a,b)
	2094	(222a,b)
trans-Rh(C≡CPh)(C=CHPh)(PPr$^i_3$)$_2$	2070^c	(273)
trans,mer-Rh(C≡CPh)$_2$(H)(PMe$_3$)$_3$	2097^g	(90)
trans,mer-Rh(C≡CPh)$_2$(H)(PBu$^n_3$)$_3$	2085^h	(86a)
Rh(C≡CPh)$_2$(SnMe$_3$)(PPh$_3$)$_2$	2073, 2082	(272a,b)
trans,mer-Rh(C≡CPh)$_2$(SnMe$_3$)(PMe$_3$)$_3$	2093	(274)
Rh(C≡CPh)(H)Cl(PPr$^i_3$)$_2$	2105^j	(270)
Rh(C≡CPh)(H)Cl(py)(PPr$^i_3$)$_2$	2100^c	(270,275)
Rh(C≡CPh)Cl(OSO$_2$CF$_3$)(CO)(PPh$_3$)$_2$	2144^b	(276)
Rh(C≡CPh){N(CH$_2$CH$_2$PPh$_2$)$_3$}	2080	(33m,88a,b)
[Rh(C≡CPh){N(CH$_2$CH$_2$PPh$_2$)$_3$}]X	2115	(33h,m,88b)
X = BF$_4$, ClO$_4$, PF$_6$, BPh$_4$		
[Rh(C≡CPh){N(CH$_2$CH$_2$PPh$_2$)$_3$}][PF$_6$]$_2$	2125	(33h)
[Rh(C≡CPh){N(CH$_2$CH$_2$PPh$_2$)$_3$}(EtOH)]X$_2$	2125	(33h)
X = PF$_6$, BF$_4$		
[Rh(C≡CPh){N(CH$_2$CH$_2$PPh$_2$)$_3$}(OCMe$_2$)][PF$_6$]$_2$	2125	(33h)
[Rh(C≡CPh)(H){N(CH$_2$CH$_2$PPh$_2$)$_3$}]X	2120	(88a,277)
X = BPh$_4$, SO$_3$CF$_3$		
[Rh(C≡CPh)(O$_2$CC$_6$H$_4$-3-Cl){N(CH$_2$CH$_2$PPh$_2$)$_3$}][BPh$_4$]	2120	(278)
[Rh(C≡CPh)(OH){N(CH$_2$CH$_2$PPh$_2$)$_3$}][BPh$_4$]	2115	(278)
Rh(C≡CPh){P(CH$_2$CH$_2$PPh$_2$)$_3$}	2070	(88a)
[Rh(C≡CPh){P(CH$_2$CH$_2$PPh$_2$)$_3$}]X	2100	(33h)
X = ClO$_4$, BF$_4$, PF$_6$		

TABLE VI (*continued*)

Compound	$\nu(C\equiv C)^a$ (cm^{-1})	Ref.
[Rh(C≡CPh){P(CH$_2$CH$_2$PPh$_2$)$_3$}(EtOH)]X$_2$	2115	(*33h*)
X = PF$_6$, BF$_4$		
[Rh(C≡CPh){P(CH$_2$CH$_2$PPh$_2$)$_3$}(OCMe$_2$)][PF$_6$]$_2$	2115	(*33h*)
[Rh(C≡CPh)(H){P(CH$_2$CH$_2$PPh$_2$)$_3$}][BPh$_4$]	2115	(*88a*)
Rh(Salen)(C≡CPh)(py)	2070^b	(*266*)
RhCp(C≡CPh)(H)(PPr$^i{}_3$)$_2$	2050^c	(*270*)
RhCp(C≡CPh)(HgCl)(PPr$^i{}_3$)	2097^c	(*279*)
RhCp(C≡CPh)(HgI)(PPr$^i{}_3$)	2096^c	(*279*)
RhCp(C≡CPh)(HgMe)(PPr$^i{}_3$)	2080^c	(*279*)
trans-Ir(C≡CPh)(CO)(PPr$^i{}_3$)$_2$	2090^q	(*280*)
trans-Ir(C≡CPh)(CO)(PPh$_3$)$_2$	2091	(*281*)
	2118	(*272*)
	2115	(*282a,b*)
Ir(C≡CPh)(CO)(PCy$_3$)$_2$	2095^e	(*271*)
Ir(C≡CPh)(CO)$_2$(PPh$_3$)$_2$	2127	(*282a*)
Ir(C≡CPh)(CO)$_2$(dppe)	2115^e	(*271*)
Ir(C≡CPh)(MeO$_2$CC≡CCO$_2$Me)(CO)(PPh$_3$)$_2$	2112	(*272a*)
	2113	(*282a*)
cis-Ir(C≡CPh)(PCy$_3$)(cod)	2090	(*93*)
Ir(C≡CPh)(cod)(PPh$_3$)$_2$	2106	(*271*)
Ir(C≡CPh)(cod)(dppe)	2105	(*271*)
Ir(C≡CPh)$_3$(PPr$^i{}_3$)$_2$	2080^c	(*283*)
mer,trans-Ir(C≡CPh)$_3$(CO)(PPr$^i{}_3$)$_2$	2112, 2157^c	(*283*)
Ir(C≡CPh)$_2$Cl(PPr$^i{}_3$)$_2$	2105^c	(*283*)
Ir(C≡CPh)$_2$(SnMe$_3$)(CO)(PPh$_3$)$_2$	2122	(*272a,b*)
Ir(C≡CPh)$_2$(SnMe$_3$)(CO)(PMe$_2$Ph)$_2$	2120	(*272a*)
cis,trans-Ir(C≡CPh)Cl$_2$(CO)(PPh$_3$)$_2$	2142	(*284*)
Ir(C≡CPh)Br$_2$(CO)(PPh$_3$)$_2$	2137	(*282a*)
Ir(C≡CPh)I$_2$(CO)(PPh$_3$)$_2$	2125^c	(*282a*)
Ir(C≡CPh)(H)$_2${MeC(CH$_2$PPh$_2$)$_3$}	2100	(*285*)
Ir(C≡CPh)(H)$_2$(CO)(dppe)	2120^e	(*271*)
Ir(C≡CPh)(SO$_4$)(CO)(PPh$_3$)$_2$	2147^c	(*282a*)
Ir(C≡CPh)(O$_2$CCF$_3$)$_2$(CO)(PPh$_3$)$_2$	2133^c	(*282a*)
Ir(C≡CPh)(O$_2$CMe)$_2$(CO)(PPh$_3$)$_2$	2138	(*282a*)
Ir(C≡CPh)(H)Cl(PPr$^i{}_3$)$_2$	2090^c	(*286a–c*)
Ir(C≡CPh)(H)Cl(PPr$^i{}_3$)$_2$(py)	2095^c	(*286c*)
mer-Ir(C≡CPh)(H)Cl(PPh$_3$)$_3$	2110	(*287*)
mer-Ir(C≡CPh)(H)Cl(PMePh$_2$)$_3$	2121	(*287*)
	2120	(*287*)
Ir(C≡CPh)(H)Cl(CO)(PPh$_3$)$_2$	2128	(*272a*)
	2119	(*287*)
	2130^l	(*282a*)
trans-Ir(C≡CPh)(H)Cl(CO)(PPr$^i{}_3$)$_2$	2120^c	(*280*)
trans-Ir(C≡CPh)(H)I(CO)(PPr$^i{}_3$)$_2$	2120^c	(*280*)
trans-Ir(C≡CPh)(H){OC(O)CF$_3$}(CO)(PPr$^i{}_3$)$_2$	2120^c	(*280*)
[Ir(C≡CPh)(H)(CO)$_2$(PEt$_3$)$_2$][BPh$_4$]	2146^l	(*282a*)
[Ir(C≡CPh)(H)(CO)$_2$(PPh$_3$)$_2$][PF$_6$]	2152	(*282a*)
Ir(C≡CPh)ClBr(CO)(PPh$_3$)$_2$	2140	(*284*)

127

(*continued*)

TABLE VI (continued)

Compound	$\nu(C\equiv C)^a$ (cm^{-1})	Ref.
$Ir(C\equiv CPh)ClI(CO)(PPh_3)_2$	2140	(284)
$Ir(C\equiv CPh)Cl(HgCl)(CO)(PPh_3)_2$	2123^e	(282a)
$Ir(C\equiv CPh)Cl(HgCCR)(CO)(PPh_3)_2$	2120^c	(288)
$Ir(C\equiv CPh)Cl(OSO_2CF_3)(CO)(PPh_3)_2$	2051^b	(276)
$Ir(C\equiv CPh)Cl(Me)(CO)(PPh_3)_2$	2132^l	(282a)
$Ir(C\equiv CPh)I(Me)(CO)(PPh_3)_2$	2120	(282a)
$Ir(C\equiv CPh)(O_2)(CO)(PPh_3)_2$	2125	(282a,b)
	2133	(281)
$Ir(C\equiv CPh)(SO_2)(CO)(PPh_3)_2$	2151^e	(282a)
$Ir(C\equiv CPh)(CO)(\mu\text{-dppm})_2CuCl$	2100^c	(94)
$Ir(C\equiv CPh)(CO)(\mu\text{-dppm})_2AgCl$	2090^c	(94)
$[Ir(C\equiv CPh)(CO)(\mu\text{-dppm})_2Ag][BPh_4]$	2090^c	(94)
$K_2[Ni(C\equiv CPh)_4]$	2062	(225)
	2066	(289)
$trans\text{-}Ni(C\equiv CPh)_2(PEt_3)_2$	2085	(290)
$trans\text{-}Ni(C\equiv CPh)_2(PBu^n_3)_2$	2090	(291)
$trans\text{-}Ni(C\equiv CPh)_2(PPh_3)_2$	2095	(289)
$Ni(C\equiv CPh)_2(PEt_2Ph)_3$	2088	(292)
$trans\text{-}Ni(C\equiv CPh)Cl(PEt_2Ph)_2$	2098^g	(292)
$trans\text{-}Ni(C\equiv CPh)(C_6Cl_5)(PMe_2Ph)$	2090	(293)
$trans\text{-}Ni(C\equiv CPh)(C_6H_2\text{-}2,4,6\text{-}Me_3)(PMe_2Ph)_2$	2070	(294)
$trans\text{-}Ni(C\equiv CPh)(C_6H_3\text{-}2,6\text{-}OMe)(PMe_2Ph)_2$	2078	(294)
$trans\text{-}Ni(C\equiv CPh)(CCl=CCl_2)(PMe_2Ph)_2$	2083	(294)
$Ni(C\equiv CPh)(NO)(PPh_3)_2$	2090^b	(295)
$Ni(C\equiv CPh)_2(\mu\text{-dppm})_2HgCl_2$	2080^e	(296)
$Ni(C\equiv CPh)_2(\mu\text{-dppm})_2HgBr_2$	2080	(296)
$Ni(C\equiv CPh)_2(\mu\text{-dppm})_2HgI_2$	2080	(296)
$[Ni(C\equiv CPh)_2(\mu\text{-dppm})_2Au]Cl$	2090	(296)
$NiCp(C\equiv CPh)(PPh_3)$	2160	$(222,^b\ 297)$
	2090^b	(298)
$trans\text{-}Pd(C\equiv CPh)_2(PMe_3)_2$	2080	(290)
$trans\text{-}Pd(C\equiv CPh)_2(PEt_3)_2$	2070	(290)
$trans\text{-}Pd(C\equiv CPh)_2(PMe_2Ph)_2$	2110^c	(299)
$trans\text{-}Pd(C\equiv CPh)_2(PPh_3)_2$	2110	$(290,299)^c$
$trans\text{-}Pd(C\equiv CPh)Cl(PPh_3)_2$	2125	(284)
$trans\text{-}Pd(C\equiv CPh)Br(PMe_3)_2$	2118	(300)
$trans\text{-}Pd(C\equiv CPh)Br(PPh_3)_2$	2120	(284)
$trans\text{-}Pd(C\equiv CPh)(Me)(PMe_3)_2$	2100^b	(301)
$trans\text{-}Pd(C\equiv CPh)(C_6F_5)(PPh_3)_2$	2113	(302)
$trans\text{-}Pd(C\equiv CPh)\{C(CO_2Me)=CH(CO_2Me)\}(PEt_3)_2$	2100	(303)
$cis\text{-}Pd(C\equiv CPh)_2(dppe)$	2112, 2101	(304)
	$2121,\ 2109^n$	
$cis\text{-}Pd(C\equiv CPh)(Me)(dppe)$	2100^b	(301)
$Pd(C\equiv CPh)_2\{PPh_2(C_{10}H_{14}O)\}_2$	2100	(305)
$Q_2[Pt(C\equiv CPh)_4]$		
$\quad Q = K$	2096^c	(306)
$\quad Q = NBu^n_4$	2100	(307)

TABLE VI (*continued*)

Compound	$\nu(C{\equiv}C)^a$ (cm^{-1})	Ref.
K$_2$[*trans*-Pt(C$\equiv$CPh)$_4$(dppe-P)$_2$]	2082	(*308*)
	2092, 2109^{h,o}	
[PMePh$_3$]$_2$[*cis*-Pt(C$\equiv$CPh)$_2$(C$_6$F$_5$)$_2$]	2083, 2096	(*302*)
[PMePh$_3$]$_2$[Pt(C$\equiv$CPh)$_3$(C$_6$F$_5$)]	2076, 2100	(*309*)
trans-Pt(C$\equiv$CPh)$_2$(PMe$_3$)$_2$	2090	(*290*)
trans-Pt(C$\equiv$CPh)$_2$(PEt$_3$)$_2$	2107	(*290,310*)
	2100	(*311*)
	2105^c	(*312*)
	2103^b	(*21*)
trans-Pt(C$\equiv$CPh)$_2$(PBu$^n_3$)$_2$	2100	(*310*)
trans-Pt(C$\equiv$CPh)$_2$(PMe$_2$Ph)$_2$	2106	(*313*)
	2116	(*314*)
	2110	(*315*)
	2100	(*316*)
trans-Pt(C$\equiv$CPh)$_2$(PMePh$_2$)$_2$	2105	(*307*)
trans-Pt(C$\equiv$CPh)$_2$(PPh$_3$)$_2$	2090	(*290*)
	2110	(*317a,b*)
	2120	(*318*)
	2100^b	(*218*)
	2130	(*319*)
trans-Pt(C$\equiv$CPh)$_2$(dppm-P)$_2$	2105	(*307*)
trans-Pt(C$\equiv$CPh)$_2${PPh$_2$CH$_2$P(O)Ph$_2$}	2100	(*307*)
trans-Pt(C$\equiv$CPh)$_2$(SbEt$_3$)$_2$	2080	(*290*)
trans-Pt(C$\equiv$CPh)Cl(PEt$_3$)$_2$	2120	(*106*)
trans-Pt(C$\equiv$CPh)Cl(PEt$_2$Ph)$_2$	2120	(*320*)
trans-Pt(C$\equiv$CPh)Cl(PMePh$_2$)$_2$	2130	(*320,321*)^c
trans-Pt(C$\equiv$CPh)Cl(PPh$_3$)$_2$	2125	(*317b*)
trans-Pt(C$\equiv$CPh)I(PPh$_3$)$_2$	2125	(*284*)
trans-Pt(C$\equiv$CPh)(Me)(PMe$_3$)$_2$	2100^b	(*301*)
trans-Pt(C$\equiv$CPh)(C$_6$H$_4$-3-F)(PEt$_3$)$_2$	2105^p	(*30*)
trans-Pt(C$\equiv$CPh)(C$_6$H$_4$-4-F)(PEt$_3$)$_2$	2110^p	(*30*)
trans-Pt(C$\equiv$CPh)(CPh$=$CH$_2$)(PPh$_3$)$_2$	2110	(*119*)
trans-Pt(C$\equiv$CPh){C(CO$_2$Me)$=$CH(CO$_2$Me)}(PEt$_3$)$_2$	2100	(*303*)
trans-Pt(C$\equiv$CPh)(CMe$=$CH$_2$)(PPh$_3$)$_2$	2102^c	(*118,128*)
trans-Pt(C$\equiv$CPh)(SnMe$_3$)(PMePh$_2$)$_2$	2110	(*272a,b*)
trans-Pt(C$\equiv$CPh)(SnMe$_3$)(PPh$_3$)$_2$	2100	(*272a,b*)
cis-Pt(C$\equiv$CPh)$_2$(PEt$_3$)$_2$	2106, 2113	(*310*)
cis-Pt(C$\equiv$CPh)$_2$(PBu$^n_3$)$_2$	2098, 2110	(*310*)
cis-Pt(C$\equiv$CPh)$_2$(PMePh$_2$)$_2$	2120, 2130^c	(*321*)
cis-Pt(C$\equiv$CPh)$_2$(PPh$_3$)$_2$	2125	(*113,317b*)
	2110^c	(*321*)
cis-Pt(C$\equiv$CPh)$_2$(dppm)	2108^c	(*322*)
cis-Pt(C$\equiv$CPh)$_2$(dppe)	2112^c	(*322*)
	2110	(*316*)
	2116	(*314*)
	2095, 2110	(*308*)
	2098, 2116^e	(*308*)
	2098, 2112^{h,o}	(*308*)

(*continued*)

TABLE VI (*continued*)

Compound	$\nu(C{\equiv}C)^a$ (cm^{-1})	Ref.
cis-Pt(C≡CPh)$_2$(cod)	2130, 2135^c	*(321)*
cis-Pt(C≡CPh)Cl(PPh$_3$)$_2$	2125	*(284)*
cis-Pt(C≡CPh)Br(PPh$_3$)$_2$	2125	*(284,323)*
cis-Pt(C≡CPh)I(PPh$_3$)$_2$	2125	*(284)*
cis-Pt(C≡CPh)Me(dppe)	2110	*(186a)*
cis-Pt(C≡CPh)(C$_6$F$_5$)(dppe)	2116	*(302)*
cis-Pt(C≡CPh)(SnPh$_3$)(PPh$_3$)$_2$	2100^b	*(324)*
Pt(C≡CPh)$_2${PPh$_2$(C$_{10}$H$_{14}$O)}$_2$	2105^c	*(305)*
Pt(C≡CPh)Cl(PEt$_3$)$_2$	2120^c	*(312)*
Pt(C≡CPh)(SnEt$_3$)(PPh$_3$)$_2$	2108	*(272a,b)*
Pt(C≡CPh)(C$_6$F$_5$)(py)(PPh$_3$)	2118	*(325)*
[Pt(C≡CPh)(AsMe$_3$)$_2$][PF$_6$]	2100^b	*(326)*
[Pt(C≡CPh)(CN-C$_6$H$_4$-4-OMe)(PMePh$_2$)][PF$_6$]	2145	*(320)*
[Pt(C≡CPh)(μ-dppm)$_2$(μ-pz-3,5-Me$_2$)Rh(CO)][PF$_6$]	2129	*(137)*
[Pt(C≡CPh)(μ-dppm)$_2$(μ-pz-4-Me$_2$)Rh(CO)][PF$_6$]	2126	*(137)*
[Pt(C≡CPh)(μ-dppm)$_2$(μ-pz)Rh(CO)][PF$_6$]	2122	*(137)*
[Pt(C≡CPh)(μ-dppm)$_2$(μ-pz-4-Br-3,5-Me$_2$)Rh(CO)][PF$_6$]	2122	*(137)*
[Pt(C≡CPh)(μ-dppm)$_2$(μ-pz-3,4,5-Br$_3$)Rh(CO)][PF$_6$]	2133	*(137)*
[Pt(C≡CPh)(μ-dppm)$_2$(μ-pz-3-Me-5-CF$_3$)Rh(CO)][PF$_6$]	2129	*(137)*
[Pt(C≡CPh)(μ-dppm)$_2$(μ-C≡CPh)Ir(SO$_2$)	2104	*(327)*
Pt(C≡CPh)$_2$(μ-dppm)$_2$(μ-SO$_2$){IrCl}	2103	*(327)*
Pt(C≡CPh)(μ-dppm)$_2$(μ-SO$_2$)(μ-C≡CPh)RhCl	2105	*(327)*
Pt$_2$(C≡CPh)(PPr$^i_2$Ph)$_2$(μ-C≡CPh)(μ-SiMe$_2$)	2108	*(145)*
Pt$_2$(C≡CPh)(PCy$_3$)$_2$(μ-C≡CPh)(μ-SiMe$_2$)	2100	*(145)*
Pt$_2$(C≡CPh)(PMeBu$^t_2$)$_2$(μ-C≡CPh)(μ-SiMe$_2$)	2104	*(145)*
[Pt(C≡CPh){C(NHBu^n)NH(C$_6$H$_4$-4-OMe)}(PMePh$_2$)][PF$_6$]	2130	*(320)*
[PMePh$_3$]$_2$[Pt(C≡CPh)(C$_6$F$_5$)(μ-C≡CPh)$_2$Pt(C$_6$F$_5$)$_2$]	2109	*(309)*
Pt(C≡CPh)(μ-dppm)$_2$(μ-C≡CPh)W(CO)$_3$	2110^c	*(132a)*
Pt(C≡CPh)(μ-dppm)$_2$(μ-C≡CPh)Mo(CO)$_3$	2120	*(132a)*
Pt(C≡CPh)(μ-dppm)$_2$(μ-C=CHPh)W(CO)$_3$	2120	*(132a)*
Pt(C≡CPh)(μ-dppm)$_2$(μ-C≡CPh)IrCl	2112	*(328)*
Pt(C≡CPh)$_2$(μ-dppm)$_2$(μ-H)Ir(H)Cl	2118	*(328)*
Pt(C≡CPh)(μ-dppm)$_2$(μ-C≡CPh)Ir(CO)Cl	2102	*(328)*
[Pt(C≡CPh)(μ-dppm)$_2$(μ-C≡CPh)Ir(CO)][BPh$_4$]	2125	*(328)*
[Pt(C≡CPh)(μ-dppm)$_2$(μ-C≡CPh)Pt(C≡CPh)][ClO$_4$]	2120	*(20a)*
[*trans*-Pt(C≡CPh)(NH=NC$_6$H$_4$-4-F)(PPh$_3$)$_2$][BF$_4$]	2124^b	*(123)*
[Pt$_2$(C≡CPh)(PEt$_3$)$_4$(μ-C=CHPh)][BF$_4$]	2100^e	*(144)*
Pt(C≡CPh)(PEt$_3$)$_2$(μ-C=CHPh){Pt(PEt$_3$)Cl}	2115^g	*(20c)*
Pt(C≡CPh)(PEt$_3$)$_2$(μ-C=CHPh){Pt(PEt$_3$)Br}	2115^g	*(20c)*
Pt(C≡CPh)(PEt$_3$)$_2$(μ-C=CHPh){Pt(PEt$_3$)I}	2115^g	*(20c)*
Pt(C≡CPh)(PEt$_3$)$_2$(μ-C=CHPh){Pt(PEt$_3$)(SPh)}	2110^g	*(20c)*
Pt(C≡CPh)(PEt$_3$)$_2$(μ-C=CHPh){Pt(PEt$_3$)(SCHMe$_2$)}	2115^g	*(20c)*
Pt(C≡CPh)(PEt$_3$)$_2$(μ-C=CHPh){Pt(PEt$_3$)(SC$_6$Cl$_5$)}	2110^g	*(20c)*
[Pt(C≡CPh)(PEt$_3$)$_2${μ-η^2,η^4-C(Ph)=CC(Ph)CH(Ph)}Pt(PEt$_3$)$_2$][BF$_4$]	2108^c	*(139)*
Pt(C≡CPh)$_2$(PEt$_3$)$_2${N(C$_6$H$_4$-4-NO$_2$)NNN(C$_6$H$_4$-4-NO$_2$)}	2154^c	*(131)*
{Pt(C≡CPh)}$_2$(μ-H)(μ-dppm)$_2$	2110^c	*(329)*
Pt$_2$(C≡CPh)$_2$(μ-dppm)$_2$	2090^c	*(329)*
{Pt(C≡CPh)}$_2$(μ-CS$_2$)(μ-dppm)$_2$	2110^c	*(329)*

TABLE VI (continued)

Compound	$\nu(C\equiv C)^a$ (cm^{-1})	Ref.
{Pt(C≡CPh)}$_2$(μ-MeO$_2$CC≡CCO$_2$Me)(μ-dppm)$_2$	2110^c	(329)
trans-{Pt(C≡CPh)$_2$}$_2$(μ-dppm)$_2$	2105	(307)
[Pt(C≡CPh)$_2$(μ-dppm)$_2$Ag][PF$_6$]	2105^c	(330)
Pt(C≡CPh)$_2$(μ-dppm)$_2$AgI	2105^c	(330)
Pt(C≡CPh)$_2$(μ-dppm)$_2$AgCl	2105^c	(330)
Pt(C≡CPh)Cl(μ-dppm)$_2$AgCl	2120^c	(330)
Pt(C≡CPh)Br(μ-dppm)$_2$AgBr	2120^c	(330)
Pt(C≡CPh)I(μ-dppm)$_2$AgI	2109^c	(330)
Pt(C≡CPh)$_2$(μ-dppm)$_2$CuCl	2110^c	(330)
[Pt(C≡CPh)$_2$(μ-dppm)$_2$Cu][BPh$_4$]	2130^c	(330)
Pt(C≡CPh)$_2$(μ-dppm)$_2$HgCl$_2$	2102, 2114	(307)
Pt(C≡CPh)(PPh$_3$)$_2$(μ-C≡CPh)Pt(C$_6$F$_5$)$_2$(CO)	2127^b	(143)
Pt(C≡CPh)(PEt$_3$)$_2$(μ-C≡CPh)Pt(C$_6$F$_5$)$_2$(CO)	2128^b	(143)
[Q][Au(C≡CPh)$_2$]		
Q = NBu$^n_4$	2105	(32)
Q = N(PPh$_3$)$_2$	2105^b	(331)
Q = PPh$_4$	2102	(32)
Q = K	2102^c	(332)
	2112^{c,o}	(332)
	2110^c	(333)
Au(C≡CPh)(PMe$_3$)	2097^b	(148)
	2102^c	(334)
Au(C≡CPh)(PPr$^i_3$)	2102^c	(334)
Au(C≡CPh)(PCy$_3$)	2113^b	(335)
Au(C≡CPh)(PPh$_3$)	2118^c	(336)
	2120^b	(148)
	2125	(32)
Au(C≡CPh){P(p-Tol)$_3$}	2120^c	(336)
Au(C≡CPh){PPh(OMe)$_2$}	2110^b	(148)

^{a} Frequencies are given in cm^{-1}. Unless noted otherwise, frequencies were determined by infrared spectroscopy for nujol mull samples.
^{b} Sample type not reported.
^{c} KBr pellet.
^{d} Halocarbon mull.
^{e} CH$_2$Cl$_2$ solution.
^{f} Cyclohexane solution.
^{g} Hexane solution.
^{h} Neat.
^{i} Toluene solution.
^{j} C$_6$H$_6$ or C$_6$D$_6$ solution.
^{k} Tetrahydrofuran solution.
^{l} CHCl$_3$ solution.
^{m} N-Methylpyrrolidone solution.
^{n} Pyridine solution.
^{o} From Raman spectroscopy.
^{p} CCl$_4$ solution.
^{q} Pentane solution.

cases where they have been studied. Nearly all $\nu(C\equiv C)$ frequencies have been measured by infrared spectroscopy.

Before discussing the distribution of $\nu(C\equiv C)$ frequencies for metal–alkynyl compounds, we make note of two caveats. First, it is clear that there are complexes in Table VI for which incorrect frequencies have been reported. A particularly egregious example is the simple compound trans-Pt(C≡CPh)$_2$(PPh$_3$)$_2$, for which $\nu(C\equiv C)$ frequencies of 2090 (290), 2100 (218), 2110 (317a,b), 2120 (318), and 2130 (319) cm^{-1} are claimed. This particular problem is easy to spot. But because for most metal–alkynyl complexes there has only been a single frequency measurement, incorrect data for them may not stand out, and the $\nu(C\equiv C)$ frequency distribution may be skewed accordingly by their inclusion in it. We have not corrected frequencies that we suspect have been misassigned, because this could only legitimately be done by inspecting the vibrational spectra, which are very rarely presented. We believe errors should be apparent by being grossly outlying points in the distribution and by comparison to frequencies of related complexes.

A second cautionary note regards the $\nu(C\equiv C)$ frequencies of complexes with two or more chemically equivalent alkynyl ligands. Depending on the symmetry of the compound, the spectra of these species may exhibit more than one $\nu(C\equiv C)$ band, because the $\nu(C\equiv C)$ modes consist of different symmetry-adapted combinations of the various C≡C stretching coordinates. An example is the compound cis-Pt(C≡CPh)$_2$(PBu$_3^n$)$_2$ (310), for which $\nu(C\equiv C)$ infrared bands at 2098 and 2110 cm^{-1} are observed. However, many such compounds are reported to exhibit only a single $\nu(C\equiv C)$ band (e.g., cis-Pt(C≡CPh)$_2$(PPh$_3$)$_2$) (113,317b,321), even when other $\nu(C\equiv C)$ modes are dipole allowed. In drawing comparisons among the frequencies of bis–alkynyl (and more highly substituted alkynyl) complexes, it is important to compare frequencies for modes of identical symmetry only; reference to Table VI reveals that $\nu(C\equiv C)$ modes of different symmetry can vary by 5–45 cm^{-1} within a single compound, and "shifts" of this magnitude could easily be misinterpreted as being the result of differences in M–CCR π-bonding. This complication would seem to be particularly acute for metal–alkynyl polymers (4a–e).

The distribution of C≡C stretching frequencies for the L$_n$M—C≡CPh complexes set out in Table VI is displayed in Fig. 5. In those cases where several different frequencies are reported for the same compound, we chose the one that was most often reported [e.g., 2120 cm^{-1} for trans-Ir(C≡CPh)(CO)(PPh$_3$)$_2$]; if a "consensus frequency" was unavailable [e.g., trans-Pt(C≡CPh)$_2$(PPh$_3$)$_2$], the compound was excluded. The C≡C frequencies of metal–phenylethynyl complexes span a roughly 180 cm^{-1} range, although 98% of them lie in the range 2030–2160 cm^{-1}.

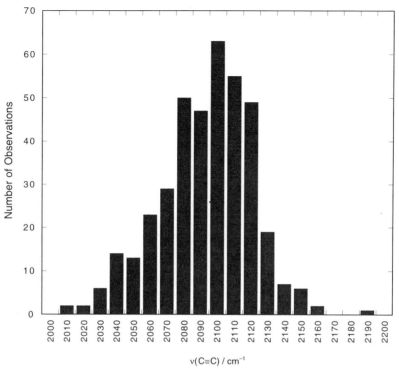

Number of Observations

$\nu(C{\equiv}C)$ / cm^{-1}

FIG. 5. Distribution of C≡C stretching frequencies for L_nM–C≡CPh complexes. Data are taken from Table VI. The frequency labels correspond to the centers of the bins.

As for the C≡C bond distances of metal–alkynyl complexes (Section III-B), the interpretation of $\nu(C{\equiv}C)$ frequencies requires a reference point; and while terminal alkynes are commonly used in this regard, they are flawed because the $\nu(C{\equiv}C)$ mode contains contributions from the terminal C–H coordinate (207). Better reference compounds are organic and main-group phenylethyne derivatives of type RC≡CPh (R ≠ H), for which $\nu(C{\equiv}C)$ frequencies range from 2150–2280 cm^{-1} (R = Me, t-Bu, CH$_2$OH, CHO, CO$_2$H, NMe$_2$, SiMe$_3$, SMe, Cl, Br, I); by comparison, $\nu(C{\equiv}C) = 2111$ cm^{-1} for phenylacetylene (207). The fact that the $\nu(C{\equiv}C)$ frequencies for L_nM—C≡CPh complexes are lower than those of RC≡CPh compounds is consistent with the observation that the C≡C bond distances of metal–alkynyl complexes are longer than those of disubstituted organic alkynes. The two possible explanations put forth previously for the C≡C bond distances (Section III-B) apply here as well, that is, that the difference between the C≡C bonds of organic alkynes and metal–alkynyl compounds reflects the effects of M $\rightarrow$ CCR π-backbonding (or RCC $\rightarrow$ M π-bonding) and/or that the M–CCR bond is polarized in a $M^{\delta+}-C^{\delta-}$ fashion—although the greater precision of the $\nu(C\equiv C)$ frequencies allows these possibilities to be addressed in more detail than was possible for the $C\equiv C$ bond distances.

Some of the trends in $\nu(C\equiv C)$ frequencies among the compounds in Table VI would appear to be consistent with M $\rightarrow$ CCR π-backbonding. Within a set of $L_nM-C\equiv CPh$ complexes of a given metal, shifts in the frequency of $\nu(C\equiv C)$ upon changing the electron donating or accepting characteristics of the ancillary ligands are typically observed to be in the expected direction, with lower frequencies being found for electron-rich metal centers; the magnitudes of these shifts are modest (<50 cm^{-1}). Some selected, relatively simple examples of this effect include the pairs of compounds and ions $[W(C\equiv CPh)(CO)_4L]^-$ (L = CO, $\nu(C\equiv C)$ = 2080 cm^{-1}; L = PPh$_3$, $\nu(C\equiv C)$ = 2065 cm^{-1}) (213), FeCp*(C$\equiv$CPh)L$_2$ (L = CO, $\nu(C\equiv C)$ = 2095 cm^{-1} (59); L = ½ dppe, $\nu(C\equiv C)$ = 2054 cm^{-1} (33o), and trans-Pd(C$\equiv$CPh)X(PMe$_3$)$_2$ (X = Me, $\nu(C\equiv C)$ = 2100 cm^{-1} (301); X = Br, $\nu(C\equiv C)$ = 2118 cm^{-1} (300). Ancillary ligands with less dramatic electronic differences than these yield smaller frequency shifts; for example, phenylethynyl complexes with ancillary phosphine ligands that differ in their R-groups commonly exhibit identical $\nu(C\equiv C)$ frequencies. To the extent that the electronic characteristics of the metal center that favor M $\rightarrow$ CCR π-backbonding tend to disfavor RCC $\rightarrow$ M π-bonding, it is less clear how to reconcile these trends in frequencies within the framework of the latter model of metal–alkynyl bonding, with the data for FeCp*(C$\equiv$CPh)L$_2$ (L = CO, ½ dppe)—the class of compounds from which theoretical (11) and photoelectron-spectroscopic (12a,b) studies yielded the RCC $\rightarrow$ M π-bonding model—standing out in this regard.

Another probe of the extent of M $\rightarrow$ CCR π-backbonding is the sensitivity of the $\nu(C\equiv C)$ frequency to the oxidation state of the metal. But because the M–CCR bond polarity should also be quite sensitive to the metal oxidation state, an unambiguous interpretation of these data cannot be devised straightforwardly. From the standpoint of M $\rightarrow$ CCR π-backbonding, the simple expectation is that the $\nu(C\equiv C)$ frequency should decrease according to $M^{(n+1)+} \geq M^{n+}$. This expectation is not always fulfilled. Consistent with this ordering, Bianchini, Zanello, and co-workers have reported tandem electrochemical and vibrational-spectroscopic studies on compounds of the general form $[M(C\equiv CR)$ $\{E(CH_2CH_2PPh_2)_3\}]^{n+}$ (M = Fe, Co, Rh; E = N, P), for which they found the $\nu(C\equiv C)$ frequencies to be ordered Co^{II} > Co^I (33j,82), Rh^{III} > Rh^{II} > Rh^I (33h,m,88a,b,277,278), and Fe^{III} $\cong$ Fe^{II} (33a,235). This was interpreted as suggesting that M $\rightarrow$ CCR π-backbonding is present in the Co^I, Rh^I, and Rh^{II} compounds but is unimportant for the iron compounds. Similarly, Bear et $al.$ concluded that M $\rightarrow$ CCH π-backbonding is present in the series of complexes $[Rh_2(C\equiv CH)(\mu\text{-}NC_5H_4\text{-}2\text{-}NPh)_4]^{-/0/+}$ based on the observation that the $\nu(C\equiv C)$ frequencies decrease by 18 and 32 cm^{-1} on one- and two-electron reduction, respectively, relative to that of the cation ($33i$). In sharp contrast to these studies are reports of two compounds for which $\nu(C\equiv C)$ $decreases$ upon an increase in metal oxidation state. Whiteley and co-workers discovered that the $\nu(C\equiv C)$ frequencies of the $[Mo(\eta^7\text{-}C_7H_7)(C\equiv CR)(dppe)]^+$ (R = Ph, t-Bu) ions are 13 cm^{-1} lower than those of the neutral compounds ($33g,n$), and independent data from Connelly et $al.$ and Whiteley et $al.$ reveal that for the pair of compounds $[FeCp^*(C\equiv CPh)(dppe)]^{0/+}$, the $\nu(C\equiv C)$ frequencies lie in the order Fe^{III} (2022 cm^{-1}) ($33d$) < Fe^{II} (2054 cm^{-1}) ($33o$). These latter results are clearly inconsistent with the HOMO's of these compounds participating in Fe $\rightarrow$ CCR π-backbonding. The frequency shifts of $[FeCp^*(C\equiv CPh)(dppe)]^{0/+}$ could be viewed as supporting the presence of RCC $\rightarrow$ Fe π-bonding, as proposed in theoretical and photoelectron-spectroscopic studies of related compounds ($11,12a,b$).

In none of the cases just described do the differences in $\nu(C\equiv C)$ frequencies among a series of redox congeners or analogous L_nM—C$\equiv$CPh complexes of a given metal exceed more than one-third of the 150-cm^{-1} or so range of frequencies exhibited in Fig. 5. This raises the additional question of the role of the nature of the metal in determining this frequency, through its potential influences on both M $\rightarrow$ CCR π-backbonding and on M–CCR bond polarity. Both of these interpretations have been invoked previously to explain trends in $\nu(C\equiv C)$ frequencies as a function of metal. In papers in which it is presumed that M $\rightarrow$ CCR π-backbonding is a general feature of metal–alkynyl compounds, differences in $\nu(C\equiv C)$ frequencies among analogous L_nM–C$\equiv$CPh complexes of different metals have been used to order the metals according to their π-basicity. Systematic differences in the frequency of $\nu(C\equiv C)$ as a function of metal are not always observed, however: the frequencies of the complexes $trans$-$M(C\equiv CPh)_2(PEt_3)_2$ stand in the order Pt (2107 cm^{-1}) > Ni (2085 cm^{-1}) > Pd (2070 cm^{-1}) (290), while those of $[M(C\equiv CPh)(CO)_5]^-$ are independent of metal (M = Cr, Mo, W; $\nu(C\equiv C)$ = 2080 cm^{-1}) (213). Moreover, the $\nu(C\equiv C)$ frequencies of the d^0 complexes $MCp_2(C\equiv CPh)_2$ are also metal-sensitive, varying according to Hf (2080 cm^{-1}) (35) > Zr (2075 cm^{-1}) (35) > Ti (2070 cm^{-1}) ($208a,b$), even though M $\rightarrow$ CCR π-backbonding must be absent. That $\nu(C\equiv C)$ frequencies cannot, in general, be interpreted exclusively within a L_nM $\rightarrow$ CCR π-backbonding framework suggests that they are also governed by bond-polarity considerations, as we noted earlier for the interpretation of C$\equiv$C bond lengths (Section III-B). Nast has noted that the bond-polarity explanation finds support among alkali-metal M—C≡CPh complexes (*1*), the $\nu(C\equiv C)$ frequencies of which range from 1990 cm^{-1} (M = Rb, Cs) to 2036 cm^{-1} (M = Li) (*337*). In interpreting these latter frequencies, however, allowance must be made for the fact

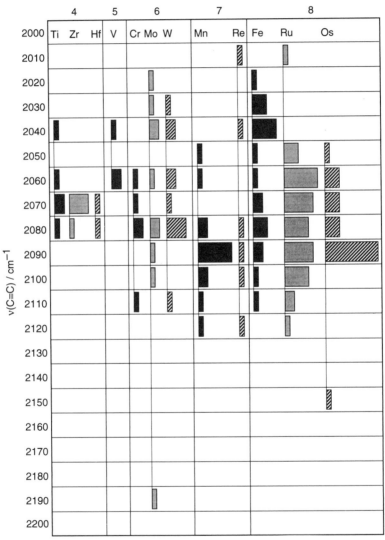

FIG. 6. Distribution of C≡C stretching frequencies for L$_n$M–C≡CPh complexes broken down according to metal. Data are taken from Table VI. The frequency labels correspond to the centers of the bins.

that the interactions between the alkali metal and alkynyl ligand in the solid state are of both the σ- and π-type, as indicated by X-ray crystallographic studies of these compounds (338).

Some insight into these issues can be gleaned from breaking down the overall distribution of $\nu(C\equiv C)$ frequencies for $L_nM—C\equiv CPh$ complexes into individual distributions for each metal. Figure 6 exhibits these distribu-

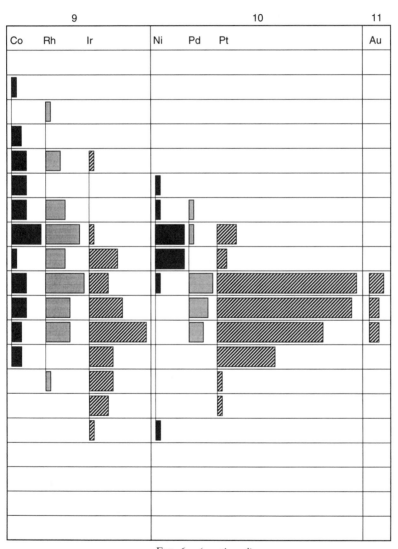

FIG. 6. (continued)

tions, arranged by group. There are too few data for the complexes of groups 4–7 to draw any definitive conclusions regarding global trends in the frequencies, but several tentative observations can be advanced. First, there appears to be a gradual increase in the frequency of $\nu(C{\equiv}C)$ from left to right across the transition series. Choosing 2100 cm^{-1} as a reference frequency, one notes that the majority of $\nu(C{\equiv}C)$ frequencies for compounds in groups 4–8 are <2100 cm^{-1}, while those for iridium, palladium, platinum, and gold are >2100 cm^{-1}. Second, the data for compounds of the groups 9 and 10 metals suggest that $\nu(C{\equiv}C)$ frequencies increase down a group. There are obviously insufficient data for the early metals to comment on the generality of this trend, nor is it clear that the data for osmium are consistent with it. On the whole, though, these trends appear to be more consistent with M–CCR bond polarity being the stronger influence on the frequency of $\nu(C{\equiv}C)$ than is M $\rightarrow$ CCR π-backbonding, because $\nu(C{\equiv}C)$ shifts to lower frequency for the more electropositive metals, with their typically lower d-electron counts. Much additional data for the early metals will be required before it is apparent whether these observations find support or whether they are merely a result of the small sample size. Nonetheless, it is clear that differences between the $\nu(C{\equiv}C)$ frequencies of compounds from different groups of the transition series should be interpreted with caution.

In closing, we emphasize that, because this discussion has been limited to complexes of the phenylethynyl ligand, it would not be appropriate to treat these observations as general for complexes of other alkynyl ligands, particularly those with R-groups whose electron donor or acceptor characteristics are significantly different from those of phenyl. Moreover, the phenyl group has a considerable electronic π-buffering capability, which may partially mask the intrinsic behavior of the C$\equiv$C bond in these data.

C. M–C and C–R Stretching Frequencies

Few M–C stretching frequencies have been reported for metal–alkynyl complexes (*15d,289,290,310,339*). The only detailed investigation has been that of Masai *et al.* on *trans*-M(C$\equiv$CR)$_2$L$_2$ (M = Ni, Pd, Pt; R = Me, CH$_2$F, Ph, CH=CH$_2$, C$\equiv$CH, C$\equiv$CPh; L = PMe$_3$, PEt$_3$, PPh$_3$, SbEt$_3$) complexes (*290*). Comparisons among the infrared spectra of these compounds suggested assignment of a band in the 535–600 cm^{-1} region as the b_{2u}-symmetry $\nu(M{-}C)$ mode. The frequency of this mode depends strongly on the nature of the alkynyl R-group and weakly on the nature of M and L. Interestingly, the squares of the M–C frequencies (ν^2) of these compounds are a linear function of the Taft σ^* constant of the R-group.

By assuming that $\nu(\text{M–C})$ contains negligible contributions from other symmetry coordinates and that ν^2 reflects the M–C bond strength (via their relationship to the force constant), these workers concluded that the M–C bond is strengthened with the increasing electronegativity of the R-group. This was interpreted as resulting from enhanced M $\rightarrow$ CCR π-backbonding in compounds with R-groups of greatest electronegativity.

The only alkynyl complexes for which $\nu(\text{C–R})$ frequencies are routinely reported are those of the parent ethynyl ligand, C≡CH [$\nu(\text{C–H})$ = 3250–3300 cm^{-1}]. Bell and Chisholm reported infrared-spectroscopic frequencies for the natural-abundance and deuterium-labeled pairs of complexes $trans$-Pt(C≡CH/D)$_2$(PMe$_2$Ph)$_2$ and $trans$-Pt(C≡CH/D)X (PMe$_2$Ph)$_2$ (X = Cl, CCl=CH$_2$/D$_2$); the alkynyl $\nu(\text{C–H})$ and $\nu(\text{C–D})$ infrared bands appear at ca. 3280 cm^{-1} and 2550 cm^{-1}, respectively, with the $\nu(\text{C≡C})$ frequencies of these isotopomers being approximately 1970 cm^{-1} for C≡CH and 1850 cm^{-1} for C≡CD (340).

V

CONCLUDING REMARKS

Viewed as a whole, it appears that the results of the structural, spectroscopic, theoretical, and other physical studies on metal–alkynyl complexes cannot be interpreted within a single, simple picture of the nature of metal–alkynyl bonding. In view of the fact that understanding the electronic-structural basis of the reactions of metal–alkynyl complexes and the physical properties of the advanced materials developed from them requires a clear picture of this bond, further work in this area would be highly desirable. Of particular value would be studies that applied a combination of the experimental techniques described above and theoretical calculations to a series of archetypal metal–alkynyl complexes from across the transition series, so as both to gauge the effect of the metal on the nature of the M–CCR bond and to clarify the benchmark parameters for each technique within different metal–alkynyl bonding regimes.

The history of routinely interpreting the structures and spectra of metal–carbonyl and –cyano complexes from a π-backbonding perspective tempts one to carry this approach over to metal-alkynyl complexes as well. But however successful this may appear to be for individual or limited series of metal–alkynyl compounds, consideration of the whole body of metrical and vibrational-frequency data associated with the M—C≡C—R linkage makes it clear that, rather than a particular datum or series of data demonstrating the presence of π-backbonding, they are probably no more than just consistent with it, the frequent, less equivocal claims in the literature notwithstanding. Different experimental techniques sometime lead to different conclusions about the nature of this bond for a single compound. The bond distances and vibrational frequencies of metal–alkynyl compounds should be interpreted with this point in mind.

VI

APPENDIX: ABBREVIATIONS USED IN THIS REVIEW

bpy	2,2'-bipyridine
cod	cycloocta-1,5-diene
Cp	η^5-C$_5$H$_5$
Cp*	η^5-C$_5$Me$_5$
Cy	cyclohexyl
dbpe	1,2-bis(di-n-butylphosphino)ethane
depe	1,2-bis(diethylphosphino)ethane
dmpe	1,2-bis(dimethylphosphino)ethane
dppe	1,2-bis(diphenylphosphino)ethane
dppf	1,1'-bis(diphenylphosphino)ferrocene
dppm	1,2-bis(diphenylphosphino)methane
dppp	1,2-bis(diphenylphosphino)propane
HOMO	highest-occupied molecular orbital
LUMO	lowest-unoccupied molecular orbital
MLCT	metal-to-ligand charge transfer
N$_4$	3,9-dimethyl-4,8-diazaundecane-2,10-dione dioxime
Pc	phthalocyaninato
phen	1,10-phenanthroline
PPN	N(PPh$_3$)$_2$
py	pyridine
pzH	pyrazole
Salen	N,N'-bis(salicylaldehdo)ethylenediamine
Tol	tolyl
Tp'	hydrotris(3,5-dimethylpyrazolyl)borato
TPP	tetraphenylporphyrin dianion

ACKNOWLEDGMENT

We are indebted to the National Science Foundation for their support of our research through Grant CHE 93-07013. M.D.H. gratefully acknowledges support through fellowships from the David and Lucile Packard Foundation and the Alfred P. Sloan Foundation. We thank John Cooper, Tara Meyer, Joe Grabowski, and colleagues at other institutions for bringing various papers to our attention. We are especially grateful to Todd Marder for providing us with preprints and unpublished crystallographic data.

REFERENCES

(1) Nast, R. *Coord. Chem. Rev.* **1982**, *47*, 89.

(2) Recent leading references to this work from various research groups include

(2a) Chiang, S.-J.; Chi, Y.: Su, P.-C.; Peng, S.-M.; Lee, G.-H. *J. Am. Chem. Soc.* **1994**, *116*, 11181.

(2b) Akita, M.; Sugimoto, S.; Takabuchi, A.; Tanaka, M.; Moro-oka, Y. *Organometallics* **1993**, *12*, 2925.

(2c) Lang, H.; Herres, M.; Zsolnai, L. *Bull. Chem. Soc. Jpn.* **1993**, *66*, 429.

(2d) Berenguer, J. R.; Falvello, L. R.; Forniés, J.; Lalinde, E.; Tomás, M. *Organometallics* **1993**, *12*, 6.

(2e) Weidmann, T.; Weinrich, V.; Wagner, B.; Robl, C.; Beck, W. *Chem. Ber.* **1991**, *124*, 1363.

(2f) Bruce, M. I.; Koutsantonis, G. A.; Tiekink, E. R. T. *J. Organomet. Chem.* **1991**, *407*, 391.

(2g) Lotz, S.; van Rooyen, P. H.; Meyer, R. *Adv. Organomet. Chem.* **1995**, *37*, 219.

(3) Werner, H. *Nachr. Chem., Tech. Lab.* **1992**, *40*, 435; Bruce M. I. *Chem. Rev.* **1991**, *91*, 197; *Pure Appl. Chem.* **1990**, *62*, 1021; Werner, H. *Angew. Chem., Int. Ed. Engl.* **1990**, *29*, 1077; Bruce, M. I. *J. Organomet. Chem.* **1990**, *400*, 321; Pombeiro, A. J. L.; Richards, R. L. *Coord. Chem. Rev.* **1990**, *104*, 13; Davies, S. G.; McNally, J. P.; Smallridge, A. J. *Adv. Organomet. Chem.* **1990**, *30*, 1; Antonova, A. B.; Ioganson, A. A. *Russ. Chem. Rev. (Engl. Transl.)* **1989**, *58*, 693; Pombeiro, A. J. L. *Polyhedron* **1989**, *8*, 1595; Consiglio, G.; Morandini, F. *Chem. Rev.* **1987**, *87*, 761; Bruce, M. I. *Pure Appl. Chem.* **1986**, *58*, 553; Silvestre, J.; Hoffmann, R. *Helv. Chim. Acta* **1985**, *68*, 1461. Bruce, M. I.; Swincer, A. G. *Adv. Organomet. Chem.* **1983**, *22*, 59.

(4a) Dick, G. S. S.; Marder, T. B.; Khan, M. S.; Kakkar, A. K.; Lewis, J. Review article in preparation.

(4b) Lang, H. *Angew. Chem., Int. Ed. Engl.* **1994**, *33*, 547.

(4c) Chisholm, M. H. *Angew. Chem., Int. Ed. Engl.* **1991**, *30*, 673.

(4d) Hanack, M.; Datz, A.; Fay, R.; Fischer, K.; Keppeler, U.; Koch, J.; Metz, J.; Mezger, M.; Schneider, O.; Schulze, H.-J. In "Handbook of Conducting Polymers"; Skotheim, T. A., Ed.; Dekker: New York, 1986; Vol. 2, pp. 133–204.

(4e) Hagihara, N.; Sonogashira, K.; Takahashi, S. *Adv. Polym. Sci.* **1980**, *41*, 149.

(5) Myers, L. K.; Langhoff, C.; Thompson, M. E. *J. Am. Chem. Soc.* **1992**, *114*, 7560; Fyfe, H. B.; Mlekuz, M.; Stringer, G.; Taylor, N. J.; Marder, T. B. In "Inorganic and Organometallic Polymers with Special Properties"; Laine, R. M., Ed.; Kluwer Academic Publishers: Dordrecht, The Netherlands, 1992; pp. 331–344; Porter, P. L.; Guha, S.; Kang, K.; Frazier, C. C. *Polymer* **1991**, *32*, 1756; Blau, W. J.; Byrne, H. J.; Cardin, D. J.; Davey, A. P. *J. Mater. Chem.* **1991**, *1*, 245; Marder, T. B.; Lesley, G.; Yuan, Z.; Fyfe, H. B.; Chow, P.; Stringer, G.; Jobe, I. R.; Taylor, N. J.; Williams, I. D.; Kurtz, S. K. *ACS Symp. Ser.* **1991**, *455*, 605; Lesley, G.; Yuan, Z.; Stringer, G.; Jobe, I. R.; Taylor, N. J.; Koch, L.; Scott, K.; Marder, T. B.; Williams, I. D.; Kurtz, S. K. In "Organic Materials for Non-Linear Optics II"; Hann, R. A.; Bloor, D., Eds.; Royal Society of Chemistry: Cambridge, UK, 1991; pp. 197–203; Davey, A. P.; Cardin, D. J.; Byrne, H. J.; Blau, W. In "Organic

Molecules for Nonlinear Optics and Photonics''; Messier, J.; Kajzar, F.; Prasad, P., Eds.; Kluwer Academic Publishers: Dordrecht, The Netherlands, 1991; pp. 391–402; Frazier, C. C.; Guha, S.; Chen, W. *PCT Int. Appl.* **1989**, *PCT Int. Appl. WO 89 01,182; Chem. Abstr.* **1989**, *111*, 105446p; Guha, S.; Frazier, C. C.; Porter, P. L.; Kang, K.; Finberg, S. E. *Opt. Lett.* **1989**, *14*, 952; Frazier, C. C.; Chauchard, E. A.; Cockerham, M. P.; Porter, P. L. *Mater. Res. Soc. Symp. Proc.* **1988**, *109*, 323; Frazier, C. C.; Guha, S.; Chen, W. P.; Cockerham, M. P.; Porter, P. L.; Chauchard, E. A.; Lee, C. H. *Polymer* **1987**, *28*, 553.

(6a) Frapper, G.; Kertesz, M. *Inorg. Chem.* **1993**, *32*, 732.

(6b) Dray, A. E.; Wittmann, F.; Friend, R. H.; Donald, A. M.; Khan, M. S.; Lewis, J.; Johnson, B. F. G. *Synth. Met.* **1991**, *41–43*, 871.

(6c) Canadell, E.; Alvarez, S. *Inorg. Chem.* **1984**, *23*, 573.

(6d) Seelig, F. F. *Z. Naturforsch., A* **1979**, *34A*, 986.

(7a) Rourke, J. P.; Bruce, D. W.; Marder, T. B. *J. Chem. Soc., Dalton Trans.* **1995**, 317.

(7b) Kaharu, T.; Matsubara, H.; Takahashi, S. *J. Mater. Chem.* **1992**, *2*, 43.

(7c) Abe, A.; Kimura, N.; Tabata, S. *Macromolecules* **1991**, *24*, 6238.

(7d) Kotani, S.; Shiina, K.; Sonogashira, K. *Appl. Organomet. Chem.* **1991**, *5*, 417.

(7e) Takahashi, S.; Takai, Y.; Morimoto, H.; Sonogashira, K. *J. Chem. Soc., Chem. Commun.* **1984**, 3.

(7f) Takahashi, S.; Takai, Y.; Morimoto H.; Sonogashira, K.; Hagihara, N. *Mol. Cryst. Liq. Cryst.* **1982**, *82*, 139.

(8) Sharpe, A. G. In "Comprehensive Coordination Chemistry"; Wilkinson, G.; Gillard, R. D.; McCleverty, J. A., Eds.; Pergamon: New York, 1987; Vol. 2, pp. 7–14.

(9) Wrackmeyer, B.; Horchler, K. *Prog. NMR Spectrosc.* **1990**, *22*, 209; Relevant papers since this review include

(9a) Wrackmeyer, B.; von Locqueghien, K. H.; Kupče, E.; Sebald, A. *Magn. Reson. Chem.* **1993**, *31*, 45.

(9b) Duer, M. J., Khan, M. S.; Kakkar, A. K. *Solid State Nucl. Magn. Reson.* **1992**, *1*, 13

(9c) Wrackmeyer, B. *Z. Naturforsch., B: Chem. Sci.* **1991**, *46B*, 35.

(10) Beck, W.; Niemer, B.; Wieser, M. *Angew. Chem., Int. Ed. Engl.* **1993**, *32*, 923.

(11) Kostić, N. M.; Fenske, R. F. *Organometallics* **1982**, *1*, 974.

(12a) Lichtenberger, D. L.; Renshaw, S. K.; Wong, A.; Tagge, C. D. *Organometallics* **1993**, *12*, 3522.

(12b) Lichtenberger, D. L.; Renshaw, S. K.; Bullock, R. M. *J. Am. Chem. Soc.* **1993**, *115*, 3276.

(13) Louwen, J. N.; Hengelmolen, R.; Grove, D. M.; Oskam, A.; DeKock, R. L. *Organometallics* **1984**, *3*, 908.

(14) Knight, E. T.; Myers, L. K.; Thompson, M. E. *Organometallics* **1992**, *11*, 3691.

(15a) Stoner, T. C.; Schaefer, W. P.; Marsh, R. E.; Hopkins, M. D. *J. Clust. Sci.* **1994**, *5*, 107.

(15b) Stoner, T. C.; Geib, S. J.; Hopkins, M. D. *Angew. Chem., Int. Ed. Engl.* **1993**, *32*, 409.

(15c) Stoner, T. C.; Geib, S. J.; Hopkins, M. D. *J. Am. Chem. Soc.* **1992**, *114*, 4201.

(15d) Stoner, T. C.; Dallinger, R. F.; Hopkins, M. D. *J. Am. Chem. Soc.* **1990**, *112*, 5651.

(15e) John, K. D.; Hopkins, M. D. Unpublished results.

(16a) Pollagi, T. P.; Manna, J.; Stoner, T. C.; Geib, S. J.; Hopkins, M. D. In "Transition Metal Carbyne Complexes"; Kreissl, F. R., Ed.; Kluwer Academic Publishers: Dordrecht, The Netherlands, 1993; pp. 71–73.

(*16b*) Manna, J.; Geib, S. J.; Hopkins, M. D. *J. Am. Chem. Soc.* **1992**, *114*, 9199.

(*16c*) Manna, J.; Johnson, J. A.; Hopkins, M. D. Unpublished results.

(*17*) Manna, J.; Geib, S. J.; Hopkins, M. D. *Angew. Chem., Int. Ed. Engl.* **1993**, *32*, 858.

(*18*) Masai, H.; Sonogashira, K.; Hagihara, N. *Bull. Chem. Soc. Jpn.* **1971**, *44*, 2226.

(*19*) Tahara, T.; Seto, K.; Takahashi, S. *Polym. J.* **1987**, *19*, 301; Takahashi, S.; Morimoto, H.; Murata, E.; Kataoka, S.; Sonogashira, K.; Hagihara, N. *J. Polym. Sci., Polym. Chem. Ed.* **1982**, *20*, 565; Takahashi, S.; Murata, E.; Sonogashira, K.; Hagihara, N. *ibid.* **1980**, *18*, 661; Takahashi, S.; Ohyama, Y.; Murata, E.; Sonogashira, K.; Hagihara, N. *ibid.*, 349; Sonogashira, K.; Kataoka, S.; Takahashi, S.; Hagihara, N. *J. Organomet. Chem.* **1978**, *160*, 319; Fujikura, Y.; Sonogashira, K.; Hagihara, N. *Chem. Lett.* **1975**, 1067.

(*20a*) Yam, V. W.-W.; Chan, L.-P.; Lai, T.-F. *Organometallics* **1993**, *12*, 2197.

(*20b*) Sacksteder, L.; Baralt, E.; DeGraff, B. A.; Lukehart, C. M.; Demas, J. N. *Inorg. Chem.* **1991**, *30*, 3955.

(*20c*) Baralt, E.; Boudreaux, E. A.; Demas, J. N.; Lenhert, P. G.; Lukehart, C. M.; McPhail, A. T.; McPhail, D. R.; Myers, J. B., Jr.; Sacksteder, L.; True, W. R. *Organometallics* **1989**, *8*, 2417.

(*21*) Sacksteder, L.; Baralt, E.; DeGraff, B. A.; Lukehart, C. M.; Demas, J. N. *Inorg. Chem.* **1991**, *30*, 2468.

(*22*) Yip, H.-K.; Lin, H.-M.; Wang, Y.; Che, C-M. *J. Chem. Soc., Dalton Trans.* **1993**, 2939.

(*23*) Lewis, J.; Khan, M. S.; Kakkar, A. K.; Johnson, B. F. G.; Marder, T. B.; Fyfe, H. B.; Wittmann, F.; Friend, R. H.; Dray, A. E. *J. Organomet. Chem.* **1992**, *425*, 165; Johnson, B. F. G.; Kakkar, A. K.; Khan, M. S.; Lewis, J.; Dray, A. E.; Friend, R. H.; Wittmann, F. *J. Mater. Chem.* **1991**, *1*, 485.

(*24*) Bryndza, H. E.; Fong, L. K.; Paciello, R. A.; Tam, W.; Bercaw, J. E. *J. Am. Chem. Soc.* **1987**, *109*, 1444.

(*25*) Bulls, A. R.; Bercaw, J. E.; Manriquez, J. M.; Thompson, M. E. *Polyhedron* **1988**, *7*, 1409; Thompson, M. E.; Baxter, S. M.; Bulls, A. R.; Burger, B. J.; Nolan, M. C.; Santarsiero, B. D.; Schaefer, W. P.; Bercaw, J. E. *J. Am. Chem. Soc.* **1987**, *109*, 203.

(*26*) Ziegler, T.; Folga, E.; Berces, A. *J. Am. Chem. Soc.* **1993**, *115*, 636; Rappé, A. K. *Organometallics* **1990**, *9*, 466.

(*27*) Birchall, T.; Myers, R. D. *Spectrosc.: Int. J.* **1983**, *2*, 22.

(*28*) Evans, D. J.; Jimenez-Tenorio, M.; Leigh, G. J. *J. Chem. Soc., Dalton Trans.* **1991**, 1785.

(*29*) Sebald, A.; Wrackmeyer, B.; Beck, W. *Z. Naturforsch., B: Anorg. Chem., Org. Chem.* **1983**, *38B*, 45.

(*30*) Parshall, G. W. *J. Am. Chem. Soc.* **1966**, *88*, 704.

(*31a*) Köhler, F. H.; Prössdorf, W.; Schubert, U. *Inorg. Chem.* **1981**, *20*, 4096.

(*31b*) Köhler, F. H.; Hofmann, P.; Prössdorf, W. *J. Am. Chem. Soc.* **1981**, *103*, 6359.

(*31c*) Köhler, F. H.; Prössdorf, W.; Schubert, U.; Neugebauer, D. *Angew. Chem., Int. Ed. Engl.* **1978**, *17*, 850.

(*31d*) Balch, A. L.; Latos-Grazyn'ski, L.; Noll, B. C.; Phillips, S. L. *Inorg. Chem.* **1993**, *32*, 1124.

(*32*) Abu-Salah, O. M.; Al-Ohaly, A.-R. A.; Al-Showiman, S. S.; Al-Najjar, I. M. *Transition Met. Chem. (Weinheim, Ger.)* **1985**, *10*, 207.

(*33a*) Bianchini, C.; Laschi, F.; Masi, D.; Ottaviani, F. M.; Pastor, A.; Peruzzini, M.; Zanello, P.; Zanobini, F. *J. Am. Chem. Soc.* **1993**, *115*, 2723.

(*33b*) Bear, J. L.; Han, B.; Huang, S. *J. Am. Chem. Soc.* **1993**, *115*, 1175.

(33c) Li, Y.; Han, B.; Kadish, K. M.; Bear, J. L. *Inorg. Chem.* **1993**, *32*, 4175.

(33d) Connelly, N. G.; Gamasa, M. P.; Gimeno, J.; Lapinte, C.; Lastra, E.; Maher, J. P.; Le Narvor, N.; Rieger, A. L.; Rieger, P. H. *J. Chem. Soc., Dalton Trans.* **1993**, 2575; Correction: p. 2981.

(33e) Field, L. D.; George, A. V.; Laschi, F.; Malouf, E. Y.; Zanello, P. *J. Organomet. Chem.* **1992**, *435*, 347.

(33f) Hills, A.; Hughes, D. L.; Kashef, N.; Lemos, M. A. N. D. A.; Pombeiro, A. J. L.; Richards, R. L. *J. Chem. Soc., Dalton Trans.* **1992**, 1775.

(33g) Beddoes, R. L.; Bitcon, C.; Whiteley, M. W. *J. Organomet Chem.* **1991**, *402*, 85.

(33h) Bianchini, C.; Meli, A.; Peruzzini, M.; Vacca, A.; Laschi, F.; Zanello, P.; Ottaviani, F. M. *Organometallics* **1990**, *9*, 360.

(33i) Yao, C. L.; Park, K. H.; Khokhar, A. R.; Jun, M.-J.; Bear, J. L. *Inorg. Chem.* **1990**, *29*, 4033.

(33j) Bianchini, C.; Innocenti, P.; Meli, A.; Peruzzini, M.; Zanobini, F.; Zanello, P. *Organometallics* **1990**, *9*, 2514.

(33k) Bianchini, C.; Laschi, F.; Ottaviani, M. F.; Peruzzini, M.; Zanello, P.; Zanobini, F. *Organometallics* **1989**, *8*, 893.

(33l) Hills, A.; Hughes, D. L.; Kashef, N.; Richards, R. L.; Lemos, M. A. N. D. A.; Pombeiro, A. J. L. *J. Organomet. Chem.* **1988**, *350*, C4.

(33m) Bianchini, C.; Laschi, F.; Ottaviani, F.; Peruzzini, M.; Zanello, P. *Organometallics* **1988**, *7*, 1660.

(33n) Adams, J. S.; Bitcon, C.; Brown, J. R.; Collison, D.; Cunningham, M.; Whiteley, M. W. *J. Chem. Soc., Dalton Trans.* **1987**, 3049.

(33o) Bitcon, C.; Whiteley, M. W. *J. Organomet. Chem.* **1987**, *336*, 385.

(33p) Chakravarty, A. R.; Cotton F. A. *Inorg. Chim. Acta* **1986**, *113*, 19.

(34) Lang, H.; Herres, M.; Zsolnai, L.; Imhof, W. *J. Organomet. Chem.* **1991**, *409*, C7.

(35) Erker, G.; Frömberg, W.; Benn, R.; Mynott, R.; Angermund K.; Krüger, C. *Organometallics* **1989**, *8*, 911.

(36) Walsh, P. J.; Hollander, F. J.; Bergman, R. G. *J. Organomet. Chem.* **1992**, *428*, 13.

(37) Wielstra Y.; Gambarotta, S.; Meetsma, A.; de Boer, J. L.; Chiang, M. Y. *Organometallics* **1989**, *8*, 2696.

(38) Evans, W. J.; Bloom, I.; Doedens, R. J. *J. Organomet. Chem.* **1984**, *265*, 249.

(39) Schubert, U.; Köhler, F. H.; Prössdorf, W. *Cryst. Struct. Commun.* **1981**, *10*, 245.

(40) Buang, N. A.; Hughes, D. L.; Kashef, N.; Richards, R. L.; Pombeiro, A. J. L. *J. Organomet. Chem.* **1987**, *323*, C47.

(41) Pombeiro, A. J. L.; Hills, A.; Hughes, D. L.; Richards, R. L. *J. Organomet. Chem.* **1990**, *398*, C15.

(42) Beddoes, R. L.; Bitcon, C.; Ricalton, A.; Whiteley, M. W. *J. Organomet. Chem.* **1989**, *367*, C21.

(43) Clark, G. R.; Nielson, A. J.; Rickard, C. E. F. *J. Chem. Soc., Chem. Commun.* **1989**, 1157.

(44) Sieber, W.; Wolfgruber, M.; Neugebauer, D.; Orama, O.; Kreissl, F. R. *Z. Naturforsch., B: Anorg. Chem., Org. Chem.* **1983**, *38B*, 67.

(45) Valín, M. L.; Moreiras, D.; Solans, X.; Miguel, D.; Riera, V. *Acta Crystallogr. Sect. C: Cryst. Struct. Commun.* **1986**, *C42*, 977.

(46) Grigsby, W. J.; Main, L.; Nicholson, B. K. *Organometallics* **1993**, *12*, 397.

(47) Goldberg, S. Z.; Duesler, E. N.; Raymond, K. N. *Inorg. Chem.* **1972**, *11*, 1397; *Chem. Commun.* **1971**, 826.

(48) Cramer, R. E.; Higa, K. T.; Gilje, J. W. *Organometallics* **1985**, *4*, 1140.

(49) Batsanov, A. S.; Struchkov, Y. T.; Zhdanovich, V. I.; Petrovskii, P. V.; Kolobova, N. E. *Metalloorg. Khim.* **1989**, *2*, 1045.
(50) Senn, D. R.; Wong, A.; Patton, A. T.; Marsi, M.; Strouse, C. E.; Gladysz, J. A. *J. Am. Chem. Soc.* **1988**, *110*, 6096.
(51) Lee, K.-W.; Pennington, W. T.; Cordes, A. W.; Brown, T. L. *J. Am. Chem. Soc.* **1985**, *107*, 631.
(52a) Field, L. D.; George, A. V.; Malouf, E. Y.; Slip, I. H. M.; Hambley, T. W. *Organometallics* **1991**, *10*, 3842.
(52b) Field, L. D.; George, A. V.; Hambley, T. W.; Malouf, E. Y.; Young, D. J. *J. Chem. Soc., Chem. Commun.* **1990**, 931.
(53) Hughes, D. L.; Leigh, G. J.; Jimenez-Tenorio, M.; Rowley, A. T. *J. Chem. Soc., Dalton Trans.* **1993**, 75; Hills, A.; Hughes, D. L.; Jimenez-Tenorio, M.; Leigh, G. J. *J. Organomet. Chem.* **1990**, *391*, C41.
(54) Field, L. D.; George, A. V.; Hambley, T. W. *Inorg. Chem.* **1990**, *29*, 4565.
(55) Hughes, D. L.; Jimenez-Tenorio, M.; Leigh, G. J.; Rowley, A. T. *J. Chem. Soc., Dalton Trans.* **1993**, 3151.
(56) Löwe, C.; Hund, H.-U.; Berke, H. *J. Organomet. Chem.* **1989**, *372*, 295.
(57) Goddard, R.; Howard, J.; Woodward, P. *J. Chem. Soc., Dalton Trans.* **1974**, 2025.
(58) Gamasa, M. P., Gimeno J.; Lastra, E. *J. Organomet. Chem.* **1991**, *405*, 333.
(59) Akita, M.; Terada, M.; Oyama, S.; Moro-oka, Y. *Organometallics* **1990**, *9*, 816.
(60) Kruger, G. J.; Ashworth, T. V.; Singleton, E. *Acta Crystallogr., Sect. A: Cryst. Phys., Differ., Theor. Geo., Crystallogr.* **1981**, *A37*, C220.
(61) Atherton, Z.; Faulkner, C. W.; Ingham, S. L.; Kakkar, A. K.; Khan, M. S.; Lewis, J.; Long, N. J.; Raithby, P. R. *J. Organomet. Chem.* **1993**, *462*, 265.
(62) Touchard, D.; Haquette, P.; Pirio, N.; Toupet, L.; Dixneuf, P. H. *Organometallics* **1993**, *12*, 3132; Haquette, P.; Pirio, N.; Touchard, D.; Toupet, L.; Dixneuf, P. H. *J. Chem. Soc., Chem. Commun.* **1993**, 163.
(63a) Sun, Y.; Taylor, N. J.; Carty, A. J. *J. Organomet. Chem.* **1992**, *423*, C43.
(63b) Sun, Y., Taylor, N. J.; Carty, A. J. *Organometallics* **1992**, *11*, 4293.
(64) Echavarren, A. M.; López, J.; Santos, A.; Romero, A.; Hermoso, J. A.; Vegas, A. *Organometallics* **1991**, *10*, 2371.
(65) Montoya, J.; Santos, A.; López, J.; Echavarren, A. M.; Ros, J.; Romero, A. *J. Organomet. Chem.* **1992**, *426*, 383.
(66a) Jia, G.; Gallucci, J. C.; Rheingold, A. L.; Haggerty, B. S.; Meek, D. W. *Organometallics* **1991**, *10*, 3459.
(66b) Jia, G.; Rheingold, A. L.; Meek, D. W. *ibid.* **1989**, *8*, 1378.
(67) Torres, M. R.; Santos, A.; Ros, J.; Solans, X. *Organometallics* **1987**, *6*, 1091.
(68) Helliwell, M.; Stell, K. M.; Mawby, R. J. *J. Organomet. Chem.* **1988**, *356*, C32.
(69) Wisner, J. M.; Bartczak, T. J.; Ibers, J. A. *Inorg. Chim. Acta* **1985**, *100*, 115.
(70) Bruce, M. I.; Humphrey, M. G.; Snow, M. R.; Tiekink, E. R. T. *J. Organomet. Chem.* **1986**, *314*, 213.
(71) Lomprey, J. R.; Selegue, J. P. *Organometallics* **1993**, *12*, 616.
(72) Bruce, M. I.; Hinterding, P.; Tiekink, E. R. T.; Skelton, B. W.; White, A. H. *J. Organomet. Chem.* **1993**, *450*, 209.
(73) Consiglio, G.; Morandini, F.; Sironi, A. *J. Organomet. Chem.* **1986**, *306*, C45.
(74) Romero, A.; Peron, D.; Dixneuf, P. H. *J. Chem. Soc., Chem. Commun.* **1990**, 1410.
(75) Matsuzaka, H.; Hirayama, Y.; Nishio, M.; Mizobe, Y.; Hidai, M. *Organometallics* **1993**, *12*, 36.
(76) Espuelas, J.; Esteruelas, M. A.; Lahoz, F. J.; Oro, L. A.; Valero, C. *Organometallics* **1993**, *12*, 663.

(77) Cherkas, A. A.; Taylor, N. J.; Carty, A. J. *J. Chem. Soc., Chem. Commun.* **1990,** 385.

(78) Klein, H.-F.; Beck, H.; Hammerschmitt, B.; Koch, U.; Koppert, S.; Cordier, G. *Z. Naturforsch., B: Chem. Sci.* **1991,** *46B,* 147.

(79) Stringer, G.; Taylor, N. J.; Marder, T. B. *Acta Crystallogr., Sect. C: Cryst. Struct. Commun.* (accepted for publication).

(80) Garcia Basallote, M.; Hughes, D. L.; Jiménez-Tenorio, M.; Leigh, G. J.; Puerta Vizcaino, M. C.; Valerga Jiménez, P. *J. Chem. Soc., Dalton Trans.* **1993,** 1841.

(81) Giese, B.; Zehnder, M.; Neuburger, M.; Trach, F. *J. Organomet. Chem.* **1991,** *412,* 415.

(82) Bianchini, C.; Peruzzini, M.; Vacca, A.; Zanobini, F. *Organometallics* **1991,** *10,* 3697.

(83) Habadie, N.; Dartiguenave, M.; Dartiguenave, Y.; Britten, J. F.; Beauchamp, A. L. *Organometallics* **1989,** *8,* 2564.

(84) Kergoat, R.; Gomes de Lima, L. C.; Jégat, C.; Le Berre, N.; Kubicki, M. M.; Guerchais, J. E. *J. Organomet. Chem.* **1990,** *389,* 71.

(85) Zargarian, D.; Chow, P.; Taylor, N. J.; Marder, T. B. *J. Chem. Soc., Chem. Commun.* **1989,** 540.

(86a) Fyfe, H. B.; Mlekuz, M.; Zargarian, D.; Taylor, N. J.; Marder, T. B. *J. Chem. Soc., Chem. Commun.* **1991,** 188.

(86b) Fyfe, H. B.; Mlekuz, M.; Zargarian, D.; Marder, T. B. In "Organic Materials for Non-Linear Optics II"; Hann, R. A.; Bloor, D., Eds.; Royal Society of Chemistry: Cambridge, UK, 1991; pp. 204–209.

(87) Bruce, M. I.; Hambley, T. W.; Snow, M. R.; Swincer, A. G. *J. Organomet. Chem.* **1982,** *235,* 105.

(88a) Bianchini, C.; Masi, D.; Meli, A.; Peruzzini, M.; Ramirez, J. A.; Vacca, A.; Zanobini, F. *Organometallics* **1989,** *8,* 2179.

(88b) Bianchini, C.; Mealli, C.; Peruzzini, M.; Vizza, F.; Zanobini, F. *J. Organomet. Chem.* **1988,** *346,* C53.

(89) Marder, T. B.; Zargarian, D.; Calabrese, J. C.; Herskovitz, T. H.; Milstein, D. *J. Chem. Soc., Chem. Commun.* **1987,** 1484.

(90) Chow, P.; Zargarian, D.; Taylor, N. J.; Marder, T. B. *J. Chem. Soc., Chem. Commun.* **1989,** 1545.

(91) Werner, H.; Baum, M.; Schneider, D.; Windmüller, B. *Organometallics* **1994,** *13,* 1089.

(92) Rappert, T.; Nürnberg, O.; Mahr, N.; Wolf, J.; Werner, H. *Organometallics* **1992,** *11,* 4156.

(93) Fernández, M. J.; Esteruelas, M. A.; Covarrubias, M.; Oro, L. A. Apreda, M.-C.; Concepcion, F.-F.; Cano, F. H. *Organometallics* **1989,** *8,* 1158.

(94) Hutton, A. T.; Pringle, P. G.; Shaw, B. L. *Organometallics* **1983,** *2,* 1889.

(95) O'Connor, J. M.; Pu, L.; Chadha, R. K. *J. Am. Chem. Soc.* **1990,** *112,* 9627.

(96) Callahan, K. P.; Strouse, C. E.; Layten, S. W.; Hawthorne, M. F. *J. Chem. Soc., Chem. Commun.* **1973,** 465.

(97) Davies, G. R.; Mais, R. H. B.; Owston, P. G. *J. Chem. Soc. A* **1967,** 1750.

(98) Spofford, W. A., III; Carfagna, P. D.; Amma, E. L. *Inorg. Chem.* **1967,** *6,* 1553.

(99) von Deuten, K.; Beyer, A.; Nast, R. *Cryst. Struct. Commun.* **1979,** *8,* 755.

(100) van der Voort, E.; Spek, A. L.; de Graaf, W. *Acta Crystallogr., Sect. C: Cryst. Struct. Commun.* **1987,** *C43,* 2311.

(101) Onitsuka, K.; Ogawa, H.; Joh, T.; Takahashi, S.; Yamamoto, Y.; Yamazaki, H. *J. Chem. Soc., Dalton Trans.* **1991,** 1531.

(102) Yasuda, T.; Kai, Y.; Yasuoka, N.; Kasai, N. *Bull. Chem. Soc. Jpn.* **1977,** *50,* 2888.
(103) Behrens, U.; Hoffmann, K. *J. Organomet. Chem.* **1977,** *129,* 273.
(104) de Graaf, W.; Harder, S.; Boersma, J.; van Koten, G.; Kanters, J. A. *J. Organomet. Chem.* **1988,** *358,* 545.
(105) Sünkel, K. *J. Organomet. Chem.* **1992,** *436,* 101.
(106) Sebald, A.; Stader, C.; Wrackmeyer, B.; Bensch, W. *J. Organomet. Chem.* **1986,** *311,* 233.
(107) Carpenter, J. P.; Lukehart, C. M. *Inorg. Chim. Acta* **1991,** *190,* 7.
(108) Chiesi Villa, A., Gaetani Manfredotti, A.; Guastini, C. *Cryst. Struct. Commun.* **1976,** *5,* 139.
(109) Cardin, C. J.; Cardin, D. J.; Lappert, M. F.; Muir, K. W. *J. Chem. Soc., Dalton Trans.* **1978,** 46; *J. Organomet. Chem.* **1973,** *60,* C70.
(110) Sünkel, K. *J. Organomet. Chem.* **1988,** *348,* C12.
(111) Chiesi Villa, A.; Gaetani Manfredotti, A.; Guastini, C.; Carusi, P.; Furlani, A.; Russo, M. V. *Cryst. Struct. Commun.* **1977,** *6,* 623.
(112) Chiesi Villa, A.; Gaetani Manfredotti, A.; Guastini, C.; Carusi, P.; Furlani, A.; Russo, M. V. *Cryst. Struct. Commun.* **1977,** *6,* 629.
(113) Furlani, A.; Licoccia, S.; Russo, M. V.; Chiesi Villa, A.; Guastini, C. *J. Chem. Soc., Dalton Trans.* **1984,** 2197.
(114) Furlani, A.; Russo, M. V.; Licoccia, S.; Guastini, C. *Inorg. Chim. Acta* **1979,** *33,* L125.
(115) Wouters, J. M. A.; Vrieze, K.; Elsevier, C. J.; Zoutberg, M. C.; Goubitz, K. *Organometallics* **1994,** *13,* 1510.
(116) Furlani, A.; Licoccia, S., Russo, M. V.; Chiesi Villa, A.; Guastini, C. *J. Chem. Soc., Dalton Trans.* **1982,** 2449.
(117) Behrens, U.; Hoffmann, K.; Kopf, J.; Moritz, J. *J. Organomet. Chem.* **1976,** *117,* 91.
(118) Kowalski, M. H.; Arif, A. M.; Stang, P. J. *Organometallics* **1988,** *7,* 1227.
(119) Furlani, A.; Russo, M. V.; Chiesi Villa, A.; Gaetani Manfredotti, A.; Guastini, C. *J. Chem. Soc., Dalton Trans.* **1977,** 2154.
(120) Chiesi Villa, A.; Gaetani Manfredotti, A.; Guastini, C. *Cryst. Struct. Commun.* **1977,** *6,* 313.
(121) Onuma, K.; Kai, Y.; Yasuoka, N.; Kasai, N. *Bull. Chem. Soc. Jpn.* **1975,** *48,* 1696.
(122) Li, X.; Lukehart, C. M.; Han, L. *Organometallics* **1992,** *11,* 3993.
(123) Croatto, U.; Toniolo, L.; Immirzi, A.; Bombieri, G. *J. Organomet. Chem.* **1975,** *102,* C31.
(124) Phillips, J. R.; Miller, G. A.; Trogler, W. C. *Acta Crystallogr., Sect. C: Cryst. Struct. Commun.* **1990,** *C46,* 1648.
(125) Bonamico, M.; Dessy, G.; Fares, V.; Russo, M. V.; Scaramuzza, L. *Cryst. Struct. Commun.* **1977,** *6,* 39.
(126) Kubota, M.; Sly, W. G.; Santarsiero, B. D.; Clifton, M. S.; Kuo, L. *Organometallics* **1987,** *6,* 1257.
(127) Baddley, W. H.; Panattoni, C.; Bandoli, G.; Clemente, D. A.; Belluco, U. *J. Am. Chem. Soc.* **1971,** *93,* 5590.
(128) Stang, P. J.; Kowalski, M. H. *J. Am. Chem. Soc.* **1989,** *111,* 3356.
(129) Weigand, W.; Robl, C. *Chem. Ber.* **1993,** *126,* 1807.
(130) Mariezcurrena, R. A.; Rasmussen, S. E. *Acta Chem. Scand.* **1973,** *27,* 2678.
(131) Geisenberger, J.; Nagel, U.; Sebald, A.; Beck, W. *Chem. Ber.* **1983,** *116,* 911.
(132a) Blagg, A.; Hutton, A. T.; Pringle, P. G.; Shaw, B. L. *J. Chem. Soc., Dalton Trans.* **1984,** 1815.

(*132b*) Blagg, A.; Hutton, A. T.; Pringle, P. G.; Shaw, B. L. *Inorg. Chim. Acta* **1983**, *76*, L265.

(*133*) Hutton, A. T.; Langrick, C. R.; McEwan, D. M.; Pringle, P. G.; Shaw, B. L. *J. Chem. Soc., Dalton Trans.* **1985**, 2121.

(*134*) Manojlović-Muir, L.; Henderson, A. N.; Treurnicht, I.; Puddephatt, R. J. *Organometallics* **1989**, *8*, 2055.

(*135*) Manojlović-Muir, L.; Muir, K. W.; Treurnicht, I.; Puddephatt, R. J. *Inorg. Chem.* **1987**, *26*, 2418; Arsenault, G. J.; Manojlović-Muir, L.; Muir, K. W.; Puddephatt, R. J.; Treurnicht, I. *Angew. Chem., Int. Ed. Engl.* **1987**, *26*, 86.

(*136*) Alcock, N. W.; Kemp, T. J.; Pringle, P. G.; Bergamini, P.; Traverso, O. *J. Chem. Soc., Dalton Trans.* **1987**, 1659.

(*137*) DePriest, J. C.; Woods, C. *Polyhedron* **1991**, *10*, 2153.

(*138*) Markham, D. P.; Shaw, B. L.; Thornton-Pett, M. *J. Chem. Soc., Chem. Commun.* **1987**, 1005.

(*139*) Baralt, E.; Lukehart, C. M.; McPhail, A. T.; McPhail, D. R. *Organometallics* **1991**, *10*, 516.

(*140*) Lukehart, C. M.; Owen, M. D. *J. Clust. Sci.* **1991**, *2*, 71.

(*141*) Smith, D. E., Welch, A. J.; Treurnicht, I.; Puddephatt, R. J. *Inorg. Chem.* **1986**, *25*, 4616.

(*142*) McDonald, W. S.; Pringle, P. G.; Shaw, B. L. *J. Chem. Soc., Chem. Commun.* **1982**, 861.

(*143*) Berenguer, J. R.; Forniés, J.; Lalinde, E.; Martinez, F.; Urriolabeitia, E.; Welch, A. J. *J. Chem. Soc., Dalton Trans.* **1994**, 1291.

(*144*) Afzal, D.; Lenhert, P. G.; Lukehart, C. M. *J. Am. Chem. Soc.* **1984**, *106*, 3050.

(*145*) Ciriano, M.; Howard, J. A. K.; Spencer, J. L.; Stone, F. G. A.; Wadepohl, H. *J. Chem. Soc., Dalton Trans.* **1979**, 1749.

(*146*) Bruce, M. I.; Duffy, D. N. *Aust. J. Chem.* **1986**, *39*, 1697.

(*147*) Corfield, P. W. R.; Shearer, H. M. M. *Acta Crystallogr.* **1967**, *23*, 156.

(*148*) Bruce, M. I.; Horn, E.; Matisons, J. G.; Snow, M. R. *Aust. J. Chem.* **1984**, *37*, 1163.

(*149*) Carriedo, G. A.; Riera, V.; Solans, X.; Solans, J. *Acta Crystallogr., Sect. C: Cryst. Struct. Commun.* **1988**, *C44*, 978.

(*150*) Jia, G.; Payne, N. C.; Vittal, J. J.; Puddephatt, R. J. *Organometallics* **1993**, *12*, 4771; Jia, G.; Puddephatt, R. J.; Vittal, J. J.; Payne, N. C. *ibid.*, 263.

(*151*) Jia, G.; Puddephatt, R. J.; Scott, J. D.; Vittal, J. J. *Organometallics* **1993**, *12*, 3565.

(*152*) Wilson, E. B., Jr.; Decius, J. C.; Cross, P. C. "Molecular Vibrations"; McGraw-Hill: New York, 1955; p 176.

(*153*) McMullan, R. K.; Kvick, Å.; Popelier, P. *Acta Crystallogr., Sect. B: Struct. Sci.* **1992**, *B48*, 726.

(*154*) Lukehart, C. M. "Fundamental Transition Metal Organometallic Chemistry"; Brooks-Cole: Monterey, CA, 1985; Chapter 2.

(*155*) The decision to include in the distribution (Figure 2) all crystallographically distinct C≡C bond lengths (e.g., eight distances for $Mo_2(CCMe)_4(PMe_3)_4$), rather than to either also double count crystallographically equivalent bond lengths (e.g., two distances of 1.206 Å for $ZrCp_2(CCMe)_2$) or to simply count the mean C≡C bond length for $ML_n(CCR)_x (x > 1)$ complexes with chemically equivalent alkynyl ligands (e.g., 1.195 Å for $Mo_2(CCMe)_4(PMe_3)_4$) was based on the presumption that the characteristics of the distribution have less to do with intrinsic chemical differences among compounds than they do with crystallographic differences. Even though different distributions are obtained under these alternate criteria, the mean C≡C bond length remains identical to that obtained by the method we adopted:

with double counting of crystallographically equivalent $C\equiv C$ bond lengths, $d(C\equiv C)_{mean}$ = 1.201(17) Å; counting only the mean $C\equiv C$ bond length for each compound with more than one chemically equivalent ligand, $d(C\equiv C)_{mean}$ = 1.201(18) Å.

(156) Fast, H.; Welsh, H. L. *J. Mol. Spectrosc.* **1972**, *41*, 203.

(157) Allen, F. H.; Kennard, O.; Watson, D. G.; Brammer, L.; Orpen, A. G.; Taylor, R. *J. Chem. Soc., Perkin Trans. 2* **1987**, S1.

(158) Botschwina, P. In "Ion and Cluster Ion Spectroscopy and Structure"; Maier, J. P., Ed.; Elsevier: Amsterdam, 1989; pp. 59–108.

(159) Ervin, K. M.; Lineberger, W. C. *J. Phys. Chem.* **1991**, *95*, 1167.

(160) Huheey, J. E.; Keiter, E. A.; Keiter, R. L. "Inorganic Chemistry: Principles of Structure and Reactivity"; Harper Collins: New York, 1993; pp. 182–199.

(161) Pauling, L. "The Nature of the Chemical Bond"; Cornell Univ. Press: Ithaca, NY, 1960; Chapters 3 and 7.

(162) Schomaker, V.; Stevenson, D. P. *J. Am. Chem. Soc.* **1941**, *63*, 37.

(163) Wells, A. F. *J. Chem. Soc.* **1949**, 55; Pearson, R. G. *J. Chem. Soc., Chem. Commun.* **1968**, 65; Barbe, J. *J. Chem. Educ.* **1983**, *60*, 640; Peter, L. *ibid.* **1986**, *63*, 123.

(164) Labinger, J. A.; Bercaw, J. E. *Organometallics* **1988**, *7*, 926.

(165) Porterfield, W. W. "Inorganic Chemistry: A Unified Approach"; Academic Press: San Diego; 1993; p. 215; Porterfield, W. W. Personal communication, 1994.

(166) Bratsch, S. G. *J. Chem. Educ.* **1988**, *65*, 34, 223.

(167) Hunter, W. E.; Hrncir, D. C.; Bynum, R. V.; Penttila, R. A.; Atwood, J. L. *Organometallics* **1983**, *2*, 750.

(168) Cotton, F. A.; Weisinger, K. J. *Inorg. Chem.* **1990**, *29*, 2594.

(169) Ressner, J. M.; Wernett, P. C.; Kraihanzel, C. S.; Rheingold, A. L. *Organometallics* **1988**, *7*, 1661.

(170) Gibson, D. H.; Owens, K.; Mandal, S. K.; Sattich, W. E.; Richardson, J. F. *Organometallics* **1990**, *9*, 424.

(171) Stille, J. K.; Smith, C.; Anderson, O. P.; Miller, M. M. *Organometallics* **1989**, *8*, 1040.

(172) Consiglio, G.; Morandini, F.; Ciani, G.; Sironi, A. *Angew. Chem., Int. Ed. Engl.* **1983**, *22*, 333.

(173) Wakatsuki, Y.; Yamazaki, H. *J. Chem. Soc. Jpn.* **1985**, 586.

(174) de Graaf, W.; Boersma, J.; Grove, D.; Spek, A. L.; van Koten, G. *Recl. Trav. Chim. Pays-Bas.* **1988**, *107*, 299.

(175) Bardi, R.; Piazzesi, A. M. *Inorg. Chim. Acta* **1981**, *47*, 249.

(176) Ito, Y.; Inouye, M.; Suginome, M.; Murakami, M. *J. Organomet. Chem.* **1988**, *342*, C41.

(177) The Pd—C bond length of PdMeBr(NMe$_2$CH$_2$C$_6$H$_4$-2-PPh$_2$) (2.142(8) Å, with methyl trans to the nitrogen atom), reported in the same paper as Pd(C≡CSiMe$_3$)Br (NMe$_2$CH$_2$C$_6$H$_4$-2-PPh$_2$) (*104*), would seem to be a perfect reference point, yielding Δd(Pd—C) = 0.20 Å. We have not used it for this purpose because it appears to be anomalously long, as seen by comparison to the Pd—C bond distances of PdMe$_2$(tmeda) (2.028[3] Å) (*174*) and *cis*-PdMe$_2$(PPh$_2$Me)$_2$ (2.090[3] Å; Wisner, J. M.; Bartczak, T. J.; Ibers, J. A. *Organometallics* **1986**, *5*, 2044).

(178) Morino, Y; Kuchitsu, K.; Fukuyama, T.; Tanimoto, M. *Acta Crystallogr., Sect. A: Cryst. Phys., Differ., Theor. Gen. Crystallogr.* **1969**, *A25*, S127; Sugié, M.; Kuchitsu, K. *J. Mol. Struct.* **1974**, *20*, 437.

(179) Gambarotta, S.; Floriani, C.; Chiesi-Villa, A.; Guastini, C. *Inorg. Chem.* **1984**, *23*, 1739.

(*180*) Staples, R. J.; Khan, M. N. I.; Wang, S.; Fackler, J. P., Jr. *Acta Crystallogr., Sect. C: Cryst. Struct. Commun.* **1992**, *C48*, 2213.

(*181*) Jones, P. G.; Lautner, J. *Acta Crystallogr., Sect. C: Cryst. Struct. Commun.* **1988**, *C44*, 2091.

(*182*) The *trans*-PtR$_2$(PR$_3'$)$_2$ (R = CN, alkynyl) comparison is valid, even though the pairs of alkynyl and cyano ligands in the two complexes do not share a common trans ligand, because the trans influences of the CN and CCR ligands are very similar (*vide infra*).

(*183*) Cowie, J.; Hamilton, E. J. M.; Laurie, J. C. V.; Welch, A. J. *J. Organomet. Chem.* **1990**, *394*, 1.

(*184*) Williams, E. A. In "The Chemistry of Organic Silicon Compounds"; Patai, S.; Rappoport, Z., Eds.; Wiley: New York, 1989; Part I, Chapter 8; Miracle, G. E.; Ball, J. L.; Powell, D. R.; West, R. *J. Am. Chem. Soc.* **1993**, *115*, 11598.

(*185*) Hartley, F. R. *Chem. Soc. Rev.* **1973**, *2*, 163. Appleton, T. G.; Clark, H. C.; Manzer, L. E. *Coord. Chem. Rev.* **1973**, *10*, 335; Burdett, J. K.; Albright, T. A. *Inorg. Chem.* **1979**, *18*, 2112.

(*186a*) Appleton, T. G.; Bennett, M. A. *Inorg. Chem.* **1978**, *17*, 738.

(*186b*) Manojlović-Muir, L. J.; Muir, K. W. *Inorg. Chim. Acta* **1974**, *10*, 47.

(*187*) Bardi, R.; Piazzesi, A. M.; Cavinato, G.; Cavoli, P.; Toniolo, L. *J. Organomet. Chem.* **1982**, *224*, 407.

(*188*) Diré, S.; Campostrini, R.; Carturan, G.; Calligaris, M.; Nardin, G. *J. Organomet. Chem.* **1990**, *390*, 267.

(*189*) Bardi, R.; Piazzesi, A. M.; del Pra, A.; Cavinato, G.; Toniolo, L. *J. Organomet. Chem.* **1982**, *234*, 107.

(*190*) Bennett, M. A.; Ho, K.-C.; Jeffery, J. C.; McLaughlin, G. M.; Robertson, G. B. *Aust. J. Chem.* **1982**, *35*, 1311.

(*191*) Graziani, R.; Cavinato, G.; Casellato, U.; Toniolo, L. *J. Organomet. Chem.* **1988**, *353*, 125.

(*192*) Kaduk, J. A.; Ibers, J. A. *J. Organomet. Chem.* **1977**, *139*, 199.

(*193*) Sen, A.; Chen, J.-T.; Vetter, W. M.; Whittle, R. R. *J. Am. Chem. Soc.* **1987**, *109*, 148.

(*194*) Wiege, M.; Brune, H. A.; Klein, H.-P.; Thewalt, U. *Z. Naturforsch., B: Anorg. Chem., Org. Chem.* **1982**, *37B*, 718.

(*195*) Bennett, M. A.; Chee, H.-K.; Robertson, G. B. *Inorg. Chem.* **1979**, *18*, 1061.

(*196*) Conzelmann, W.; Koola, J. D.; Kunze, U.; Strähle, J. *Inorg. Chim. Acta* **1984**, *89*, 147.

(*197*) Solans, X.; Miravitlles, C.; Arrieta, J. M.; Germain, G.; Declercq, J. P. *Acta Crystallogr., Sect. B: Struct. Crystallogr. Cryst. Chem.* **1982**, *B38*, 1812.

(*198*) del Pra, A.; Zanotti, G.; Bombieri, G.; Ros, R. *Inorg. Chim. Acta* **1979**, *36*, 121.

(*199*) Davies, J. A.; Pinkerton, A. A.; Staples, R. J. *Acta Crystallogr., Sect. C: Cryst. Struct. Commun.* **1990**, *C46*, 48.

(*200*) Phang, L.-T.; Au-Yeung, S. C. F.; Hor, T. S. A.; Khoo, S. B.; Zhou, Z.-Y.; Mak, T. C. W. *J. Chem. Soc., Dalton Trans.* **1993**, 165.

(*201*) Montgomery, C. D.; Payne, N. C.; Willis, C. J. *Inorg. Chem.* **1987**, *26*, 519.

(*202*) Riley, P. E.; Davis, R. E. *J. Organomet. Chem.* **1980**, *192*, 283.

(*203*) Barron, P. F.; Engelhardt, L. M.; Healy, P. C.; Oddy, J.; White, A. H. *Aust. J. Chem.* **1987**, *40*, 1545.

(*204*) Ahrland, S.; Dreisch, K.; Norén, B.; Oskarsson, Å *Acta Chem. Scand., Ser. A* **1987**, *A41*, 173.

(*205*) Baenziger, N. C.; Bennett, W. E.; Soboroff, D. M. *Acta Crystallogr., Sect. B: Struct. Crystallogr. Cryst. Chem.* **1976**, *B32*, 962.

(*206*) Di Vaira, M.; Midollini, S.; Sacconi, L. *Inorg. Chem.* **1981**, *20*, 3430.

(207) Grindley, T. B.; Johnson, K. F.; Katritzky, A. R.; Keogh, H. J.; Topsom, R. D. *J. Chem. Soc., Perkin Trans. 2* **1974,** 273; Grindley, T. B.; Johnson, K. F.; Katritzky, A. R.; Keogh, H. J.; Thirkettle, C.; Brownlee, R. T. C.; Munday, J. A.; Topsom, R. D. *ibid.,* 276; Grindley, T. B.; Johnson, K. F.; Katritzky, A. R.; Keogh, H. J.; Thirkettle, C.; Topsom, R. D. *ibid.,* 282. See also the references cited in these papers.

(208a) Köpf, H.; Schmidt, M. *J. Organomet. Chem.* **1967,** *10,* 383.

(208b) Teuben, J. H.; de Liefde Miejer, H. J. *J. Organomet. Chem.* **1969,** *17,* 87.

(209a) Jimenez, R.; Barral, M. C.; Moreno, V.; Santos, A. *J. Organomet. Chem.* **1979,** *174,* 281.

(209b) Lang, H.; Seyferth, D. *Z. Naturforsch., B: Chem. Sci.* **1990,** *45B,* 212.

(210) Jenkins, A. D.; Lappert, M. F.; Srivastava, R. C. *J. Organomet. Chem.* **1970,** *23,* 165.

(211) Barral, M. C.; Jimenez, R.; Santos, A. *Inorg. Chim. Acta* **1982,** *63,* 257.

(212) Lang, H.; Seyferth, D. *Appl. Organomet. Chem.* **1990,** *4,* 599.

(213) Schlientz, W. J.; Ruff, J. K. *J. Chem. Soc. A* **1971,** 1139.

(214) Nast, R.; Köhl, H. *Z. Anorg. Allg. Chem.* **1963,** *320,* 135.

(215) Taube, R.; Drevs, H.; Marx, G. *Z. Anorg. Allg. Chem.* **1977,** *436,* 5.

(216) Nesmeyanov, A. N.; Makarova, L. G.; Vinogradova, V. N.; Korneva, V. N.; Ustynyuk, N. A. *J. Organomet, Chem.* **1979,** *166,* 217.

(217) Nast, R.; Köhl, H. *Chem. Ber.* **1964,** *97,* 207.

(218) Villemin, D.; Schigeko, E. *J. Organomet. Chem.* **1988,** *346,* C24.

(219) Green, M. L. H.; Mole, T. *J. Organomet. Chem.* **1968,** *12,* 404.

(220) Benfield, F. W. S.; Green, M. L. H. *J. Chem. Soc., Dalton Trans.* **1974,** 1324.

(221) Birdwhistell, K. R.; Templeton, J. L. *Organometallics* **1985,** *4,* 2062.

(222) Bruce, M. I.; Humphrey, M. G.; Matisons, J. G.; Roy, S. K.; Swincer, A. G. *Aust. J. Chem.* **1984,** *37,* 1955.

(223) Eberl, K.; Uedelhoven, W.; Wolfgruber, M.; Kreissl, F. R. *Chem. Ber.* **1982,** *115,* 504.

(224) Feng, S. G.; White, P. S.; Templeton, J. L. *J. Am. Chem. Soc.* **1992,** *114,* 2951.

(225) Nast R., Müller, H.-P. *Chem. Ber.* **1978,** *111,* 415.

(226) Ustynyuk, N. A.; Vinogradova, V. N.; Petrovskii, P. V. *Izv. Akad. Nauk SSSR, Ser. Khim.* **1982,** 680.

(227a) Carriedo, G. A.; Riera, V.; Miguel, D.; Manotti Lanfredi, A. M.; Tiripicchio, A. *J. Organomet. Chem.* **1984,** *272,* C17.

(227b) Miguel, D.; Riera, V. *J. Organomet. Chem.* **1985,** *293,* 379.

(228) Bruce, M. I.; Harbourne, D. A.; Waugh, F.; Stone, F. G. A. *J. Chem. Soc. A* **1968,** 356.

(229) Carvalho, M. F. N. N.; Herrmann, R.; Pombeiro, A. J. L. *Monatsh. Chem.* **1993,** *124,* 739.

(230) Johnson, B. F. G.; Kakkar, A. K.; Khan, M. S.; Lewis, J. *J. Organomet. Chem.* **1991,** *409,* C12.

(231) Lewis, J.; Khan, M. S.; Kakkar, A. K.; Raithby, P. R.; Fuhrmann, K.; Friend, R. H. *J. Organomet. Chem.* **1992,** *433,* 135.

(232) Ittel, S. D.; Tolman, C. A.; English, A. D.; Jesson, J. P. *J. Am. Chem. Soc.* **1978,** *100,* 7577.

(233) Ikariya, T.; Yamamoto, A. *J. Organomet. Chem.* **1976,** *118,* 65.

(234) Nast, R.; Krüger, K. W.; Beck, G. *Z. Anorg. Allg. Chem.* **1967,** *350,* 177.

(235) Bianchini, C.; Meli, A.; Peruzzini, M.; Frediani, P.; Bohanna, C.; Esteruelas, M. A.; Oro, L. A. *Organometallics* **1992,** *11,* 138; Bianchini, C.; Meli, A.; Peruzzini, M.; Vizza, F.; Zanobini, F.; Frediani, P. *ibid.* **1989,** *8,* 2080.

(236) Hanack, M.; Knecht, S.; Schulze, H.-J. *J. Organomet. Chem.* **1993,** *445,* 157.

(237) Taube, R.; Drevs, H. *Z. Anorg. Allg. Chem.* **1977,** *429,* 5.

(238) Steinborn, D.; Sedlak, U.; Dargatz, M. *J. Organomet. Chem.* **1991**, *415*, 407.
(239) Gotzig, J.; Werner, R.; Werner, H. *J. Organomet. Chem.* **1985**, *285*, 99.
(240) Davies, S. J.; Johnson, B. F. G.; Lewis, J.; Raithby, P. R. *J. Organomet. Chem.* **1991**, *414*, C51.
(241) Werner, H.; Meyer, U.; Esteruelas, M. A.; Sola, E.; Oro, L. A. *J. Organomet. Chem.* **1989**, *366*, 187.
(242) Wakatsuki, Y.; Yamazaki, H.; Kumegawa, N.; Satoh, T.; Satoh, J. Y. *J. Am. Chem. Soc.* **1991**, *113*, 9604.
(243) Werner, H.; Esteruelas, M. A.; Otto, H. *Organometallics* **1986**, *5*, 2295.
(244) Romero, A.; Santos, A.; Vegas, A. *Organometallics* **1988**, *7*, 1988.
(245) Santos, A.; López, J.; Matas, L.; Ros, J.; Galán, A.; Echavarren, A. M. *Organometallics* **1993**, *12*, 4215.
(246) Critchlow, P. B., Robinson, S. D. *Inorg. Chem.* **1978**, *17*, 1902.
(247) Kohlmann, W.; Werner, H. *Z. Naturforsch., B: Chem. Sci.* **1993**, *48B*, 1499.
(248) Jia, G.; Meek, D. W. *Organometallics* **1991**, *10*, 1444.
(249) Bianchini, C.; Bohanna, C.; Esteruelas, M. A.; Frediani, P.; Meli, A.; Oro, L. A.; Peruzzini, M. *Organometallics* **1992**, *11*, 3837.
(250) Bruce, M. I.; Wong, F. S.; Skelton, B. W.; White, A. H. *J. Chem. Soc., Dalton Trans.* **1982**, 2203.
(251a) Bruce, M. I.; Hameister, C.; Swincer, A. G.; Wallis, R. C. *Inorg. Synth.* **1982**, *21*, 82.
(251b) Bruce, M. I.; Wallis, R. C. *Aust. J. Chem.* **1979**, *32*, 1471.
(252) Abu Salah, O. M.; Bruce, M. I. *J. Chem. Soc., Dalton Trans.* **1975**, 2311.
(253) Bruce, M. I.; Cifuentes, M. P.; Snow, M. R.; Tiekink, E. R. T. *J. Organomet. Chem.* **1989**, *359*, 379.
(254) Bruce, M. I.; Humphrey, M. G. *Aust. J. Chem.* **1989**, *42*, 1067.
(255) Sato, M.; Sekino, M. *J. Organomet. Chem.* **1993**, *444*, 185.
(256) Dev, S.; Mizobe, Y.; Hidai, M. *Inorg. Chem.* **1990**, *29*, 4797.
(257) Le Bozec, H.; Ouzzine, K.; Dixneuf, P. H. *Organometallics* **1991**, *10*, 2768.
(258) Esteruelas, M. A.; Sola, E.; Oro, L. A.; Werner, H.; Meyer, U. *J. Mol. Catal.* **1988**, *45*, 1.
(259) Bruce, M. I.; Swincer, A. G. *Aust. J. Chem.* **1980**, *33*, 1471.
(260) Pourreau, D. B.; Geoffroy, G. L.; Rheingold, A. L.; Geib, S. J. *Organometallics* **1986**, *5*, 1337.
(261) Knaup, W.; Werner, H. *J. Organomet. Chem.* **1991**, *411*, 471.
(262) Werner, H.; Weinand, R.; Knaup, W.; Peters, K.; von Schnering, H. G. *Organometallics* **1991**, *10*, 3967; Weinand, R.; Werner, H. *J. Chem. Soc., Chem. Commun.* **1985**, 1145.
(263) Nast, R.; Fock, K. *Chem. Ber.* **1976**, *109*, 455.
(264) Nast, R.; Fock, K. *Chem. Ber.* **1977**, *110*, 280.
(265) Klein, H.-F.; Karsch, H. H. *Chem. Ber.* **1975**, *108*, 944.
(266) van den Bergen, A. M.; Elliott, R. L.; Lyons, C. J.; MacKinnon, K. P.; West, B. O. *J. Organomet. Chem.* **1985**, *297*, 361.
(267) Mestroni, G.; Zassinovich, G.; Camus, A.; Costa, G. *J. Organomet. Chem.* **1975**, *92*, C35.
(268) Khan, M. S.; Pasha, N. A.; Kakkar, A. K.; Raithby, P. R.; Lewis, J.; Fuhrmann, K.; Friend, R. H. *J. Mater. Chem.* **1992**, *2*, 759.
(269) Werner, H.; Schneider, D.; Schulz, M. *J. Organomet. Chem.* **1993**, *451*, 175.
(270) Werner, H.; Wolf, J.; Garcia Alonso, F. J.; Ziegler, M. L.; Serhadli, O. *J. Organomet. Chem.* **1987**, *336*, 397.
(271) Fernández, M. J.; Esteruelas, M. A.; Covarrubias, M.; Oro, L. A. *J. Organomet. Chem.* **1990**, *381*, 275.

(272a) Cetinkaya, B.; Lappert, M. F.; McMeeking, J.; Palmer, D. E. *J. Chem. Soc., Dalton Trans.* **1973**, 1202.

(272b) Cetinkaya, B.; Lappert, M. F.; McMeeking, J.; Palmer, D. *J. Organomet. Chem.* **1972**, *34*, C37.

(273) Schäfer, M.; Wolf, J.; Werner, H. *J. Chem. Soc., Chem. Commun.* **1991**, 1341.

(274) Khan, M. S.; Davies, S. J.; Kakkar, A. K.; Schwartz, D.; Lin, B.; Johnson, B. F. G.; Lewis, J. *J. Organomet. Chem.* **1992**, *424*, 87; Davies, S. J.; Johnson, B. F. G.; Khan, M. S.; Lewis, J. *J. Chem. Soc., Chem. Commun.* **1991**, 187.

(275) Wolf, J.; Werner, H.; Serhadli, O.; Ziegler, M. L. *Angew. Chem., Int. Ed. Engl.* **1983**, *22*, 414.

(276) Stang, P. J.; Crittell, C. M. *Organometallics* **1990**, *9*, 3191.

(277) Bianchini, C.; Meli, A.; Peruzzini, M.; Zanobini, F. *J. Chem. Soc., Chem. Commun.* **1987**, 971.

(278) Bianchini, C.; Meli, A.; Peruzzini, M.; Zanobini, F.; Zanello, P. *Organometallics* **1990**, *9*, 241.

(279) Brekau, U.; Werner, H. *Organometallics* **1990**, *9*, 1067.

(280) Werner, H.; Höhn, A. *Z. Naturforsch., B: Anorg. Chem., Org. Chem.* **1984**, *39B*, 1505.

(281) Nast, R.; Dahlenburg, L. *Chem. Ber.* **1972**, *105*, 1456.

(282a) Walter, R. H.; Johnson, B. F. G. *J. Chem. Soc., Dalton Trans.* **1978**, 381.

(282b) Reed, C. A.; Roper, W. R. *J. Chem. Soc., Dalton Trans.* **1973**, 1370.

(283) Werner, H.; Höhn, A.; Schulz, M. *J. Chem. Soc., Dalton Trans.* **1991**, 777.

(284) Burgess, J.; Howden, M. E.; Kemmitt, R. D. W.; Sridhara, N. S. *J. Chem. Soc., Dalton Trans.* **1978**, 1577.

(285) Bianchini, C.; Barbaro, P.; Meli, A.; Peruzzini, M.; Vacca, A.; Vizza, F. *Organometallics* **1993**, *12*, 2505.

(286a) Höhn, A.; Werner, H. *J. Organomet. Chem.* **1990**, *382*, 255.

(286b) Garcia Alonso, F. J.; Höhn, A.; Wolf, J.; Otto, H.; Werner, H. *Angew. Chem., Int. Ed. Engl.* **1985**, *24*, 406.

(286c) Werner, H.; Höhn, A. *J. Organomet. Chem.* **1984**, *272*, 105.

(287) Bennett, M. A.; Charles, R.; Fraser, P. J. *Aust. J. Chem.* **1977**, *30*, 1213.

(288) Collman, J. P.; Kang, J. W. *J. Am. Chem. Soc.* **1967**, *89*, 844.

(289) Ballester, L.; Cano, M.; Santos, A. *J. Organomet. Chem.* **1982**, *229*, 101.

(290) Masai, H.; Sonogashira, K.; Hagihara, N. *J. Organomet. Chem.* **1971**, *26*, 271.

(291) Carusi, P.; Furlani, A. *Gazz. Chim. Ital.* **1980**, *110*, 7.

(292) Chatt, J.; Shaw, B. L. *J. Chem. Soc.* **1960**, 1718.

(293) Oguro, K.; Wada, M.; Okawara, R. *J. Organomet. Chem.* **1978**, *159*, 417.

(294) Wada, M.; Oguro, K.; Kawasaki, Y. *J. Organomet. Chem.* **1979**, *178*, 261.

(295) Seidel, W.; Genitz, D. *Z. Chem.* **1979**, *19*, 413.

(296) Fontaine, X. L. R.; Higgins, S. J.; Langrick, C. R.; Shaw, B. L. *J. Chem. Soc., Dalton Trans.* **1987**, 777.

(297) Yamazaki, H.; Hagihara, N. *Bull. Chem. Soc. Jpn.* **1964**, *37*, 907.

(298) Yamazaki, H.; Nishido, T.; Matsumoto, Y.; Sumida, S.; Hagihara, N. *J. Organomet. Chem.* **1966**, *6*, 86.

(299) Nelson, J. H.; Verstuyft, A. W.; Kelly, J. D.; Jonassen, H. B. *Inorg. Chem.* **1974**, *13*, 27.

(300) Klein, H.-F.; Zettel, B.; Flörke, U.; Haupt, H.-J. *Chem. Ber.* **1992**, *125*, 9.

(301) Kim, Y.-J.; Osakada, K.; Yamamoto, A. *J. Organomet. Chem.* **1993**, *452*, 247.

(302) Espinet, P.; Forniés, J.; Martínez, F.; Sotes, M.; Lalinde, E.; Moreno, M. T.; Ruiz, A.; Welch, A. J. *J. Organomet. Chem.* **1991**, *403*, 253.

(*303*) Tohda, Y.; Sonogashira, K.; Hagihara, N. *J. Organomet. Chem.* **1976**, *110*, C53.
(*304*) Nast, R.; Müller, H.-P.; Pank, V. *Chem. Ber.* **1978**, *111*, 1627.
(*305*) Perera, S. D.; Shaw, B. L. *J. Organomet. Chem.* **1991**, *402*, 133.
(*306*) Nast, R.; Heinz, W.-D. *Chem. Ber.* **1962**, *95*, 1478.
(*307*) Langrick, C. R.; McEwan, D. M.; Pringle, P. G.; Shaw, B. L. *J. Chem. Soc., Dalton Trans.* **1983**, 2487.
(*308*) Nast, R.; Voss, J.; Kramolowsky, R. *Chem. Ber.* **1975**, *108*, 1511.
(*309*) Forniés, J.; Gómez-Saso, M. A.; Lalinde, E.; Martínez, F.; Moreno, M. T. *Organometallics* **1992**, *11*, 2873.
(*310*) Sonogashira, K.; Fujikura, Y.; Yatake, T.; Toyoshima, N.; Takahashi, S.; Hagihara, N. *J. Organomet. Chem.* **1978**, *145*, 101.
(*311*) Chatt, J.; Shaw, B. L. *J. Chem. Soc.* **1959**, 4020.
(*312*) Glockling, F.; Hooton, K. A. *J. Chem. Soc. A* **1967**, 1066.
(*313*) Yamazaki, S. *Polyhedron* **1992**, *11*, 1983.
(*314*) Yamazaki, S.; Deeming, A. J. *J. Chem. Soc., Dalton Trans.* **1993**, 3051.
(*315*) Bell, R. A.; Chisholm, M. H.; Couch, D. A.; Rankel, L. A. *Inorg. Chem.* **1977**, *16*, 677.
(*316*) Almeida, J. F.; Pidcock, A. *J. Organomet, Chem.* **1981**, *209*, 415.
(*317a*) Abu Salah, O. M.; Bruce, M. I. *Aust. J. Chem.* **1976**, *29*, 73.
(*317b*) Collamati, I.; Furlani, A. *J. Organomet. Chem.* **1969**, *17*, 457.
(*318*) Russo, M. V.; Furlani, A. *J. Organomet. Chem.* **1979**, *165*, 101.
(*319*) Blake, D. M.; Nyman, C. J. *J. Am. Chem. Soc.* **1970**, *92*, 5359.
(*320*) Cardin, C. J.; Cardin, D. J.; Lappert, M. F. *J. Chem. Soc., Dalton Trans.* **1977**, 767.
(*321*) Cross, R. J.; Davidson, M. F. *J. Chem. Soc., Dalton Trans.* **1986**, 1987.
(*322*) Anderson, G. K.; Lumetta, G. J. *J. Organomet. Chem.* **1985**, *295*, 257.
(*323*) Cook, C. D.; Jauhal, G. S. *Can. J. Chem.* **1967**, *45*, 301.
(*324*) Butler, G.; Eaborn, C.; Pidcock, A. *J. Organomet. Chem.* **1981**, *210*, 403.
(*325*) Berenguer, J. R.; Forniés, J.; Martínez, F.; Cubero, J. C.; Lalinde, E.; Moreno, M. T.; Welch, A. J. *Polyhedron* **1993**, *12*, 1797.
(*326*) Chisholm, M. H.; Clark, H. C. *Chem. Commun.* **1970**, 763.
(*327*) Schenk, W. A.; Hilpert, G. H. *J. Chem. Ber.* **1991**, *124*, 433.
(*328*) McEwan, D. M.; Markham, D. P.; Pringle, P. G.; Shaw, B. L. *J. Chem. Soc., Dalton Trans.* **1986**, 1809.
(*329*) Langrick, C. R.; Pringle, P. G.; Shaw, B. L. *J. Chem. Soc., Dalton Trans.* **1985**, 1015.
(*330*) Cooper, G. R.; Hutton, A. T.; Langrick, C. R.; McEwan, D. M.; Pringle, P. G.; Shaw, B. L. *J. Chem. Soc., Dalton Trans.* **1984**, 855.
(*331*) Abu-Salah, O. M.; Al-Ohaly, A. R. *J. Organomet. Chem.* **1983**, *255*, C39.
(*332*) Nast, R.; Schneller, P.; Hengefeld, A. *J. Organomet. Chem.* **1981**, *214*, 273.
(*333*) Nast, R.; Kirner, U. *Z. Anorg. Allg. Chem.* **1964**, *330*, 311.
(*334*) Werner, H.; Otto, H.; Ngo-Khac, T.; Burschka, C. *J. Organomet. Chem.* **1984**, *262*, 123.
(*335*) Bailey, J. *J. Inorg. Nucl. Chem.* **1973**, *35*, 1921.
(*336*) Cross, R. J.; Davidson, M. F. *J. Chem. Soc., Dalton Trans.* **1986**, 411.
(*337*) Nast, R.; Gremm, J. *Z. Anorg. Allg. Chem.* **1963**, *325*, 62.
(*338*) Weiss, E.; Plass, H. *Chem. Ber.* **1968**, *101*, 2947.
(*339*) Klein, H.-F.; Beck-Hemetsberger, H.; Reitzel, L.; Rodenhäuser, B.; Cordier, G. *Chem. Ber.* **1989**, *122*, 43; Barral, M. C.; Jimenez, R.; Royer, E.; Moreno, V.; Santos, A. *Inorg. Chim. Acta* **1978**, *31*, 165.
(*340*) Bell, R. A.; Chisholm, M. H. *Inorg. Chem.* **1977**, *16*, 687.

Organotransition-Metal Chemistry and Homogeneous Catalysis in Aqueous Solution

D. M. ROUNDHILL

Department of Chemistry
Tulane University
New Orleans, Louisiana

I. Introduction 155
II. Metal Alkyls 156
   A. Alkylchromium(III) Complexes 156
   B. Alkylcobalt(III) Complexes 160
   C. Alkylplatinum(III) Complexes 162
   D. Metal–Carbonyl Complexes 163
III. Tertiary Phosphines 164
   A. Synthesis 164
   B. Metal Complexes 166
IV. Homogeneous Catalysis 167
   A. Transition-Metal Halides in Catalysis 167
   B. Wacker Oxidation of Alkenes 171
   C. Cobalt–Cyanide Complexes as Alkene Hydrogenation Catalysts . . 171
   D. Tertiary Phosphine Complexes in Catalytic Hydroformylation and Hydrogenation Reactions 172
   E. Asymmetric Catalysis with Tertiary Phosphine Complexes 179
   F. Supramolecular-Medium Effects in Catalytic Reactions 181
   G. Other Catalyzed Reactions 182
   References 184

I

INTRODUCTION

Although it is generally perceived that organometallic complexes are only stable in the absence of water, there is a growing interest in the use of water as both a reagent and a solvent for reactions involving such compounds. Whereas in some cases it is clear that this solvent acts destructively on organometallic complexes, there are other cases where the complexes are stable. For transition metals, the organometallic compound methylcobalamin has long been known to be stable in an aqueous environment (1), and for the main group elements both methylmercury and methyllead compounds are stable (2). In general for transition-metal systems, water-soluble organometallics are known for the kinetically inert metal ions.

155

More recently there has been a strong growth of interest in the use of transition metals as homogeneous catalysts in aqueous solution. These processes involve the intermediacy of organometallic compounds for cases where the catalysts are simple transition-metal halide salts or where the metal complexes contain water-soluble tertiary phosphines as ligands (3,4).

Water as a solvent offers a number of variations to those offered by organic solvents. Because of its high dielectric constant and its ability to hydrate strongly both cations and anions, it is a solvent that strongly favors ionic reactions. Structurally, water has only O–H bonds. The enthalpy of this bond is 436 kJ mol^{-1}, which is a very high value for a single bond. A consequence of this high bond enthalpy is that water is an excellent solvent for carrying out free-radical reactions because in order for a radical to abstract a hydrogen atom from water, the bond enthalpy of the bond formed between the new heteroatom and hydrogen must be greater than that of the O–H bond.

In this review we consider both the organometallic compounds that can be obtained in aqueous solution and the catalytic systems that can be used in this solvent. An attempt is made to correlate the two topics such that parallels can be drawn between the reactions that are observed in aqueous solution and the catalytic processes that can be used.

II

METAL ALKYLS

A. Alkylchromium(III) Complexes

A series of alkylchromium(III) complexes have been prepared in aqueous solution. These complexes can be synthesized by reaction of the chromium(II) precursor with alkyl halides (Eq. 1) or alkyl radicals (Eq. 2) (5,6). Both of these schematic routes have been used to synthesize the

$$2\ Cr(II)\ +\ RX\ \rightarrow\ Cr(III)X\ +\ Cr(III)R \tag{1}$$

$$Cr(II)\ +\ R^{\cdot}\ \rightarrow\ Cr(III)R \tag{2}$$

alkyl complexes $CrR(H_2O)_5^{2+}$. In the first case, with RX, the reaction sequence is a stepwise one, where the initial step involves halogen atom abstraction from the alkyl halide, followed by the trapping of the alkyl radical by a second molecule of the chromium(II) complex (Scheme I). The formation of alkyl complexes $CrR(H_2O)_5^{2+}$ by trapping of the chromium(II)

$$Cr(II) + RX \rightarrow Cr(III)X + R^{\cdot}$$

$$Cr(II) + R^{\cdot} \rightarrow Cr(III)R$$

SCHEME 1.

hexaquo complex (Eq. 3) with alkyl radicals occurs for a wide range of R-groups. The complexes $CrR(H_2O)_5{}^{2+}$ are identified by absorption bands at 390–405 nm ($\varepsilon = 200$–400 L mol^{-1} cm^{-1}), 270–290 nm ($\varepsilon = 2$–4×10^3 L mol^{-1} cm^{-1}) and 520–570 nm ($\varepsilon = 10$–40 L mol^{-1} cm^{-1}). The log of the

$$Cr(H_2O)_6^{2+} + R^{\cdot} \rightarrow CrR(H_2O)_5^{2+} + H_2O \tag{3}$$

rate constant (L mol^{-1} sec^{-1}) for the different R-groups covers the range between 7.5 and 8.2 (6). The reaction shows a small positive volume of activation, which supports a pathway where chromium–carbon sigma-bond formation is controlled by solvent exchange on $Cr(H_2O)_6{}^{2+}$. The reaction therefore follows an I_d mechanism (7). Macrocyclic Cr(II) complexes $CrL(H_2O)_2{}^{2+}$ (L = [14]aneN$_4$) (Fig. 1) undergo similar reactions with alkyl halides RX to give $CrXL(H_2O)_2{}^{2+}$ and $CrRL(H_2O)^{2+}$ (Eq. 4).

$$2 \, CrL(H_2O)_2^{2+} + RX \longrightarrow CrRL(H_2O)^{2+} + CrXL(H_2O)^{2+} + 2H_2O \tag{4}$$

Although there is a linear correlation between the reduction potentials of RX and the rate constants for their reaction with $CrL(H_2O)_2{}^{2+}$, the pathway involves atom transfer and not electron transfer (8). An electron transfer pathway initially to give X^- and a kinetically inert Cr(III) complex would not result in subsequent substitution to give the product $CrXL(H_2O)^{2+}$.

The alkyl complexes $CrR(H_2O)_5{}^{2+}$ undergo substitution reactions with either neutral ligands L or anions X^-. For the case of ethylenediamine as a neutral ligand, substitution leads to the formation of the complex $CrR(en)_2{}^{2+}$ (Eq. 5) (9). For the case of the thiocyanate ion, substitution

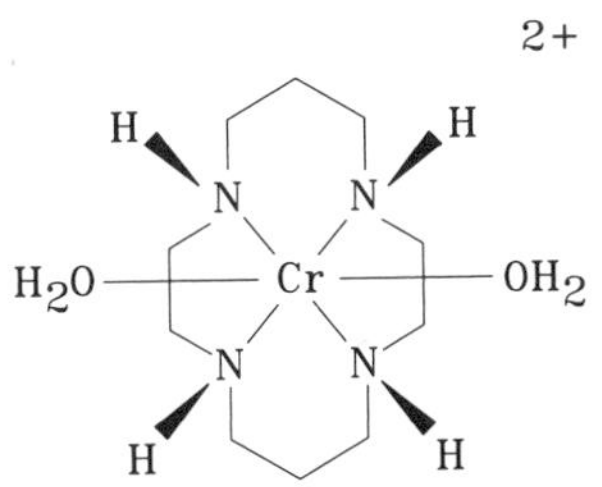

FIG. 1.

$$CrR(H_2O)_5^{2+} + 2\ en \rightarrow CrR(en)_2^{2+} + 5\ H_2O \qquad (5)$$

gives the complex $trans$-$CrR(NCS)(H_2O)_4^{2+}$ (Eq. 6) (10). The complexes $CrR(H_2O)_5^{2+}$ generally decompose by the reverse of reaction 3. Since the high strength of the Cr–R bond frequently leads to this homolysis being

$$CrR(H_2O)_5^{2+} + NCS^- \rightarrow trans\text{-}CrR(NCS)(H_2O)_4^{2+} + H_2O \qquad (6)$$

thermodynamically unfavorable, the reaction may occur only upon the addition of radical scavengers. The rates for this homolysis reaction are dependent on the nature of the R-group, with aralkyl and secondary alkyl derivatives being the most reactive. The primary alkyl complexes do not undergo significant homolysis. For the homolysis of the benzyl derivatives $Cr(CH_2C_6H_4X\text{-}4)(H_2O)_5^{2+}$, the reaction rate is lower when the substituent X is an electron-withdrawing group (11). Pulse radiolysis has been used effectively to determine the equilibrium constants for the homolytic cleavage of metal–carbon bonds. This kinetic method gives $\Delta H^{\neq}$ for the homolysis reaction. Taking this value as being a measure of the metal–carbon bond enthalpy, however, assumes that the radical recombination occurs at a rate close to the diffusion-controlled limit and has a negligible activation energy (12–14). The equilibrium constants for the Cr–R bond homolysis in $CrR(H_2O)_5^{2+}$ for different R-groups are given in Table 1. These values for K do not correlate either with the redox potential of the group R or with the activation energy for the homolytic cleavage of the corresponding substituted ethanes.

The complexes $CrR(H_2O)_5^{2+}$ can also decompose by a heterolytic pathway (Eq. 7). The reaction rate for this pathway has terms that are both

$$CrR(H_2O)_5^{2+} + H_2O \rightarrow Cr(OH)(H_2O)_5^{2+} + RH \qquad (7)$$

first-order and independent of $[H^+]$. In addition, the less-reactive alkyl complexes have heterolysis pathways that involve catalysis by sulfate

TABLE I

EQUILIBRIUM CONSTANTS FOR THE CR–R
HOMOLYSIS IN $CrR(H_2O)_5^{2+}$

R	K (mol dm^{-3})
CH_2OH	2.3×10^{-13}
$CH(Me)OH$	1.1×10^{-11}
$C(Me)_2OH$	2.5×10^{-9}
$CH(Me)OEt$	5.9×10^{-11}

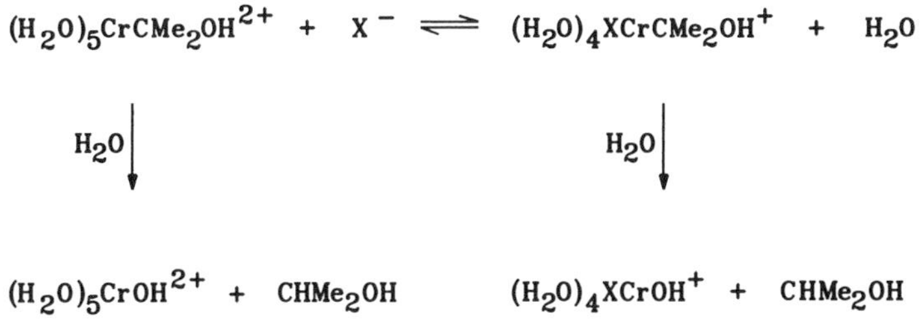

$$(H_2O)_5CrCMe_2OH^{2+} + X^- \rightleftharpoons (H_2O)_4XCrCMe_2OH^+ + H_2O$$

$$H_2O \downarrow \qquad\qquad\qquad H_2O \downarrow$$

$$(H_2O)_5CrOH^{2+} + CHMe_2OH \qquad (H_2O)_4XCrOH^+ + CHMe_2OH$$

SCHEME 2.

ion. The uncatalyzed heterolysis pathway for a range of alkyl groups have $\Delta H^{\neq} = 21 \pm 2$ kcal mol^{-1} and $\Delta S^{\neq} = -13.5 \pm 5$ cal mol^{-1} K^{-1} (15). Acetate and similar anions catalyze this heterolysis reaction by *trans* labilization of the Cr–R bond (16–18). The presence of acetate ion also decreases the rate of reaction between $Cr(H_2O)_5^{2+}$ and alkyl radicals (R) to give alkylchromium(III) complexes $CrR(H_2O)_5^{2+}$. This decrease corresponds with the reactive monomeric chromium(II) aquo complexes being converted by acetate ion into higher aggregates. Under high pressure, dissociation to monomers occurs and the reaction rate with alkyl radicals increases correspondingly (19). The heterolysis reaction can, however, become more dominant by the addition of aqueous chromium(II), which suppresses the homolysis of the Cr–R bond.

For the anion-catalyzed heterolysis reaction with the complex $(H_2O)_5Cr(CMe_2OH)^{2+}$, the catalytic effect increases with basicity of the anion. The activation volumes are all positive, in support of a dissociatively activated heterolysis mechanism. Overall, the decomposition mechanism involves parallel pathways whereby hydrolytic cleavage occurs at both $(H_2O)_5Cr(CMe_2OH)^{2+}$ and the more-reactive substituted complex $(H_2O)_4XCr(CMe_2OH)^+$ (Scheme 2).

Alkylchromium(III) complexes having alkyl groups that contain β-hydrogen and hydroxyl groups can decompose by an acid-catalyzed β-hydrogen transfer pathway. For the complex $(H_2O)_5Cr\{CH(R)CH_2OH\}^{2+}$ (R = H, Me) in acidic solution, the decomposition products are $Cr(H_2O)_6^{3+}$ and the alkene $RCH{=}CH_2$, which are formed via the π-bonded alkene complex (Eq. 8) (20). The replacement of a hydrogen substituent by a methyl

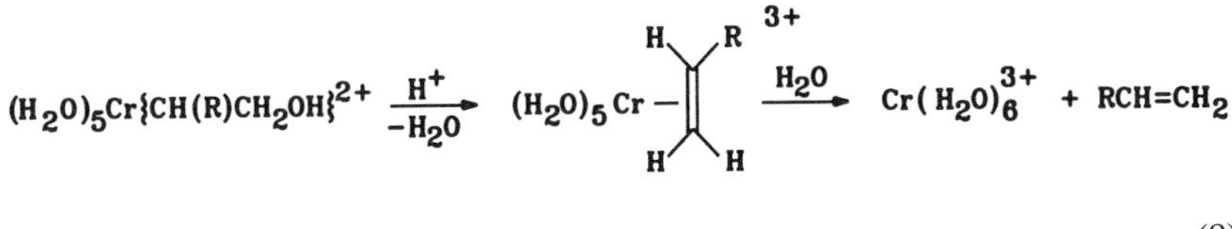

$$(H_2O)_5Cr\{CH(R)CH_2OH\}^{2+} \xrightarrow[-H_2O]{H^+} (H_2O)_5Cr\underset{H \quad H}{\overset{H \quad R}{-\big\Vert}}{}^{3+} \xrightarrow{H_2O} Cr(H_2O)_6^{3+} + RCH{=}CH_2$$

$$(R = H, Me)$$

(8)

group results in a decrease in the rate of this reaction. Since similar studies on alkyl–copper and –cobalt complexes lead to different effects (21,22); therefore the origin of this retardation effect in the reaction rate is uncertain.

Organochromium complexes undergo a series of chemical reactions with electrophiles that result in cleavage of the Cr–R bond. The examples of such reactions shown in Scheme 3 result in transfer of the alkyl group to the electrophile and formation of $Cr(H_2O)_6^{3+}$. The protonation reaction to give RH is slower than that with halogens or mercuric ions (23). These organochromium(III) complexes also react with other radicals, such as molecular oxygen. The reaction with oxygen follows a chain mechanism. In the mechanism shown for this reaction in Scheme 4 with the 2-propyl complex $Cr(CHMe_2)(H_2O)_5^{2+}$, the oxygen molecule acts to trap any free 2-propyl radicals and convert them into hydroperoxyl radicals (24). Subsequent reaction of these hydroperoxyl radicals with $Cr(CHMe_2)(H_2O)_5^{2+}$ gives $Cr(OOCHMe_2)(H_2O)_5^{2+}$ and 2-propyl radicals. The final organic products are 2-propanol and acetone, with the radicals $^\bullet CHMe_2$ and $^\bullet OO$—$CHMe_2$ being the chain-carrying intermediates. The inorganic product is $Cr(H_2O)_6^{3+}$, which is formed from the reaction of $Cr(H_2O)_5^{2+}$ with oxygen in water.

B. Alkylcobalt(III) Complexes

Alkylcobalt(III) complexes can also be synthesized in aqueous solution. Two of the best-known systems are methylcobalamin and a group of related cobaloximes, and alkylcobalt(III) complexes having ancillary cyanide ligands. As with the chromium(III) system, alkyl cobalt(III) complexes having dimethylglyoxime (DMG) or cyanide ligands can be synthesized by reaction of the cobalt(II) precursor with alkyl halides (Scheme

$$Cr(H_2O)_6^{3+} + HgR^+$$

$$\uparrow Hg^{2+}$$

$$Cr(H_2O)_6^{3+} + RH \xleftarrow{H_3O^+} CrR(H_2O)_5^{2+} \xrightarrow{X_2} Cr(H_2O)_6^{3+} + RX + X^-$$

$$\downarrow RHg^+$$

$$Cr(H_2O)_6^{3+} + HgR_2$$

SCHEME 3.

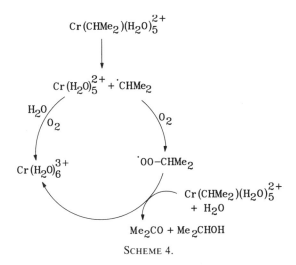

$$\text{Scheme 4.}$$

5) (25). The reaction of the cobalt(II) complex $Co(CN)_5^{3-}$ with RX follows second-order kinetics, with the rate decreasing in the sequence RI > RBr $\gg$ RCl. This sequence correlates with the bond enthalpies of the alkyl halides. The free-radical pathway for the reaction of $Co(CN)_5^{3-}$ with alkyl halides is shown in Scheme 6. The first step in Scheme 6 is the slow step; the second step is the fast one. For alkyl iodides the reaction rates follow the sequence $PhCH_2I < Me_3CI > Me_2CHI > EtI > MeI$. Again, this ordering corresponds with the highest rates being observed for compounds having the weakest carbon–iodine bond enthalpies (25).

Alkylcobalt(III) cobalamin-type complexes can be prepared in aqueous solution by reaction between the reduced vitamin B_{12r} and organic halides (Eq. 9) (26,27). The reaction of vitamin B_{12r} with organic iodides in aqueous

$$2\ B_{12r}\ +\ RX\ \longrightarrow\ R\text{-}B_{12}\ +\ B_{12a}\ +\ X^- \tag{9}$$

solution, while exhibiting the same stoichiometry as in methanol, shows different kinetic behavior. In methanol as solvent, a second-order rate

$$Co(II)L_5\ +\ RX\ \rightarrow\ Co(III)XL_5\ +\ R$$

$$Co(II)L_5\ +\ R\ \rightarrow\ Co(III)RL_5$$

$$2\ Co(II)L_5\ +\ RX\ \rightarrow\ Co(III)XL_5\ +\ Co(III)RL_5$$

$$\text{Scheme 5.}$$

$$Co(CN)_5^{3-} + RX \rightarrow CoX(CN)_5^{3-} + R^{\cdot}$$

$$Co(CN)_5^{3-} + R^{\cdot} \rightarrow CoR(CN)_5^{3-}$$

$$2\ Co(CN)_5^{3-} + RX \rightarrow CoX(CN)_5^{3-} + CoR(CN)_5^{3-}$$

SCHEME 6.

law (Eq. 10) is observed; in aqueous solution, a third-order rate law (Eq. 11) is found. This third-order rate law has been explained on the basis of

$$-d[B_{12r}]/dt = 2k[B_{12r}][RX] \tag{10}$$

$$-d[B_{12r}]/dt = 2k[B_{12r}]^2[RX] \tag{11}$$

the mechanism shown in Scheme 7. This pathway involves a pre-equilibrium step where the initially formed adduct B_{12r}.RI reacts with a second molecule of B_{12r} to give the observed products.

Pulse radiolysis has been used to study the transient formation and decomposition of cobalt–alkyl bonds in aqueous solution in the same manner as it has been used for chromium alkyls. And as for chromium alkyls, bond homolysis is a major decomposition pathway (28). For bond formation reactions, pulse radiolysis shows that they are assisted by increases in pressure. This feature results from the homolysis having a larger activation volume than the bond formation reaction, resulting in a significantly negative overall reaction volume for the process (29). In general for all of these metal–alkyl bond homolysis reactions of the aquo complexes, steric hindrance facilitates the reaction. Ligand effects also play a role, but the factors involved are more subtle.

C. Alkylplatinum(III) Complexes

Stable water-soluble alkyldiplatinum(III) complexes are known where the bridging ligand is a pyrophosphite. The iodomethyl diplatinum(III) complex has been prepared by the addition of methyl iodide to the tetrakis

$$B_{12r} + RI \rightleftharpoons B_{12r}.RI$$

$$B_{12r}.RI + B_{12r} \longrightarrow R\text{-}B_{12} + B_{12a} + I^-$$

SCHEME 7.

(μ-pyrophosphito-P,P'-)diplatinum(II) tetraanion $Pt_2(pop)_4^{4-}$ (Eq. 12).

$$\text{Pt-Pt}^{4-} + \text{MeI} \longrightarrow \text{Me-Pt-Pt-I}^{4-} \tag{12}$$

$$\left(\frown = \begin{array}{c} HO \quad O \quad OH \\ \diagdown P \diagdown \diagup P \diagup \\ O \quad \_ \quad \_ \quad O \end{array} \right)$$

This reaction can even be carried out in strongly acidic solution, since the iodomethyl complex is stable in such conditions. No evidence is found for protolysis of the platinum–methyl bond. For higher-homolog alkyl groups, the stability of the iodoalkyl complex is lower, and the reaction of $Pt_2(pop)_4^{4-}$ with higher alkyl iodides may give the diiodo diplatinum(III) complex rather than the iodoalkyl complex (Eq. 13). These reactions

$$\text{Pt-Pt} + 2\,RI \longrightarrow I-\text{Pt-Pt}-I + R_2 \tag{13}$$

resemble the reaction of $Co(CN)_5^{3-}$ with alkyl halides, where a one-electron oxidation occurs at each metal center (30).

D. Metal–Carbonyl Complexes

Hydrophilic functional groups have been attached to cyclopentadienyl groups, and these compounds used as η^5-ligands toward transition-metal carbonyls. Two complexes that have been prepared with such ligands are $[(\eta^5\text{-}C_5H_4CO_2H)W(CO)_3]_2$ and $[(\eta^5\text{-}C_5H_4CH_2CH_2NH_3^+)Mo(CO)_3]_2$ (31,32). Photolysis of these complexes in aqueous solution in the presence of added ligands L results in photodisproportionation of the complex and formation of 19-electron complexes $(\eta^5\text{-}cp)M(CO)_3L$ (M = Mo, cp = $C_5H_4CH_2CH_2NH_3^+$. M = W, cp = $C_5H_4CO_2H$) (Eq. 14). These complexes act as reductants to $Fe(CN)_6^{3-}$, methyl viologen, and cytochrome c.

$$\tfrac{1}{2}\,[(\eta^5\text{-}cp)M(CO)_3]_2 \xrightarrow{h\nu} (\eta^5\text{-}cp)M(CO)_3 \xrightarrow{L} (\eta^5\text{-}cp)M(CO)_3L \tag{14}$$

$$(M = Mo,\ cp = C_5H_4CH_2CH_2NH_3^+.\ M = W,\ cp = C_5H_4CO_2H)$$

Metal–carbonyl complexes of the sodium salt of trisulfonated triphenylphosphine TPPTS (Section III-A) have also been synthesized. The preparative route involves photolysis of $M(CO)_6$ (M = Mo, W) with TPPTS in a coordinating solvent such as methanol (33). The complexes are stable, with the thermodynamics for TPPTS dissociation being comparable with PPh_3 in nonaqueous solvents. The dissociation of TPPTS can, however, be accelerated by encapsulating the sodium ion of the ligand in a cryptand (34).

III

TERTIARY PHOSPHINES

Since many organometallic compounds are stabilized by tertiary phosphines and phosphites, then if organometallic chemistry is to find major applications in aqueous solution it is necessary that these complexes be made compatible with such a solvent system. Water-soluble phosphines can be prepared by incorporating sulfonate or phosphonate groups into the molecule. Phosphites, however, are not a particularly good choice of ligand for reactions to be carried out in aqueous solution because they undergo hydrolysis to give phosphonites (Eq. 15). The coordination and

$$P(OR)_3 + H_2O \longrightarrow PH(O)(OR)_2 + ROH \tag{15}$$

organometallic chemistry of phosphinites has been reviewed elsewhere (35,36), as has the use of tertiary phosphine complexes in aqueous solution as homogeneous catalysts (37–39).

A. Synthesis

A group of water-soluble tertiary phosphines that can be used as ligands to synthesize organometallic complexes soluble in aqueous solution are shown in Fig. 2. These individual phosphines can be synthesized by the sulfonation of triphenylphosphine, 1 (Eq. 16) (40); the reaction between

$$PPh_3 \xrightarrow{\;H_2SO_4\;} P{-}\left(\!\!\left\langle \bigcirc \right\rangle\!\!\right)_3 \quad SO_3^- Na^+ \tag{16}$$

lithium diphenylphosphide and 2-chloroethyldimethylamine, 2 (Eq. 17) (41); the reaction between diphenylphosphide and sodium 2-bromoethyl-sulfonate, 3 (Eq. 18), or 2-bromoethyl-diethyl phosphonate, 4 (Eq. 19)

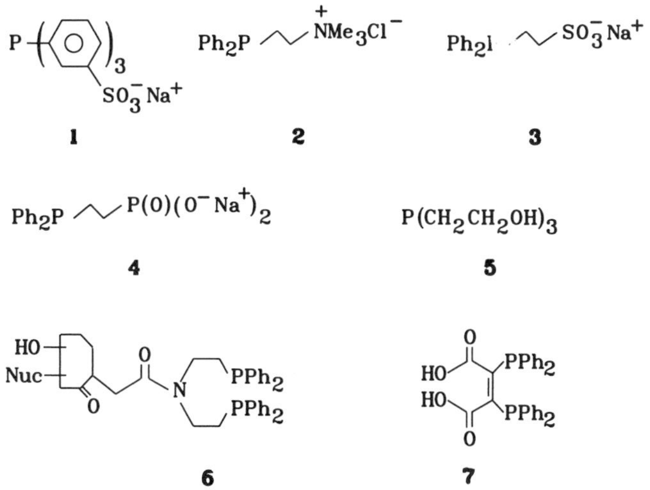

$$P{-}{\left(\bigcirc\right)}_3 \quad Ph_2P{\wedge}{\overset{+}{N}Me_3Cl^-} \quad Ph_2I{\wedge}SO_3^-Na^+$$
$$SO_3^-Na^+$$

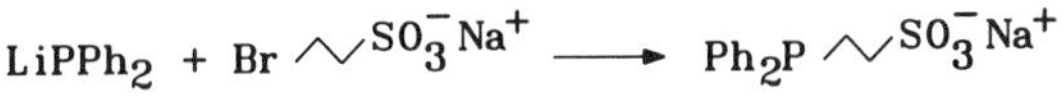

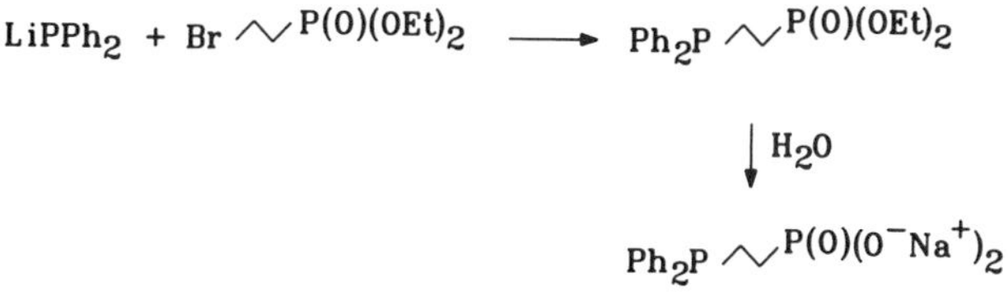

1   **2**   **3**

$$Ph_2P{\wedge}P(O)(O^-Na^+)_2 \qquad P(CH_2CH_2OH)_3$$

4   **5**

(structures for **6** and **7**)

6   **7**

FIG. 2.

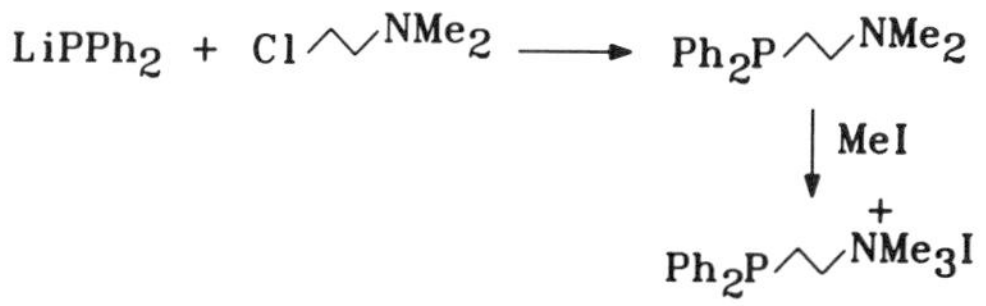

$$LiPPh_2 + Cl{\wedge}NMe_2 \longrightarrow Ph_2P{\wedge}NMe_2$$
$$\downarrow MeI$$
$$Ph_2P{\wedge}\overset{+}{N}Me_3I \qquad (17)$$

$$LiPPh_2 + Br{\wedge}SO_3^-Na^+ \longrightarrow Ph_2P{\wedge}SO_3^-Na^+ \qquad (18)$$

$$LiPPh_2 + Br{\wedge}P(O)(OEt)_2 \longrightarrow Ph_2P{\wedge}P(O)(OEt)_2$$
$$\downarrow H_2O \qquad (19)$$
$$Ph_2P{\wedge}P(O)(O^-Na^+)_2$$

(42,43); the reaction between phosphine and formaldehyde, **5** (Eq. 20) (44); the acylation of bis[2-(diphenylphosphino)ethyl] amine followed by

$$PH_3 + 3\,CH_2O \longrightarrow P(CH_2OH)_3 \qquad (20)$$

reaction with a nucleophile, **6** (Eq. 21) (45); and the oxidative hydrolysis of an anhydride, **7** (Eq. 22) (46). The alkylphosphines $P[(CH_2)_x(C_6H_5)]_3$ (x = 1, 2, 3, 6) undergo sulfonation at both the *ortho* and *para* positions

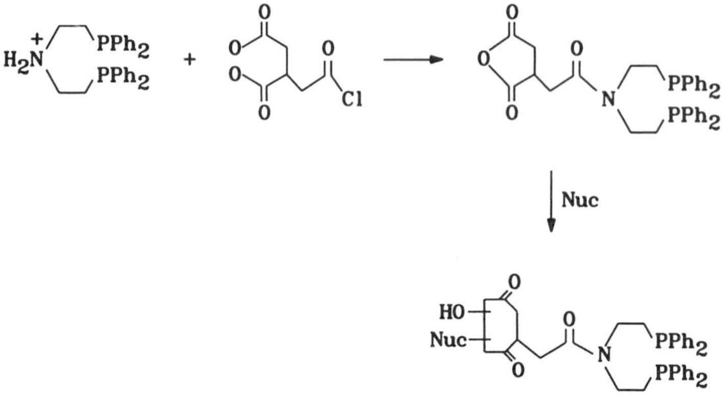

(21)

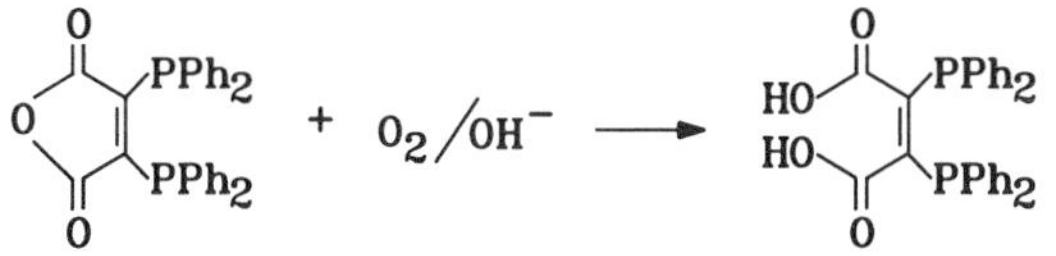

(22)

to give water-soluble derivatives. The pure *para*-substituted compounds $P[(CH_2)_X(C_6H_4\text{-}p\text{-}SO_3Na)]_3$ can be subsequently isolated (47).

These compounds all act as regular tertiary phosphines and coordinate to metal ions via their electron pair on phosphorus. These phosphines find particular application in phase-transfer or homogeneously catalyzed reactions (Section IV-D).

B. *Metal Complexes*

Water-soluble transition-metal complexes of these tertiary phosphines have been synthesized by procedures analogous to those used for phosphine complexes in nonaqueous solution. For oxygen-sensitive complexes, the high solubility of oxygen in water makes it particularly important to purge the solvent thoroughly with an inert gas prior to use. Since the tertiary phosphorus atom coordinates to form very stable complexes, only for the early transition metals must one be concerned that complexation will occur via the sulfonate group rather than via phosphorus. In essentially all cases, transition-metal complexes of these water-soluble tertiary phosphines have been synthesized for catalytic applications where water-soluble organotransition-metal complexes are important intermediates.

IV

HOMOGENEOUS CATALYSIS

A. *Transition-Metal Halides in Catalysis*

Salts of $PtCl_4^{2-}$ in acidified aqueous solution catalyze hydrogen–deuterium exchange in arenes *(48)*. Temperatures in the 80–100°C range are used. For the exchange reaction of benzene in D_2O/CH_3CO_2D, the rate is proportional to the concentration of both $PtCl_4^{2-}$ and benzene, and inversely proportional to the concentration of Cl^-. The reaction is not catalyzed by the platinum(IV) complex $PtCl_6^{2-}$ *(49,50)*. Because substituted benzenes show an *ortho* deactivating effect, and because substituents such as nitro groups do not significantly affect the exchange rate, the dissociative mechanism shown in Scheme 8 is favored *(51–53)*. This platinum catalyst system also catalyzes the exchange of hydrogen and deuterium in alkanes *(54)*. The maximum rate of exchange in the alkane pentane is found for a solution containing 50 mol% acetic acid in water *(55)*. The exchange mechanism for alkanes parallels that found for aromatics *(56)*. The hydrogen–deuterium exchange rate increases with an increase in the carbon chain length of the *n*-alkane, with a linear correlation being observed between the logarithm of the exchange rate and the ionization potential of the alkane *(55)*. The rate decreases with increased chain branching in the alkane, with the reactivity decreasing in the order primary > secondary > tertiary. For cyclic alkanes, the rates do not correlate with the ionization potentials; instead, steric factors are more important.

Alkanes and arenes can also be activated to other reactions by platinum complexes in aqueous solution *(57,58)*. For arenes in the presence of H_2PtCl_6, reduction from Pt(IV) to Pt(II) occurs and the arene undergoes chlorination. The reaction is catalyzed by platinum(II) *(59)*. Similarly, if a platinum(IV) catalyst such as H_2PtCl_6 is used, chloroalkanes are formed from alkanes. As an example, chloromethane is formed from methane (Eq. 23) *(60–62)*. Linear alkanes preferentially substitute at the methyl

$$CH_4 \xrightarrow{H_2PtCl_6} CH_3Cl + HCl \tag{23}$$

group. For cyclohexanes, however, the major products are arenes, which are likely formed by dehydrochlorination of intermediate chloroalkanes. The system has been made catalytic either by the addition of Cu^{2+} and air or by the deposition of the catalyst on silica *(63,64)*. It has been suggested that the reaction pathway involves the formation of intermediate

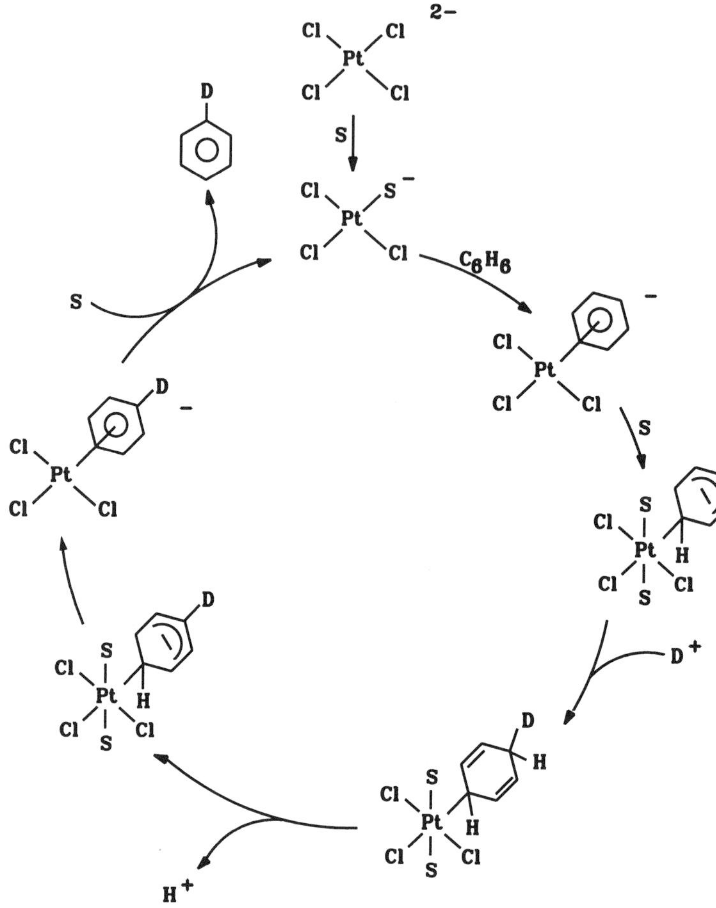

SCHEME 8.

alkylplatinum(II) complexes, which subsequently react with $PtCl_6^{2-}$ to form a carbon–chlorine bond (65).

Shilov has suggested that these catalytic reactions of alkanes involve initial addition of a C–H bond to platinum(II) to give alkyl-hydride–platinum(IV) species, which then undergo subsequent proton loss to give an alkylplatinum(II) intermediate (66). A related observation is the photolysis of a mixture of $PtCl_6^{2-}$ and hexane in aqueous trifluoroacetic acid, leading to the formation of the 1-hexene–platinum(II) complex $[PtCl_2(C_6H_{12})]_2$ (Eq. 24) (67). It has been proposed that this 1-hexene complex results from β-elimination in an alkylplatinum(IV) intermediate. The reaction is

$$PtCl_6^{2-} + C_6H_{14} \longrightarrow 1/2 \ (PtCl_2(1\text{-}C_6H_{12}))_2 \tag{24}$$

accompanied by the formation of small quantities of chlorohexane (68). An alternative explanation for these results is that the photochemical reaction follows the pathway shown in Scheme 9, which is supported by the early work of Taube showing that $PtCl_6^{2-}$ can undergo photocatalytic redox reactions (69).

At 120°C, a mixture of $PtCl_6^{2-}$ and $PtCl_4^{2-}$ in the presence of oxygen can be used for the oxidation of alkanes to alcohols (70). The substrate p-toluenesulfonic acid is oxidized sequentially at the side-chain functionality, first to the alcohol and then to the aldehyde (Eq. 25). For ethylbenzene,

$$ArCH_3 \longrightarrow ArCH_2OH \longrightarrow ArCHO$$
$$(Ar = p\text{-}HO_3SC_6H_4) \tag{25}$$

oxidation occurs both at the β-position, to give $ArCH_2CH_2OH$, and at the α-position, to give $ArCH(OH)CH_3$. This observation argues against a free-radical mechanism. When ethanol is used as substrate, ethylene glycol, acetic acid, and carbon dioxide are among the products formed.

A mixture of $PtCl_4^{2-}$ and metallic Pt with oxygen in aqueous media can be used to oxidize ethane, propane, ethers, and esters. This combination acts sequentially, whereby the initial cleavage of an unactivated C–H bond is induced by reaction with $PtCl_4^{2-}$. The subsequent oxidation step with oxygen is catalyzed by the metallic Pt. For ethane this sequence of

$$PtCl_6^{2-} \xrightarrow{\ h\nu\ } PtCl_5^{2-} + Cl^\cdot$$

$$PtCl_5^{2-} + C_6H_{14} \longrightarrow PtHCl_5^{2-} + C_6H_{13}$$

$$PtCl_6^{2-} + C_6H_{13} \longrightarrow PtCl_5^{2-} + C_6H_{12} + HCl$$

$$PtCl_6^{2-} + C_6H_{13} \longrightarrow PtCl_5^{2-} + C_6H_{13}Cl$$

$$PtHCl_5^{2-} \longrightarrow PtCl_4^{2-} + HCl$$

$$2\,PtCl_5^{2-} \longrightarrow PtCl_4^{2-} + PtCl_6^{2-}$$

$$PtCl_4^{2-} + C_6H_{12} \longrightarrow \frac{1}{2}\,(PtCl_2(1\text{-}C_6H_{12}))_2 + 4\,Cl^-$$

SCHEME 9.

steps results in the initial formation of ethyl alcohol, which is subsequently oxidized to acetic acid (71). In the absence of metallic Pt and O_2, the oxidation of unactivated C–H bonds by $PtCl_4^{2-}$ is still observed. For substrates with oxygen heteroatoms, the C–H bonds that are in α positions to an oxygen atom are activated and catalytically oxidized by metallic Pt and O_2 (72). This reaction occurs even in the absence of $PtCl_4^{2-}$. The selective activation and oxidation of C–H bonds in ethers by $PtCl_4^{2-}$ follow the reactivity order $\alpha-C-H < \beta-C-H < \gamma-C-H$ from the oxygen heteroatom. For the oxidation of ethers with $PtCl_4^{2-}$, the observation of an induction period makes it likely that the pathway involves the initial reduction to elemental platinum, followed by the initial C–H activation on a platinum surface rather than in aqueous solution. By judicious choice of platinum systems having either single or combinations of reagents, a series of selective hydrocarbon oxidations can be carried out, to give different products.

Palladium(II) catalysts, such as $PdCl_4^{2-}$ in trifluoroacetic acid solvent, have been used for the conversion of methane to methanol (73). The system uses hydrogen peroxide as oxidant; the overall reaction involving the addition of one mole of water is shown in Eq. 26. The function of the

$$CH_4 + H_2O + Pd^{2+} \longrightarrow CH_3OH + 2H^+ + Pd(0) \tag{26}$$

hydrogen peroxide is to reoxidize the palladium(0) back to palladium(II). The reaction sequence involves the initial formation of methyl trifluoroacetate, which is subsequently hydrolyzed to give methanol. The reaction occurs with both alkanes and arenes ($74,75$). These reactions show the characteristics of an electrophilic rather than a radical mechanism (76). Palladium halides such as $PdCl_4^{2-}$ also act as catalysts for the carbonylation of aryl iodides in aqueous solution. The reaction can be used to convert aryl iodides into arene carboxylic acids under a 1-atm pressure of CO at ambient temperature (77).

Hydrated $RuCl_3$ and $OsCl_3$ in aqueous solution are effective catalysts for ring-opening metathesis polymerization (ROMP). An example of such a reaction is shown in Eq. 27 (78). These halides are catalyst precursors,

$$\tag{27}$$

since they undergo hydrolysis to give the aquo complex $Ru(H_2O)_6^{2+}$, which then coordinates the double bond of the alkene to give the active complex

$Ru(H_2O)_5(alkene)^{2+}$. In the absence of water these metal halides act as ROMP catalysts only after long induction periods.

B. Wacker Oxidation of Alkenes

The Wacker process is an example of a homogeneously catalyzed reaction that occurs in water. The overall reaction involves the conversion of ethylene into acetaldehyde, or terminal alkenes into ketones (Eq. 28). The

$$\text{(28)}$$

reaction uses a catalyst system comprised of $PdCl_2$, $CuCl_2$, and molecular oxygen. The mechanism involves initial coordination of the alkene to palladium(II), followed by nucleophilic attack by water at this complexed alkene. Although the details of the mechanism have not been fully elucidated, the pathway shown in Scheme 10 for the conversion of ethylene into acetaldehyde provides a reasonable description of the steps that are involved (79,80). The catalytic reaction is inhibited by Cl^-, which competes with coordination of the alkene. Following formation of the C–O bond, the organic moiety rearranges to give acetaldehyde. The pathway also involves reduction of Pd(II) to Pd(0), followed by reoxidation back to Pd(II) by Cu(II). The molecular oxygen present converts the reduced Cu(I) back to Cu(II). Labeling studies show that the oxygen atom in acetaldehyde or ketones originates from water rather than molecular oxygen. Since carrying out the reaction with ethylene in D_2O solvent results in no incorporation of deuterium into acetaldehyde, it is apparent that no free vinyl alcohol is formed during the reaction. Instead, the vinyl alcohol β-elimination product must rearrange within the coordination sphere of Pd(II) before it has time to exchange with the solvent.

C. Cobalt–Cyanide Complexes as Alkene Hydrogenation Catalysts

The cobalt(II) cyanide complex $Co(CN)_5^{3-}$ in aqueous solution acts as a homogeneous catalyst for the selective hydrogenation of conjugated dienes to monoenes (81). The initial step in this catalysis is the reaction

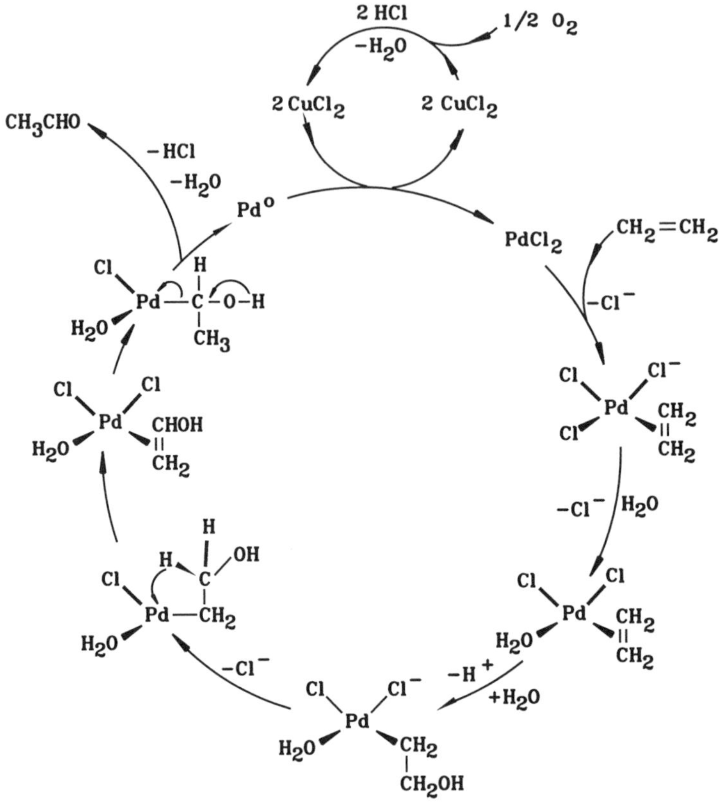

SCHEME 10.

of $Co(CN)_5^{3-}$ with hydrogen to give $CoH(CN)_5^{3-}$ (82,83). The catalytic system can be used over a temperature range of 0–125°C and a hydrogen pressure range of 1–220 atm. The reduction of carbon–carbon double and triple bonds occurs only when they are in a conjugated system. This system differs from other hydrogenation catalysts in that the pathway does not involve prior coordination of the alkene to cobalt (84). Although the detailed mechanism is complex, the major pathway is that shown in Scheme 11.

D. Tertiary Phosphine Complexes in Catalytic Hydroformylation and Hydrogenation Reactions

Many of the water-soluble phosphine complexes of the late transition metals that have been synthesized are designed for use as homogeneous

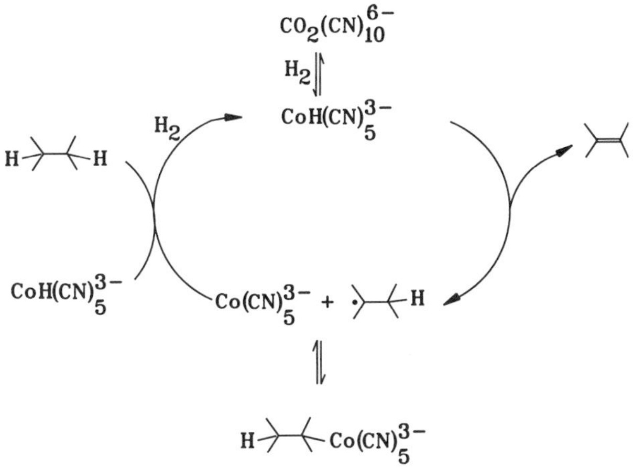

SCHEME 11.

hydroformylation and hydrogenation catalysts. This direction was taken in order that hydroformylation catalysts could be separated readily from the organic products of the reaction by subsequent extraction of the catalyst into an aqueous phase. A commonly used approach for incorporating water solubility into a transition-metal complex is to incorporate one or more sulfonate groups onto the organic functionalities of the phosphine. The two most frequently used water-soluble phosphines for catalytic applications are the mono- and trisulfonated triphenylphosphines, TPPMS and TPPTS, respectively (Fig. 3). Several features of ligands such as TPPTS are important when they are being considered for use in the synthesis of homogeneous catalysts. One factor is that the cone angle of TPPTS is larger than that of triphenylphosphine, which results in the formation

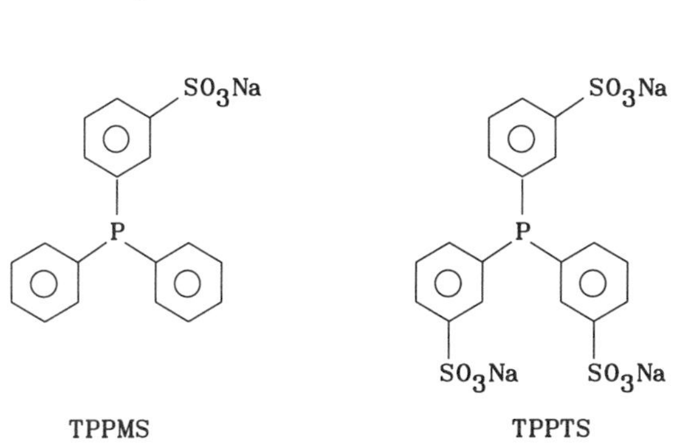

TPPMS   TPPTS

FIG. 3.

of complexes having lower coordination numbers. A second important feature, especially for solutions having a pH of 7 or above, is that hydroxy complexes may be obtained in solution. Such complexes are relatively uncommon in nonaqueous media, but in aqueous solution they are readily formed by the replacement of a coordinated Cl^- ligand by OH^- (85).

1. *TPPMS Complexes*

Transition-metal complexes of TPPMS have been used as catalysts for hydrogenation, hydroformylation, and carbonylation reactions. As an example of one such system, the complex $RhCl(TPPMS)_3$ catalyzes the hydrogenation of alkenes in neutral or slightly acidic solution (86). For this system it has been found that the *cis*-isomer maleic acid is hydrogenated more slowly than is the *trans*-isomer fumaric acid (87). This unexpected selectivity of $RhCl(TPPMS)_3$ may be due to differences in its chemistry with hydrogen as compared to that of $RhCl(PPh_3)_3$, a frequently used homogeneous hydrogenation catalyst in nonaqueous solution. Thus, whereas $RhCl(PPh_3)_3$ reacts with hydrogen to give $RhH_2Cl(PPh_3)_3$, the complex $RhCl(TPPMS)_3$ reacts to give $RhH(TPPMS)_3$ (Eq. 29) (88). A

$$RhCl(TPPMS)_3 + H_2 \rightleftharpoons RhH(TPPMS)_3 + HCl \tag{29}$$

possible pathway for the formation of $RhH(TPPMS)_3$ involves the sequence of reactions in Scheme 12, where the final hydride complex is formed by reaction of hydrogen with the hydroxy complex. The predominance of $RhH(TPPMS)_3$ can cause significant changes in the hydrogenation mechanism. For a monohydride the final reduction step cannot involve an intramolecular reaction of an alkylhydride intermediate. Instead, the alkyl intermediate must react with either hydrogen, a molecule of $RhH(TPMS)_3$, or a proton. Support for the protonation route comes from the observation of extensive deuterium incorporation from the solvent D_2O in the final alkane (89). Nevertheless, this observation does not verify such a pathway, since $H\backslash D$ exchange between the hydride ligand in $RhH(TPPMS)_3$ and D^+ may also be facile.

Aromatic and aliphatic aldehydes can be reduced to the corresponding alcohols by hydrogen transfer from formate using TPPMS complexes of

$$RhCl(TPPMS)_3 + H_2O \rightleftharpoons Rh(OH)(TPPMS)_3 + HCl$$

$$Rh(OH)(TPPMS)_3 + H_2 \rightleftharpoons RhH(TPPMS)_3 + H_2O$$

SCHEME 12.

TABLE II

CATALYTIC REDUCTION OF BENZALDEHYDE BY FORMATE AT 80°C

Catalyst	Time (min)	Conversion (%)
$RuCl_2(TPPMS)_2$	20	100
$RuHCl(CO)(TPPMS)_3$	60	80
$RuCl_2(CO)_2(TPPMS)_2$	60	0
$IrCl(CO)(TPPMS)_3$	130	11
$RhCl(TPPMS)_3$	72	22

Ru(II), Ru(I) and Ir(I) in an aqueous medium (90). Data for the different complexes are shown in Table II. The highest conversion is observed with the ruthenium(II) complexes, which correlates with the facility of Ru complexes to catalyze other hydrogen transfer reactions. The complex $PdCl_2(TPPMS)_2$ has been used as a two-phase aqueous-organic catalyst for the carbonylation of allylic chlorides (Eq. 30) (91). The reaction pro-

$$\text{Cl} + CO + ROH \longrightarrow \text{CO}_2H + HCl \qquad (30)$$

$$(R = H, alkyl)$$

ceeds smoothly under a 1-atm pressure of CO. A similar reaction of allyl chlorides and acetates with formate (Eq. 31) is also catalyzed by

$$C_6H_{13}\text{—Cl} + HCOONa \longrightarrow C_7H_{15} + C_6H_{13} \qquad (31)$$

$PdCl_2(TPPMS)_2$ (92). In this system the water-soluble complex transports the substrate into the aqueous phase and causes it to be reduced by sodium formate. The product is a mixture of the 1- and the 2-alkene.

2. *TPPTS Complexes as Hydroformylation Catalysts*

More extensive use has been made of TPPTS than of TPPMS as a ligand for preparing water-soluble homogeneous catalysts. The major reason for this is that the presence of three sulfonate groups on TPPTS causes it to have a greater solubility in aqueous solution than does TPPMS. The principal application of TPPTS has been as a ligand for rhodium in catalyzed hydroformylation reactions. In the hydroformylation of propene with such catalyst systems, the reaction conditions use an equimolar mixture of CO and H_2 at 40-bar pressure and 125°C in an aqueous solution of pH 6.0

($40,93$). Excess TPPTS ligand stabilizes the system against decomposition (94). By comparison, this catalyst system is less susceptible to poisoning than is its rhodium triphenylphosphine analog. The yield of butanals is 99%, with the relative amounts of linear isomer to branched isomer being in the ratio of $94:6$ (95). This selectivity ratio can be increased to $97:3$ by using the sulfonated ligand shown in Fig. 4 (96). Extremely high normal : branched selectivities are also shown for the binuclear rhodium complex $[Rh(\mu\text{-SR})(CO)TPPTS]_2$. Thus, for 1-hexene, a linear : branched aldehyde ratio of $96:4$ is observed ($97–101$). This highest selectivity ratio is observed when R = $tert$-Bu, and falls to $92:8$ when R = C_6F_5 for the series of alkyl groups: R = $tert$-Bu, CH_2Ph, Me, Ph, C_6F_5. Initially, it was believed that a cooperativity effect between the two Rh centers was playing an important role, that is, one of the steps in the hydroformylation pathway was being facilitated by the presence of a second metal center. Since for hydroformylation catalysts in nonaqueous solvent the final elimination of aldehyde involves an intermolecular reaction between rhodium acyl and hydride intermediates, there was a realistic possibility that it is this step that is accelerated by the presence of a proximal rhodium center. A problem with interpreting these results with both 1-hexene and 1-octene, however, is that under hydroformylation conditions in the presence of water the binuclear complex $[Rh(\mu\text{-SBu}^t)(TPPTS)]_2$ is converted into $RhH(CO)(TPPTS)_3$. Any $[Rh(\mu\text{-SBu}^t)(CO)_2]_2$ that is not converted into $RhH(CO)(TPPTS)_3$ migrates into the organic phase. The introduction of the water-soluble thiol $HS(CH_2)_3NMe_2$ gives the ionic complex $[Rh(\mu\text{-}S(CH_2)_3NHMe_2)(CO)(TPPTS)]_2^{2+}$ that does remain in the aqueous phase. This complex has a low activity level as a homogeneous hydroformylation catalyst (102).

The complex $PtCl_2(TPPTS)_2$, when immobilized by controlled-pore glass, has been used in conjunction with $SnCl_2$ as a hydroformyla-

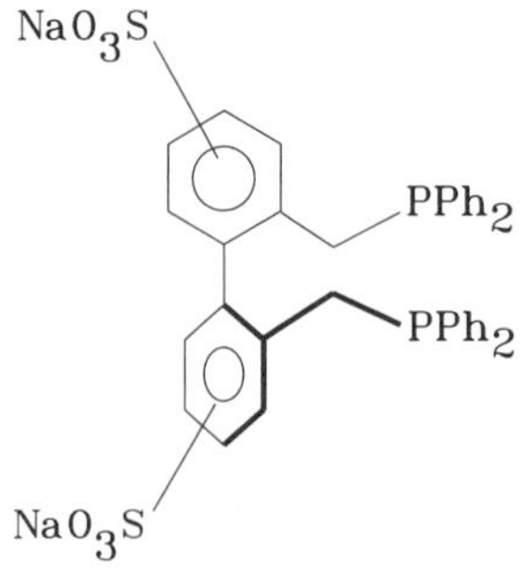

FIG. 4.

tion catalyst for 1-hexene (103). The catalytically active species is $PtCl(SnCl_3)(TPPTS)_2$, which is adsorbed in the presence of a small amount of water. A normal : branched aldehyde ratio of 11.5 : 1 has been achieved for this hydroformylation when the catalytic reaction is carried out at a temperature of 100°C and a hydrogen pressure of 1000 psig.

Hydroformylations can be carried out in aqueous solution in the absence of hydrogen. Such a situation arises because these water-soluble phosphine-substituted rhodium complexes also catalyze the water-gas shift reaction (Eq. 32). Since carbon monoxide is a reagent in both the hydrofor-

$$CO + H_2O \rightleftharpoons CO_2 + H_2 \tag{32}$$

mylation and the water-gas shift reaction, the initial involvement of CO in the water-gas shift reaction leads to the formation of hydrogen, which is then consumed along with CO in the subsequent hydroformylation step (100).

A comparison of homogeneous rhodium hydroformylation catalysts having either TPPTS or PPh_3 as supporting ligands shows that the hydrophilic TPPTS gives higher linear : branched isomer ratios but lower overall activities than does PPh_3 (104). The linear : branched isomer ratio reflects the relative amounts of linear : branched alkyl and acyl complexes that are formed in the insertion steps (Scheme 13). Higher steric requirements favor the linear isomer. Under reaction conditions using $RhH(CO)L_3$ (L = PPh_3, TPPTS) and a 200-atm pressure of CO and H_2 in a 1 : 1 ratio, one difference between the two complexes is that with triphenylphosphine the substitution product $RhH(CO)_2(PPh_3)_2$ is formed, whereas with TPPTS the analog complex $RhH(CO)_2(TPPTS)_2$ is not observed. This difference in the chemistries is reflected in the relative activation energies for phosphine dissociation, which is 19 kcal mol^{-1} for PPh_3 and 30 kcal mol^{-1} for TPPTS.

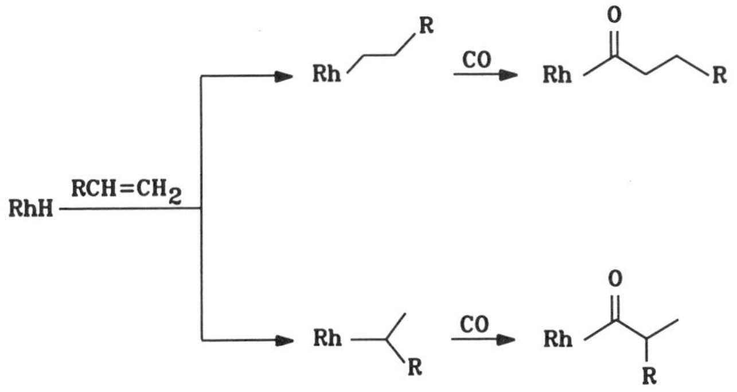

SCHEME 13.

Since these complexes are precursors to the coordinately unsaturated intermediate $RhH(CO)_2L$ (L = PPh_3, TPPTS), the presence in solution of complexes having a higher phosphine : carbon monoxide ratio for TPPTS than for PPh_3, along with the greater cone angle for TPPTS, may be the cause of the higher linear : branched ratios being found for the TPPTS rhodium hydroformylation system. In general, the SO_3^- substituents on TPPMS and TPPTS have only minimal electronic influence on the metal–phosphorus bond (105–107). The steric differences between TPPTS and PPh_3, however, are significant and can cause differences in the chemistry (33,34).

3. TPTS Complexes as Hydrogenation Catalysts

Water-soluble phosphine complexes of the late transition metals can also be used as catalysts for the hydrogenation of C=C and C=O double bonds. Thus the complex $RhCl(TPPTS)_3$ is an effective catalyst in aqueous solution for the hydrogenation of alkenes. A problem with using this complex, however, is that the TPPTS ligand is oxidized more readily than is PPh_3 in $RhCl(PPh_3)_3$. This situation occurs because Rh(III) complexes catalyze the oxygen transfer reaction from water to TPPTS to give the oxide OTPPTS (108). This observation has been confirmed by the use of $^{18}OH_2$. Nevertheless, both the Rh(I) and the Ru(II) complexes of TPPTS have been used as catalysts for the homogeneous hydrogenation of the unsaturated aldehydes RCH=CHCHO (109). These homogeneous catalysts can be used for the selective reduction of α,β-unsaturated aldehydes to unsaturated alcohols (Eq. 33).

$$\text{(33)}$$

The catalytic hydrogenation of propionaldehyde in water also occurs with the series of complexes $[RuCl_2L_2]_2$, $RuHClL_3$, $RuH(OAc)L_3$, RuH_2L_4, $RuHIL_3$, $RuCl_2(CO)_2L_2$, and $[Ru(OAc)(CO)_2L]_2$ (L = TPPTS) (110). The noncarbonylated complexes in this group all show comparable hydrogenation rates at 100°C. In all cases, a series of equilibria lead to the formation of $RuH_2L_3(H_2O)$, which suggests that RuH_2L_3 may be the catalytically active species.

The carbonylated complexes are less active as catalysts. The addition of alkali or alkaline earth metal ions or NH_4^+ dramatically increases the activity without any loss of selectivity. For cations the rate increases in the order $NR_4^+ < Na^+ < Li^+ < K^+ < Mg^{2+} < Ca^{2+}$, and for anions the rate increases in the order $SiF_6^{2-} < NO_3^- < Cl^- < Br^- < I^-$. The cation effect is believed to be due to stabilization of the C-bonded alkoxide intermediate by the binding of the cation to the β-oxygen (Eq. 34).

$$\text{(34)}$$

Other water-soluble tertiary phosphines can be used as ligands for the synthesis of hydrogenation catalysts. An example of such a phosphine is 1,3,5-triaza-7-phosphaadamantane, which has been prepared by reacting tris(hydroxymethyl) phosphine, formaldehyde, and hexamethylenetetramine (Eq. 35) (*111,112*). The ruthenium(II) complex of 1,3,5-triaza-7 phos-

$$\text{(35)}$$

phaadamantane is an effective catalyst for the catalytic hydrogenation of alkenes and aldehydes (*113*), and for the conversion of unsaturated aldehydes into alcohols using sodium formate as the source of hydrogen (*114*).

Rhodium complexes of the (2-diphenylphosphinoethyl)trimethylammonium cation have been used as both hydroformylation and hydrogenation catalysts (*41*). An advantage of these complexes is that they are soluble in water, and also that the ligand periphery has a cationic alkylammonium group. These two properties allow the complexes to be absorbed onto an ion exchange resin. Virtually no leaching of rhodium into the alkene-containing phases occurs with these catalysts, although the rhodium can ultimately be recovered from the resin by elution with acid (*115,116*).

E. Asymmetric Catalysis with Tertiary Phosphine Complexes

Chiral-sulfonated phosphines have also been synthesized and used in the preparation of water-soluble late transition-metal complexes. These complexes, like their analogs in nonaqueous solution, can be used as homogeneous catalysts for the hydrogenation of prochiral alkenes. When the phosphine ligand has a chiral center, the alkane product can potentially be obtained in enantiomeric excess. A variety of different approaches have been used to prepare chiral phosphines that are soluble in aqueous

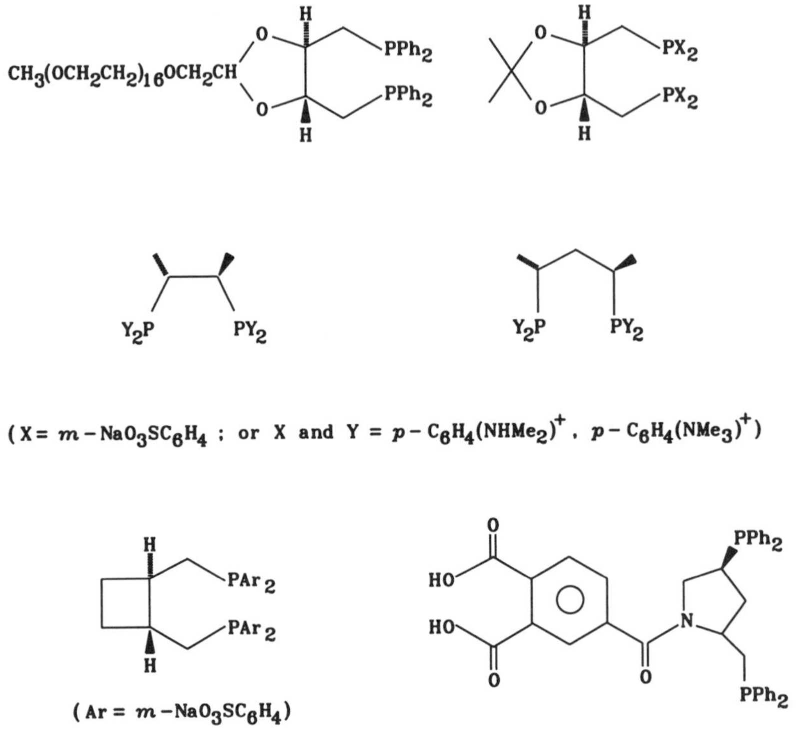

CH3(OCH2CH2)16OCH2CH
PPh2
PPh2
PX2
PX2
Y2P
PY2
(X = m−NaO3SC6H4 ; or X and Y = p−C6H4(NHMe2)+ , p−C6H4(NMe3)+)
PAr2
PAr2
(Ar = m−NaO3SC6H4)
HO
HO
PPh2
PPh2

FIG. 5.

solution. As with other homogeneous hydrogenations in aqueous solution, the enantioselective hydrogenation of prochiral alkenes frequently is slow. Also, the optical yields as measured by the enantiomeric excess (ee) in the chiral alkane often are low. A series of phosphines that have been used in the synthesis of enantioselective rhodium hydrogenation catalysts are shown in Fig. 5 (*45,117–121*). In the absence of hydrogen as a reagent, the enantioselective reduction of alkenecarboxylic acids can be carried out with formate acting as the hydrogen transfer substrate (*122*).

The low enantioselectivity of chiral hydrogenation catalysts in aqueous solution has been considered to be a consequence of the effect of solvent on the key step that determines the chirality in the product alkane. For homogeneous rhodium catalysts this step is the addition of hydrogen to rhodium(I) to give a rhodium(III) species (*123*). A large difference between the transition-state energies for this hydrogen addition to the pair of diastereomeric Rh(I) complexes leads to the observation of a large ee in the product (Fig. 6). The low ee values found in aqueous solution can therefore

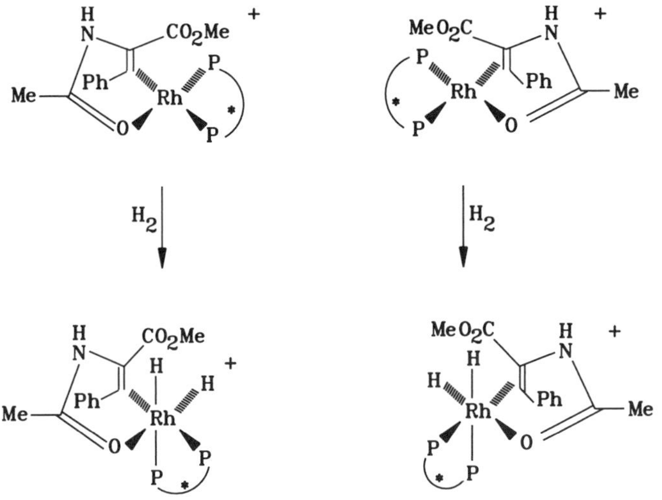

FIG. 6.

result from a small difference in these transition-state energies (124). This premise is supported by considerations that relate the observed ee with the hydrophobic and hydrophilic nature of the solvent ($125,126$).

Such generalizations may, however, be premature, since a comparison has also been made of the ee obtained from the hydrogenation of 2-acetamidoacrylic acid with the rhodium complexes of R-Binap(SO_3Na)$_4$ and its unsulfonated analog. The ee's obtained in ethanol for the R-Binap(SO_3Na)$_4$ complex ($\sim69\%$) are closely comparable with those obtained for the rhodium complex of the unsulfonated R-Binap (67.0%) (127). Further evidence against such a generalization in the values of the ee's obtained in aqueous solution is also found in the observation that rhodium(I) catalysts obtained with sulfonated diphosphines yield ee's of up to 88% in the hydrogenation of carbon–carbon double bonds in a two-phase system with both an aqueous and an organic solvent system (128).

F. Supramolecular-Medium Effects in Catalytic Reactions

Catalytic reactions in aqueous solution can be modified by changes in the supramolecular state of the medium in which they are conducted. One approach to using medium changes to influence catalytic reactions is to introduce host molecules, such as cyclodextrins, into the reaction mixtures. Cyclodextrins have inner lipophilic cavities enclosed by hydrophilic surfaces. A cyclodextrin can bind a hydrophilic terminal alkene into its cavity, where the alkyl chain is encapsulated and the alkene functionality is free. Such a cavity will preferentially encapsulate "rodlike" molecules; therefore, a cyclodextrin host incorporated into a hydroformylation catalyst will favor linear aldehyde formation (129).

Metal complexes of TPPMS and TPPTS have amphophilic character because of the presence of both hydrophilic sulfonate groups and hydrophobic phenyl groups in the ligand structure. This feature allows the complex to transfer readily between the aqueous and organic phases in a biphasic system. Furthermore, these complexes can aggregate to form micelles, or surface-active compounds. This property may be a particularly important one when the properties and selectivities of catalysts formed by such phosphines are being considered (130).

Supported aqueous-phase catalysts can also be used to advantage. These supported catalysts have a thin aqueous film adhering to silica gel that contains the water-soluble complex (131). These catalysts are particularly useful for the hydroformylation of substrates such as oleyl alcohol (132). Since these catalytic reactions occur at the phase boundary, characteristics such as the water content can cause changes both in the reactivity and in the linear : branched chain ratio of the product aldehyde.

G. *Other Catalyzed Reactions*

Water-soluble palladium(0) complexes have also been used as homogeneous catalysts in aqueous-solution alkylation reactions. The particular complex that has been used is $Pd(TPPMS)_3$. Aryl or heteroaromatic halides can be coupled with aryl or vinyl boronic acids, alkynes, alkenes, or dialkyl phosphites with this palladium(0) complex. This complex in aqueous solution can also be used for the coupling of alkynes with unprotected iodonucleotides, iodonucleosides, and iodoamino acids (133).

Palladium(II) complexes in aqueous THF have been used as catalysts for the addition of water to the alkene diethyl maleate to give the alcohol diethyl malate. The two catalytic systems that have been used for this reaction are the dimer $[Pd(\mu\text{-OH})dppe]_2$ and a catalyst mixture composed of $PdCl_4^{2-}$ and $CuCl_2$ (134). The proposed pathway for the $[Pd(\mu\text{-OH})dppe]_2$ catalyzed reaction is that shown in Fig. 7. Any free dppe in the reaction mixture catalyzes the isomerization of diethyl maleate to diethyl fumarate. A second product of this catalytic reaction that is formed in large amounts with a $PdCl_4^{2-}$ and $CuCl_2$ catalyst system is the ketoester diethyl oxaloacetate. The distribution between diethyl maleate and diethyl oxaloacetate depends on the competition for an intermediate alkylpalladium(II) complex between reaction with water to give alcohol and an intramolecular β-elimination reaction to give the ketone (Fig. 8).

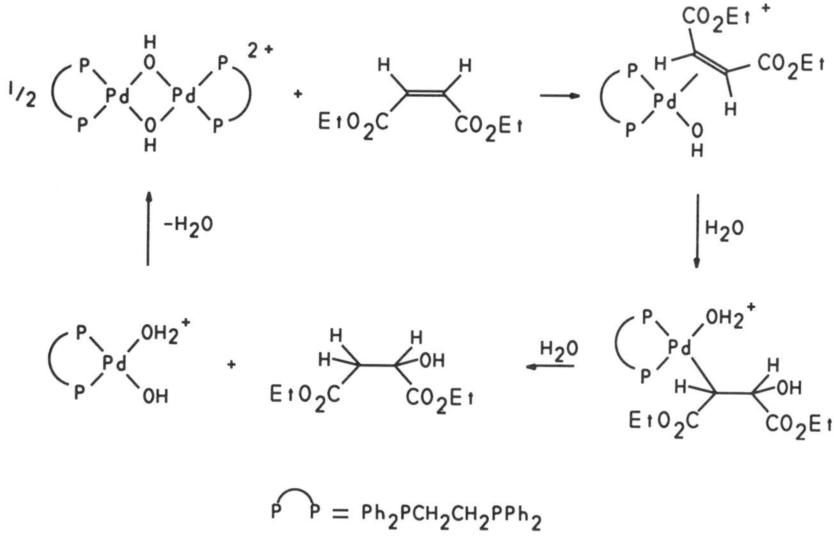

$$\stackrel{\frown}{P \quad P} = Ph_2PCH_2CH_2PPh_2$$

Fig. 7.

The water-soluble complex cis-PtCl$_2$(TPPTS)$_2$ has been used as a catalyst for the 1,4-addition of diethylamine to isoprene (Eq. 36). Water-

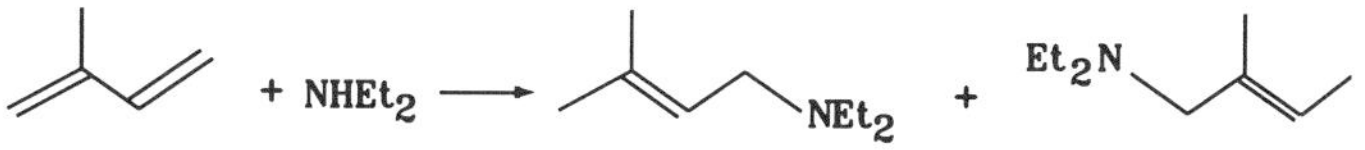

(36)

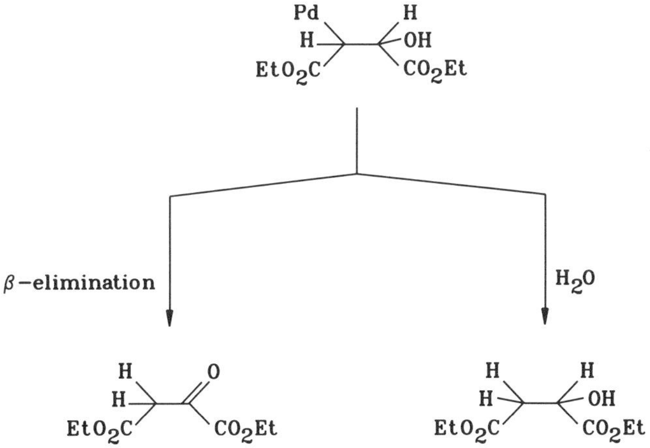

Fig. 8.

soluble rhodium complexes of TPPTS in a biphasic mixture of water and triethylamine are active catalysts for the hydrogenation of carbon dioxide to formic acid (Eq. 37) (135). The amine functions in this system to stabilize

$$H_2 + CO \rightleftharpoons HCO_2H \tag{37}$$

the formic acid that is formed. Metal complexes of 2,2-bipyridine-5 sulfonic acid have been prepared as possible catalysts for the water-gas shift reaction (136).

The water-soluble zerovalent complexes $M\{P(CH_2OH)_3\}_4$ (M = Ni, Pd, Pt) have been used as homogeneous catalysts for the addition of PH_3 to CH_2O to give $P(CH_2OH)_3$. No mechanistic study of the reaction has been carried out. But since aqueous solutions of $Pt\{P(CH_2OH)_3\}_4$ are in equilibrium with the hydride complex $[PtH\{P(CH_2OH)_3\}_4]OH$, it is possible that $Pt\{P(CH_2OH)_3\}_4$ is sufficiently basic to be protonated by PH_3 in one of the early steps in the catalytic cycle (137,138).

Supercritical water has the potential to be an interesting solvent system for organotransition metal systems. An example of this potential is found with the complex $(\pi\text{-}^5cp)Co(CO)_2$, which in organic solvents catalyzes the cyclotrimerization of hexyne-1, phenylacetylene, and butyne-2. At 140°C in aqueous media these reactions show poor selectivity to benzenes, but at 374°C in supercritical water the selectivity is again high (139).

REFERENCES

(1) Brown, D. G. Prog. Inorg. Chem. 1973, 18, 177.
(2) Craig, P. J. (Ed.). "Organometallic Compounds in the Environment"; Wiley: New York, 1986.
(3) Barton, M.; Atwood, J. D. J. Coord. Chem. 1991, 24, 43.
(4) Hermann, W. A.; Kohlpaintner, C. W. Angew. Chem., Int. Ed. Engl. 1993, 32, 1524.
(5) Kochi, J. K. "Organometallic Mechanisms and Catalysis"; Academic Press: New York, 1978; p. 139.
(6) Espenson, J. H. Acc. Chem. Res. 1992, 25, 222.
(7) van Eldik, R.; Gaede, W.; Cohen, H.; Meyerstein, D. Inorg. Chem. 1992, 31, 3695.
(8) Bakac, A.; Butkovic, V.; Espenson, J. H.; Orhanovic, M. Inorg. Chem. 1993, 32, 5886.
(9) Kochi, J. K.; Powers, J. W. J. Am. Chem. Soc. 1970, 92, 137.
(10) Bushey, W. R.; Espenson, J. H. Inorg. Chem. 1977, 16, 2772.
(11) Nohr, R. S.; Espenson, J. E. J. Am. Chem. Soc. 1975, 97, 3392.
(12) Meyerstein, D. Pure Appl. Chem. 1982, 61, 885.
(13) Halpern, J. Acct. Chem. Res. 1983, 15, 238.
(14) Kirker, G. W.; Bakac, A.; Espenson, J. M. J. Am. Chem. Soc. 1982, 104, 1249.
(15) Zhang, Z.; Jordan, R. B. Inorg. Chem. 1993, 32, 5472.
(16) Rotman, A.; Cohen, H.; Meyerstein, D. Inorg. Chem. 1985, 24, 4158.
(17) Cohen, H.; Meyerstein, D. Inorg. Chem. 1984, 23, 84.

(18) Cohen, H.; Gaede, W.; Gerhard, A.; Meyerstein, D.; van Eldik, R. *Inorg. Chem.* **1992**, *31*, 3805.

(19) Gaede, W.; Gerhard, A.; van Eldik, R.; Cohen, H.; Meyerstein, D. *J. Chem. Soc., Dalton Trans.* **1993**, 2065.

(20) Cohen, A.; Feldman, A.; Ish-Shalom, R.; Meyerstein, D. *J. Am. Chem. Soc.* **1991**, *113*, 5292.

(21) Cohen, H.; Meyerstein, D. *J. Chem. Soc., Faraday Trans. 1* **1988**, *84*, 4157.

(22) Sorek, Y.; Cohen, H.; Meyerstein, D. *J. Chem. Soc. Faraday Trans. 1* **1989**, *85*, 1169.

(23) Kirker, G. W.; Bakac, A.; Espenson, J. H. *J. Am. Chem. Soc.* **1982**, *104*, 1249.

(24) Ryan, D. A.; Espenson, J. H. *J. Am. Chem. Soc.* **1982**, *104*, 704.

(25) Chock, P. B.; Halpern, J. *J. Am. Chem. Soc.* **1969**, *91*, 582.

(26) Blaser, H.-U.; Halpern, J. *J. Am. Chem. Soc.* **1980**, *102*, 1684.

(27) Halpern, J.; Phelan, P. F. *J. Am. Chem. Soc.* **1972**, *94*, 1881.

(28) Meyerstein, D.; Schwarz, H. A. *J. Chem. Soc. Faraday Trans. 1* **1988**, *84*, 2933.

(29) van Eldik, R.; Cohen, H.; Meyerstein, D. *Angew. Chem. Int. Ed. Engl.* **1991**, *30*, 1158.

(30) Roundhill, D. M.; Gray, H.B.; Che, C.-M. *Acc. Chem. Res.* **1989**, *22*, 55.

(31) Avey, A.; Tenhaeff, S. C.; Weakley, T. J. R.; Tyler, D. R. *Organometallics* **1991**, *10*, 3607.

(32) Avey, A.; Tyler, D. R. *Organometallics* **1992**, *11*, 3856.

(33) Darensbourg, D. J.; Bischoff, C. J.; Reibenspies, J. M. *Inorg. Chem.* **1991**, *30*, 1144.

(34) Darensbourg, D. J.; Bischoff, C. J. *Inorg. Chem.* **1993**, *32*, 47.

(35) Roundhill, D. M.; Sperline, R. P.; Beaulieu, W. B. *Coord. Chem. Rev.* **1978**, *26*, 263.

(36) Landon, S.; Brill, T. B. *Chem. Rev.* **1984**, *84*, 577.

(37) Barton, M.; Atwood, J. D. *J. Coord. Chem.* **1991**, *24*, 43.

(38) Herrmann, W. A.; Kohlpaintner, C. W. *Angew. Chem., Int. Ed. Engl.* **1993**, *32*, 1524.

(39) Kalck, P.; Monteil, F. *Adv. Organomet. Chem.* **1992**, *34*, 219.

(40) Kuntz, E.G. *CHEMTECH* **1987**, *17*, 570.

(41) Smith, R. T.; Baird, M. C. *Inorg. Chim. Acta* **1982**, *62*, 135.

(42) Ganguly, S.; Mague, J. T.; Roundhill, D. M. *Inorg. Chem.* **1992**, *31*, 3500.

(43) Murray, R. E.; Wenzel, T. T. *Prepr. Div. Petrol. Chem., Am. Chem. Soc. Meet.* Miami, **1989**, *34*, 599.

(44) Hoye, P. A. T.; Pringle, P. G.; Smith, M. B., Worboys, K. *J. Chem. Soc. Dalton Trans.* **1993**, 269.

(45) Nuzzo, R. G.; Haynie, S. L.; Wilson, M. E.; Whitesides, G. M. *J. Org. Chem.* **1981**, *46*, 2861.

(46) Avey, A.; Schut, D. M.; Weakley, T. J. R., Tyler, D. R. *Inorg. Chem.* **1993**, *32*, 233.

(47) Bartik, T.; Bartik, B.; Hanson, B. E.; Guo, I.; Toth, I. *Organometallics* **1993**, *12*, 164.

(48) Webster, D. E. *Adv. Organomet. Chem.* **1977**, *15*, 147.

(49) Garnett, J. L.; Hodges, R. J. *J. Am. Chem. Soc.* **1967**, *87*, 4546.

(50) Hodges, R. J., Garnett, J. L. *J. Phys. Chem.* **1968**, *72*, 1673.

(51) Hodges, R. J.; Garnett, J. L. *J. Catal.* **1969**, *13*, 83.

(52) Garnett, J. L. *Catal. Rev.* **1972**, *5*, 229.

(53) Blackett, L.; Gold, V.; Reuben, D. M. E. *J. Chem. Soc., Perkin Trans. 2* **1974**, 1869.

(54) Gol'dshleger, N. F.; Tyabin, M. B.; Shilov, A. E.; Shteinman, A. A. *Zh. Fiz. Khim.* **1969**, *43*, 2174.

(55) Hodges, R. J.; Webster, D. E.; Wells, P.B. *J. Chem. Soc. A.* **1971**, 3230.

(56) Gol'dshleger, N. F.; Moiseev, I. I.; Khidekel, M. L.; Shteinman, A. A. *Dokl. Akad Nauk SSSR* **1972**, *206*, 106.

(57) Shilov, A. E. "Activation of Saturated Hydrocarbons by Transition Metal Complexes"; Reidel: Dordrecht, The Netherlands, 1984.

(58) Crabtree, R. H. *Chem. Rev.* **1985,** *85,* 245.
(59) Hodges, R. J.; Garnett, J. L. *J. Phys. Chem.* **1968,** *72,* 1673.
(60) Gol'dsleger, N. G.; Eskova, V. A.; Shilov, A. E.; Shteinman, A. A. *Zh. Fiz. Khim.* **1972,** *46,* 1353.
(61) Saunders, J. R.; Webster, D. E.; Wells, P. B. *J. Chem. Soc., Dalton Trans.* **1975,** 1191.
(62) Eskova, V. V.; Shilov, A. E.; Shteinman, A. A. *Kinet. Katal.* **1972,** *13,* 534.
(63) Gol'dsleger, N. F.; Lavrushko, V. V.; Krushch, A. P.; Shteinman, A. A. *Izv. Akad. Nauk SSSR, Ser. Khim.* **1976,** 2174.
(64) Trateyakov, V. P.: Zimtseva, C. P.: Rudakov, E.S.; Osetskii, A. N. *React. Catal. Lett.* **1979,** *12,* 543.
(65) Trateyakov, V. P.; Rudakov, E. S.: Bogdanov, A. V.; Zimtseva, G. P.; Kozhevina, L. I., *Dokl. Akad. Nauk SSSR* **1979,** *249,* 878.
(66) Khrushch, L. A.; Lavrushko, V. V.; Misharin, Y. S.; Moravsky, A. P.; Shilov, A. E. *Nouv. J. Chim.* **1983,** *7,* 729.
(67) Shulpin, G. B.; Nizova, G. V.; Shilov, A. E. *J. Chem. Soc., Chem. Commun.* **1983,** 761.
(68) D. M. Roundhill, Unpublished observations.
(69) Rich, R. L.; Taube, H. *J. Am. Chem. Soc.* **1954,** *76,* 2608.
(70) Labinger, J. A.; Herring, A. M.; Bercaw, J. E. *J. Am. Chem. Soc.* **1990,** *112,* 5628.
(71) Sen, A.; Lin, M. *J. Chem. Soc., Chem. Commun.* **1992,** 508.
(72) Sen, A.; Lin, M.; Kao, L.-C., Hutson, A. C. *J. Am. Chem. Soc.* **1992,** *114,* 6385.
(73) Sen, A. *Platinum Met. Rev.* **1991,** *35,* 126.
(74) Gretz, E.; Oliver, T. F.; Sen, A. *J. Am. Chem. Soc.* **1987,** *109,* 8109.
(75) Sen, A.; Gretz, E.; Oliver, T. F.; Jiang, Z. *New J. Chem.* **1989,** *13,* 755.
(76) Kao, L.-C.; Hutson, A. C.; Sen, A. *J. Am. Chem. Soc.* **1991,** *113,* 700.
(77) Bumagin, N. A.; Nikitin, K. V.; Beletskaya, I. P. *J. Organomet. Chem.* **1988,** *358,* 563.
(78) Novak, B. M.; Grubbs, R. H. *J. Am. Chem. Soc.* **1988,** *110,* 7542.
(79) Crabtree, R. H. "The Organometallic Chemistry of the Transition Metals"; Wiley: New York, 1987; pp. 173–176.
(80) Elschenbroich, C.; Salzer, A. "Organometallics"; VCH: New York, 1989; pp. 425–427.
(81) James, B. R. "Homogeneous Hydrogenation"; Wiley: New York, 1973.
(82) De Vries, B. *J. Catal.* **1962** *1,* 489.
(83) Halpern, J.; Pribanic, M. *Inorg. Chem.* **1970,** *9,* 2616.
(84) Halpern, J.; Wong, L. Y. *J. Am. Chem. Soc.* **1968,** *90,* 6665.
(85) Herrmann, W. A.; Kulpe, J. A.; Kellner, J.; Riepl, H.; Bahrmann, H.; Konkol, W. *Angew Chem., Int. Ed. Engl.* **1990,** *29,* 391.
(86) Borowski, A. F.; Cole-Hamilton, D. J.; Wilkinson, G. *Nouv. J. Chem.* **1978,** *2,* 137.
(87) Joo, F.; Somsak, L.; Beck, M. T. *J. Mol. Catal.* **1984,** *24,* 71.
(88) Joo, F.; Csiba, P.; Benyei, A. *J. Chem. Soc., Chem Commun.* **1993,** 1602.
(89) Laghmari, M.; Sinou, D. *J. Mol. Catal.* **1991,** *66,* L15.
(90) Bengei, A.; Joo, F. *J. Mol. Catal.* **1980,** *58,* 151.
(91) Kiji, J.; Okano, T.; Nishuiri, W.; Konishi, H. *Chem. Lett.* **1988,** 957.
(92) Okano, T.; Moriyama, Y.; Konishi, H.; Kiji, J. *Chem. Lett.* **1986,** 1463.
(93) Cornils, B.; Hibbel, J.; Konkol, W.; Lieder, B.; Much, J.; Schmidt, V.; Wiebus, E. Eur. Pat. 0103810, 09.7, 1983.
(94) Rhone-Poulenc, Fr. Pat 2478 078 03.12, 1980.
(95) Kuntz, E. J. Fr. Patent 2314910, 20.06, 1975.
(96) Herrmann, W. A.; Kohlpainter, C. W.; Bahrmann, H.; Konkol, W. *J. Mol. Catal.* **1992,** *73,* 191.

(97) Escaffre, P.; Thorez, A.; Kalck, P. *J. Chem. Soc., Chem. Commun.* **1987,** 146.

(98) Escaffre, P.; Thorez, A.; Kalck, P. *New J. Chem.* **1987,** *11,* 601.

(99) Kalck, P.; Escaffre, P.; Serein Spirau, F.; Thorez, A. *New J. Chem.* **1988,** *12,* 687.

(100) Kalck, P. *Pure Appl. Chem.* **1989,** *61,* 967.

(101) Kalck, P. *Polyhedron* **1988,** *7,* 2441.

(102) Monteil, F.; Queau, R.; Kalck, R. *J. Organomet. Chem.* **1994,** *480,* 177.

(103) Guo, I.; Hanson, B. E.; Toth, I. *J. Mol. Catal.* **1991,** *70,* 363.

(104) Horvath, I. T.; Kastrup, R. V.; Oswald, A. A.; Mozeleski, E. J. *Catal. Lett.* **1989,** *2,* 85.

(105) Horvath, I.; Kastrup, R.; Oswald, A.; Mozeleski, E. *Catal. Lett.* **1989,** *2,* 85.

(106) Herrmann, W. A.; Kellner, J.; Riepl, M. *J. Organomet. Chem.* **1990,** *389,* 103.

(107) Larpent, C.; Patin, M. *Appl. Organomet. Chem.* **1987,** *1,* 529.

(108) Larpent, C.; Dabard, R.; Patin, H. *Inorg. Chem.* **1987,** *26,* 2922.

(109) Grosselin, J. M.; Mercier, C.; Allmang, C.; Graes, F. *Organometallics* **1991,** *10,* 2126.

(110) Fache, E.; Santini, C. Senocq, F.; Basset, J. M. *J. Mol. Catal.* **1992,** *72,* 331, 337.

(111) Darensbourg, M. Y.; Daigle, D. *Inorg. Chem.* **1975,** *14,* 1217.

(112) Fisher, K. J.; Alyea, E. C.; Shahnazarian, N. *Phosphorus, Sulfur, Silicon Relat. Elem.* **1990,** *48,* 37.

(113) Darensbourg, D. J.; Joo, F.; Kannisto, M.; Katho, A.; Reibenspies, J. H. *Organometallics* **1992,** *11,* 1990.

(114) Darensbourg, D. J.; Joo, F.; Kannisto, M.; Katho, A.; Reibenspies, J. M.; Daigle, D. J. *Inorg. Chem.* **1994,** *33,* 200.

(115) Smith, R. T.; Ungar, R. K.; Baird, M. C. *Transition Met. Chem. (Weinheim, Ger.)* **1982,** *7,* 288.

(116) Smith, R. T.; Ungar, R. K.; Sanderson, L. J.; Baird, M.C. *Organometallics* **1983,** *2,* 1138.

(117) Amrani, Y.; Sinou, D. *J. Mol. Catal.* **1984,** *24,* 231.

(118) Toth, I.; Hanson, B. E. *Tetrahedron: Asymmetry* **1990** *1,* 895.

(119) Toth, I.; Hanson, B. E.; Davis, M. E. *Tetrahedron: Asymmetry* **1990,** *1,* 913.

(120) Laghmari, M.; Sinou, D. *J. Mol. Catal.* **1991** *66,* L15.

(121) Alario, F.; Amrani, Y. *J. Chem. Soc. Chem. Commun.* **1986,** 202.

(122) Sinou, D.; Safi, M.; Claver, C.; Masdeu, A. *J. Mol. Catal.* **1991,** *68,* L9.

(123) Chan, A. S. C.; Pluth, J. J.; Halpern, J. *J. Am. Chem. Soc.* **1986,** *102,* 5952.

(124) Amrani, Y.; Sinou, D. *J. Mol. Catal.* **1986,** *36,* 319.

(125) Lecomte, L.; Sinou, D.; Bakos, J.; Toth, I.; Heil, B. *J. Organomet. Chem.* **1989,** *370,* 277.

(126) Abraham, M. H.; Grellier, P. L.; McGill, R. A. *J. Chem. Soc., Perkin Trans. 2* **1988,** 339.

(127) Wan, K.; Davis, M. E. *J. Chem. Soc. Chem. Commun.* **1993,** 1262.

(128) Amari, Y.; Lecomte, L.; Sinou, D.; Bakos J.; Toth, I.; Heil, B. *Organometallics* **1989,** *8,* 542.

(129) Anderson, J. R.; Campi, E. M.; Jackson, W. R. *Catal. Lett* **1991,** *9,* 55.

(130) Hanson, B. E.; Ding. H. *46th Am. Chem. Soc. Reg. Meet.* Birmingham, AL, *1994,* INOR 212.

(131) Arhancet, J. P.; Davis, M. E.; Merola, J. S.; Hanson, B. E. *Nature (London)* **1989,** *339,* 454.

(132) Arhancet, J. P.; Davis, M. E.; Merola, J. S.; Hanson, B. E. *J. Catal.* **1990,** *121,* 327.

(133) Casalnuova A. L.; Calabrese, J. C. *J. Am. Chem. Soc.* **1990,** *112,* 4324.

(134) Ganguly, S.; Roundhill, D. M. *Organometallics* **1193,** *12,* 4825.

(135) Gassner, F.; Leitner, W. *J. Chem. Soc., Chem. Commun.* **1993,** 1465.
(136) Herrmann, W. A.; Thiel, W. R.; Kuchler, J. G.; Behm, J.; Herdtweck, *Chem. Ber.* **1990,** *123,* 1963.
(137) Harrison, K. N.; Hoye P. A. T.; Orpen, A. G.; Pringle, P. G.; Smith, M. B. *J. Chem. Soc., Chem. Commun.* **1989,** 1096.
(138) Ellis, J. W.; Harrison, K. N.; Hoye, P. A. T.; Orpen, A. G.; Pringle, P. G.; Smith, M. B. *Inorg. Chem.* **1992,** *31,* 3026.
(139) Jerome, K. S.; Parsons, E. J. *Organometallics* **1993,** *12,* 2991.

Structure/Property Relationships of Polystannanes

LAWRENCE R. SITA

Searle Chemistry Laboratory
Department of Chemistry
University of Chicago
Chicago, Illinois

I

INTRODUCTION

Due to proposals that adjacent d and p orbitals can overlap to form delocalized band-type orbitals either in the ground state or the excited state (1–4), there has been much interest in the synthesis and properties of structures whose skeletal frameworks consist of covalently bonded metal atoms. However, early attempts at the rational synthesis of oligomers and polymers containing a mix of d- and p-block metals were frustrated by the general lack of synthetic methodology that can be used to prepare metal–metal bonds under relatively mild conditions, and by the propensity of difunctional reagents to form cyclic rather than linear structures in potential polymerization processes (5). As a result, our knowledge concerning fully characterized examples of such catenated metal chains is still limited to a very small number of short oligomers containing only three to five metal atoms (5–8). In contrast, compounds whose molecular backbones consist entirely of σ-bonded Group 14 elements (Si, Ge, and Sn) have a rich history that extends over a wide range of structural diversity that now includes linear, cyclic, and polycyclic frameworks (9). Of these, high molecular weight polysilanes with the general formula X–$(R_2Si)_n$–X (R = alkyl or aryl) display a number of interesting phenomena, which include efficient photoconduction, large third-order nonlinear optical coefficients (X^3), reversible thermochromic behavior associated with structural phase transitions, and a high degree of radiation sensitivity that makes them suitable as photoresists for microlithography (10). Accordingly, more recent attention has focused on the development of refined models that can be used to reconcile these experimental observations with theory, with the concept of "σ-delocalization," as championed by Michl and others (11), gaining wide acceptance.

Although possessing intriguing properties, it can be argued that, in terms of localized molecular orbital pictures of electronic structure, polysilanes closely resemble their organic counterparts. On the other hand, polystannanes could be anticipated to display properties that are decidedly more "metal-like" in nature, possibly as a result of several factors that include, to name just a few, a greater reluctance to engage in formal sp hybridization, and a greater accessibility of d orbitals. A successful interpretation of the origins of these properties in terms of a unified molecular orbital theory, which takes into account the relativistic effects associated with heavy-atom systems, would then contribute greatly to our understanding of the molecular and electronic structure of compounds containing elements beyond the second row. Thus, it is the goal of this article to provide a comprehensive review of the extent to which theoretical and experimental investigations have probed the structure/property relationships of polystannanes. As provided in detail, this review will be limited to only those structures whose discrete skeletal frameworks consist entirely of σ-bonded tetracoordinate tin atoms (12).

II

DISTANNANES

Distannane, H_3Sn–SnH_3 (1), which can be considered a heavy-atom analog of ethane, is the simplest chemical structure that falls under the preceding definition for the class of compounds known as polystannanes.

Experimentally, distannane itself is highly reactive, and attempts to obtain and study it in pure form have so far all been thwarted (13). However, it has long been known that partial or full replacement of the hydrogen atoms in 1 with organic substituents, either alkyl or aryl, imparts stability to the system; and through this mechanism then, a large number of distannane derivatives of the general formula $R_3Sn–SnR_3$ have been prepared and their structures and properties investigated. As the discussion that follows reveals, this database of results concerning distannanes has proven to be an invaluable source of information regarding the nature of the Sn–Sn bonds in polystannane structures.

A. Synthesis

The general synthetic methodology that can be utilized for the construction of Sn–Sn bonds in molecular compounds is summarized in Scheme 1 (14). Of these procedures, it is important to point out that the hydrostannolysis reaction, which is thought to occur through metathetical exchange between an organotin hydride and an organotin amide according to Eq. (1), is a powerful tool for the synthesis of unsymmetric distannanes of

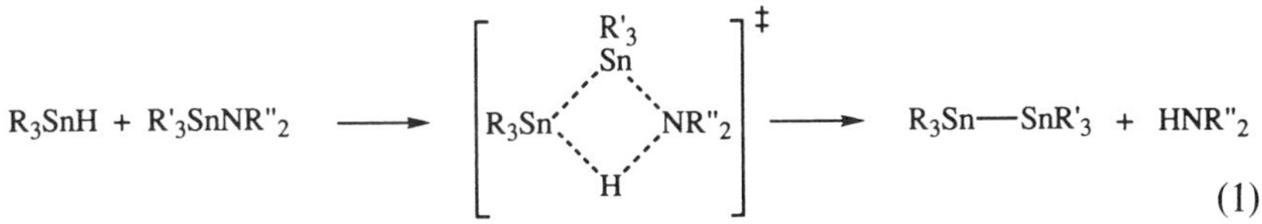

$$R_3SnH + R'_3SnNR''_2 \longrightarrow \left[R_3Sn \underset{H}{\overset{\overset{R'_3}{\underset{|}{Sn}}}{\diagup\diagdown}} NR''_2 \right]^{\ddagger} \longrightarrow R_3Sn-SnR'_3 + HNR''_2 \tag{1}$$

the general formula $R_3Sn–SnR'_3$ (15). For this class of distannane, the other methods typically provide statistical mixtures of all possible products. However, if the desired compound is a symmetric distannane, then it is often simplest to carry out either the Wurtz-type coupling of organotin halides with alkaline metals or the catalyzed dehydrogenative coupling of organotin hydrides according to Scheme 1. Both of these methods are

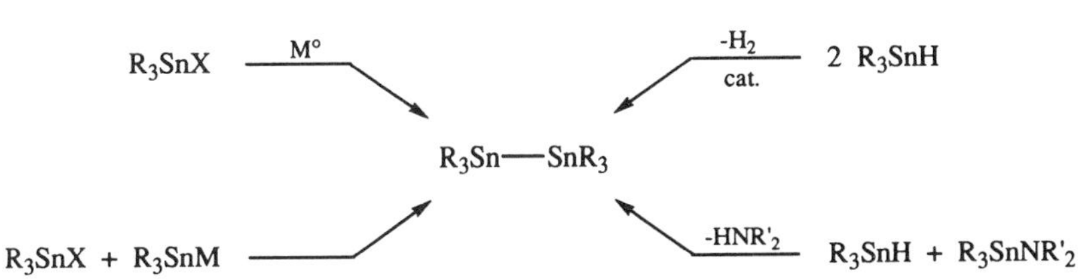

$$R_3SnX \xrightarrow{\ M^\circ\ } \searrow \qquad \swarrow \xrightarrow[\text{cat.}]{-H_2} 2\ R_3SnH$$

$$R_3Sn-SnR_3$$

$$R_3SnX + R_3SnM \longrightarrow \nearrow \qquad \searrow \xrightarrow{-HNR'_2} R_3SnH + R_3SnNR'_2$$

X = Cl, Br, I; M = Li, Na, K; cat. = pyridine, Pd, R$_3$N

SCHEME 1

also preferable for preparing distannanes that possess substituents that are sterically demanding. Finally, it is interesting to note that the dehydrogenative coupling reaction has been successfully used to prepare the first and, to date, only example of an optically active distannane, $[Ph(PhCMe_2CH_2)MeSn]_2$ (2), from the corresponding organotin hydride, which was prepared in a stereoselective manner and which is configurationally stable at the stereogenic tin center (16).

In addition to the general methods shown in Scheme 1 for the construction of Sn–Sn bonds, a number of distannanes have been obtained by a variety of other routes, most of which were initially discovered serendipitously (Scheme 2). For instance, in an attempt to prepare tetra(t-butyl)stannane (3) from tri(t-butyl)tin chloride (4), Kandil and Allred (17) found

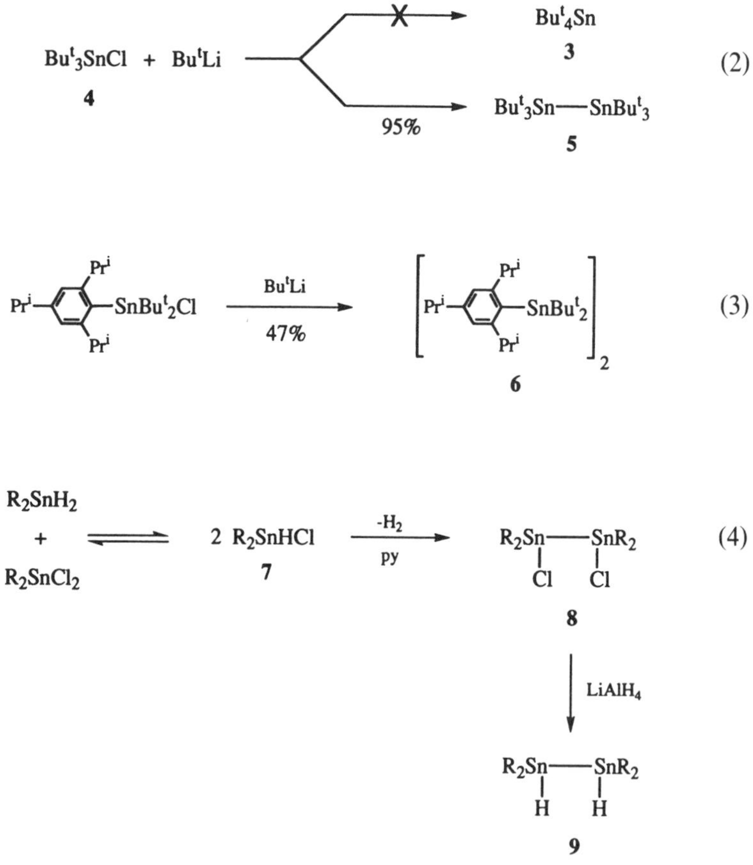

SCHEME 2

that, rather than substituting, t-butyllithium served to reduce **4** to provide hexa(t-butyl)distannane (**5**), as shown in Scheme 2 [Eq. (2)]. As determined by these authors, this side-reaction becomes the dominant pathway when the initial organotin halide is sterically encumbered, and a number of hindered distannanes have now been prepared by this method, including, more recently, 1,2-bis(2,4,6-triisopropylphenyl)-1,1,2,2-tetra(t-butyl)distannane **6** according to Eq. (3) (Scheme 2) (*18*). Taking advantage of the rapid equilibrium that is known to occur between organotin dihydrides and organotin dihalides to produce organotin hydrido halides R_2SnHCl (**7**) in solution, Neumann and co-workers (*19*) developed a convenient "one-pot" synthesis of 1,2-dichlorodistannanes **8** via the facile dehydrogenative coupling of **7**, which is catalyzed by pyridine [Eq. (4), Scheme 2]. These 1,2-dichloro-1,1,2,2-tetraalkyldistannanes are valuable starting materials for the synthesis of cyclic compounds containing distannane fragments (*20*) and for the synthesis of the 1,1,2,2-tetraalkyldistannanes **9** (R = Bu and i-Bu) via reaction with lithium aluminum hydride (*19*). Regarding hindered 1,2-dihalodistannanes, there now exists several different routes to 1,2-dichloro- and 1,2-dibromo-1,1,2,2-tetrakis(2,4,6-triisopropylphenyl)distannane **10**. These include the reductive coupling of the corresponding diorganotin dichloride **11** with lithium naphthalenide (*21*) or directly from SnX_4 (X = Cl or Br) and the corresponding aryllithium reagent (*22*) (Scheme 3). Compound **10** has also been prepared from the

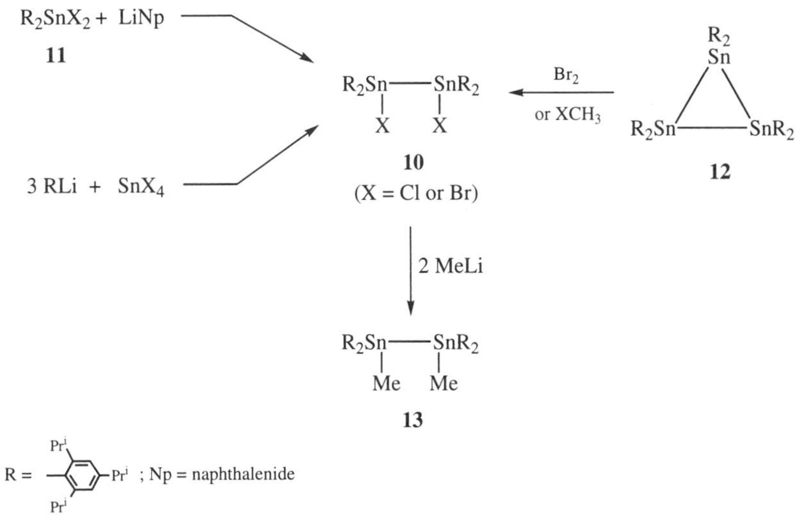

SCHEME 3

halogenation of hexakis(2,4,6,-triisopropyl-phenyl)cyclotristannane **12**
(*21*) using either bromine (*23*) or the respective haloforms (*24*). With alkyllithium reagents of small steric bulk, compound **10** (X = Cl) has in turn
been used to prepare 1,2-dimethyl-1,1,2,2-tetrakis(2,4,6-triisopropylphe-
nyl)distannane **13**, as shown in Scheme 3 (*24*). Finally, it must be mentioned that a few distannanes have been isolated as either by-products or
products of decomposition pathways. These include the isolation of 1,2-
dichloro-1,1,2,2-tetrakis[bis(trimethylsilyl)methyl]distannane **14** from the
reaction of the stable stannylene derivative, bis[bis(trimethylsilyl)methyl]-
tin **15** (*25*), with the Pt^{II} substrate, $\{PtCl(\mu\text{-}Cl)(PEt_3)\}_2$ [Eq. (5)] (*26*), and

$$R_2Sn \; + \; \{PtCl(\mu\text{-}Cl)(PEt_3)\}_2 \longrightarrow \underset{\underset{Cl \quad\; Cl}{|\quad\;\; |}}{R_2Sn\!\!-\!\!-\!\!SnR_2} \; + \text{other products} \quad (5)$$

$$\mathbf{15} \qquad\qquad\qquad\qquad\qquad\qquad\qquad \mathbf{14}$$

$$R = \underset{SiMe_3}{\overset{SiMe_3}{-\!\!\!\diagdown}}$$

the isolation of hexakis(2,6-diethylphenyl)distannane **16** from the thermolysis of hexakis(2,6-diethylphenyl)cyclotristannane **17** (*27*) either in the
solid state or in solution (*28*) [Eq. (6)].

$$\underset{\underset{\mathbf{17}}{R_2Sn\!\!-\!\!-\!\!SnR_2}}{\overset{\overset{R_2}{Sn}}{\triangle}} \quad \overset{\Delta}{\longrightarrow} \quad \underset{\mathbf{16}}{R_3Sn\!\!-\!\!SnR_3} \; + \text{other products} \quad (6)$$

$$R = \underset{Et}{\overset{Et}{-\!\!\!\diagdown\!\!\!\bigcirc}}$$

B. *Structural Studies*

1. *Peraryl Derivatives*

As predicted by theoretical calculations, the force constant for Sn–Sn
bond stretching in distannane is quite small (*29*), and thus, for synthetic
derivatives that possess severe nonbonded steric interactions between
substituents, elongation of the Sn–Sn bond is observed in the set of solid-
state structural parameters for these compounds. Given this, as Table I
indicates, there has been much current interest in probing the limit to

TABLE I

SELECTED BOND LENGTHS AND TORSION ANGLES FOR DISTANNANES

	Bond lengths (Å)		Torsion angles (°)		
Compound	Sn–Sn	Sn–C (avg)	C–Sn–Sn–C	Sn–Sn–C_{Ph}–C_{Ph}	Ref.
$R_3Sn–SnR_3$ *Peraryl*					
R = ⬡ (18)	2.770	2.180	60	13, 23, 45	(30)
R = ⬡–Me (19)	2.777	2.150	87, 33	42, 58	(33a,b)
R = ⬡ (20) Me	2.883	2.168	60	53, 54, 56	(34)
R = Et ⬡ Et (16)	3.052	2.243	63, 61, 56	48, 51, 55	(28)
Peralkyl					
R = benzyl (21)	2.823	2.191	60	—	(24)
R = t-butyl (5)	2.894	2.238	75, 44	—	(24)
RR′R″Sn–SnRR′R″					
R = R′ = Me; R′ = Cl (8)	2.770	2.14	130.1, 13.5	—	(36)
R = R′ = SiMe₃ / SiMe₃ (14) R″ = Cl	2.844	2.19	~60	—	(26)
R = R′ = iPr ⬡ iPr iPr (10) R″ = Br	2.841	2.191	~60	—	(22,24)
R = R′ = iPr ⬡ iPr iPr (13) R″ = Me	2.829	2.22, 2.18	77, 43	—	(18)
R = iPr ⬡ iPr iPr (6) R′ = R″ = t-Bu	3.034	2.259	5.1, 5.6, 21.1	—	(18)

which this bond length distortion can be forced and still maintain a stable distannane structure. In this regard, the Sn–Sn bond length of 2.770 Å determined for the crystal structure of hexaphenyldistannane (18) is often taken as the reference point, since this value is fairly close to the one calculated for the parent compound (cf. 2.804 Å) (29,30). Another feature of the structure of 18 is how the nonbonded interactions between the phenyl groups dictate this compound's symmetry properties in the solid state. As pointed out by Mislow and co-workers (31), 18 can potentially exist in two distinct conformations that are defined by the helical sense of the "propeller-like" Ph_3Sn moieties when viewed along the threefold axis of the molecule. One of these conformations is achiral (meso), with point group symmetry S_6, in which the two moieties are heterochiral, i.e., have opposite helicities, while the other conformation is chiral, with point group symmetry D_3, in which the two moieties are homochiral, i.e., have the same helicity. For 18, only a staggered S_6 conformation is observed in the solid state, either as a result of its being the lowest-energy structure or as a result of crystal packing forces that enforce it over a D_3 conformation. Evidence that, due to the calculated small barrier to rotation about the Sn–Sn bond in the parent distannane 1 (32), these latter solid-state interactions can play a significant role in influencing the geometical parameters of distannanes is provided by the crystal structure of hexa(4-methylphenyl)distannane 19 (33a,b). Thus, as Fig. 1a reveals, the molecular structure of 19 is still one in which the opposing R_3Sn moities have helicities of opposite chirality, as in 18; but now instead of a similar fully staggered conformation, 19 adopts one in which the torsion angles are 87.33° and 32.67°, respectively. Interestingly, it appears that by increasing the tilt angle of the phenyl groups from 20° to 55°, the expected increase in nonbonded interactions brought about by the closer proximity of the phenyl groups on the adjacent tin atoms in this conformation can be accommodated without significantly distorting bond lengths [cf. Sn–Sn, 2.777 Å; Sn–C, 2.150 Å for 19 (Table I)].

In compound 19, the steric influence of the 4-methyl groups on the aromatic substituents should be, and apparently are, negligible. However, the placement of groups at ortho positions should have profound effects; and indeed, this assertion is supported by the structural parameters shown in Table I and Fig. 1b for the isomeric compound hexa(2-methylphenyl)distannane (20) (34). Once more, only a staggered S_6 structure is observed in the solid state for compound 20. However, now due to severe nonbonded interactions between substituents that cannot be relieved by further tilting of the phenyl groups, the Sn–Sn bond length has been substantially elongated to 2.883 Å. Furthermore, from an analysis of this structure, it is

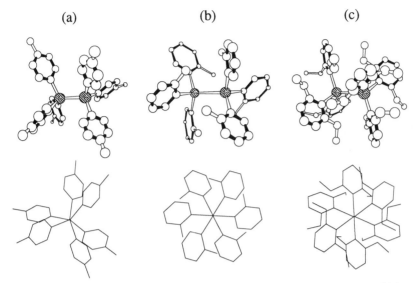

FIG. 1. Molecular structures of: (a) hexakis(4-methylphenyl)distannane (**19**), (b) hexakis(2-methylphenyl)distannane (**20**), and (c) hexakis(2,6-diethylphenyl)distannane (**16**). Structures redrawn from coordinates provided in references *33, 34,* and *28,* respectively.

clear that a D_3 conformation for **20** would be even more strained, due to an increase in nonbonded interactions.

A good example of how far the Sn–Sn bond of distannanes can be elongated and still obtain a stable structure is provided by the crystal structure of hexakis(2,6-diethylphenyl) distannane **16** (*35*). As Table I and Fig. 1c reveal, both the Sn–Sn and the mean Sn–C bond lengths, at 3.052 and 2.243 Å, respectively, are now dramatically elongated relative to **18**. Not surprisingly, to relieve best the strong nonbonded interactions present in this compound, the structure is the fully staggered S_6 conformation observed previously. In addition, it becomes clear that a 55–57° tilt of the phenyl groups is likely to be the maximum that can be accommodated in sterically encumbered peraryl distannane derivatives.

2. Peralkyl Derivatives

The body of structural data for peralkyl distannanes is a little more sparse than that for peraryl derivatives, with only two crystal structures having been determined. However, through a comparison of these two structures, a similar trend regarding the impact of nonbonded steric interactions on Sn–Sn bond length can be observed. Thus, the solid-state structure of hexabenzyldistannane 21 reveals a Sn–Sn bond length of 2.823 Å for a fully staggered conformation (Table I) (24). In contrast, the molecular structure of hexa(t-butyl)distannane 5 possesses a Sn–Sn bond that has now been elongated to 2.894 Å by virtue of the strong steric interactions between the branched alkyl substituents on adjacent tin atoms (Table I) (24). In this latter structure it is interesting to note that a more eclisped conformation is adopted with torsional angles of 77° and 45°, respectively. The molecular structures of compounds 5 and 21 are shown in Fig. 2.

3. *Mixed Derivatives*

In addition to the peraryl and peralkyl derivatives already described, the crystal structures for a number of distannanes with mixed types of substituents, including halogens, have been determined (Table I). For the most part, the geometric parameters and solid-state conformations of these compounds are in keeping with what has been presented so far. However,

(a)   (b)

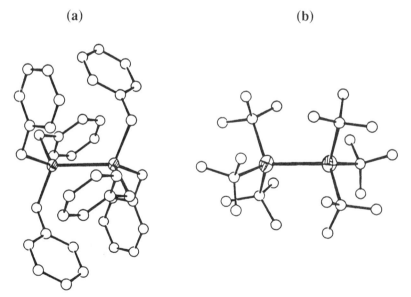

FIG. 2. Molecular structures of: (a) hexabenzyldistannane (21) and (b) hexa(t-butyl)distannane (5). Reprinted with permission from Puff *et al.* (24).

some unique features of these structures deserve comment. To begin with, the molecular structure of 1,2-dichloro-1,1,2,2-tetramethyldistannane **8** (R = Me) displays a normal Sn–Sn bond length of 2.770 Å, but due to strong Sn---Cl intermolecular interactions (Sn---Cl, 3.240 Å vs Sn–Cl, 2.448 Å), this compound is polymeric in the solid state. As a result, the methyl substituents on the adjacent tin atoms are nearly eclipsed (smallest torsion angle, 13.5°) (*36*). In contrast to what is observed for **8**, by increasing the steric bulk of the organic substituents, intermolecular interactions between tin and a halogen atom can be prevented, and as a result these distannanes remain monomeric in the solid state. Examples of distannanes that fall into this category are provided by 1,2-dichloro-1,1,2-2-tetrakis[bis-(trimethylsilyl)methyl]distannane **14** (*26*) and 1,2-dibromo-1,1,2,2-tetrakis(2,4,6-triisopropylphenyl)distannane **10** (*22,24*). Both of these compounds adopt nearly staggered conformations, with Sn–Sn bond lengths that are only moderately distorted (Sn–Sn, 2.844 Å for **14** and 2.841 Å for **10**).

A final class of distannanes to consider are those that possess a mixture of aryl and alkyl substituents. In this regard, the structures of two derivatives are known. One of these is represented by 1,2-dimethyl-1,1,2,2-tetrakis(2,4,6-triisopropylphenyl)distannane **13**, which is conveniently prepared from the corresponding 1,2-dichloro derivative (*24*). This compound adopts a solid-state conformation that is more eclipsed (mean torsion angles, 77° and 43°, respectively) and that has a Sn–Sn bond length of 2.829 Å (Table I). The other compound that falls into this class is 1,2-bis(2,4,6-triisopropylphenyl)-1,1,2,2-tetra(*t*-butyl)distannane **6**. This compound surprisingly adopts a nearly eclipsed conformation, with a very long Sn–Sn bond length of 3.034 Å, which is nearly as long as that of the distannane **16** (Table I and Fig. 3) (*18*). Apparently, it is this conformation that can best relieve the strong steric interactions between the two aryl substituents.

C. Properties

1. Electronic and Vibrational Spectroscopy

It has long been known that the electronic spectrum of hexaphenyldistannane (**18**) exhibits a strong absorption maximum (λ_{max} 247 nm; ε_{max} 40,400), which has been attributed to the presence of the Sn–Sn bond in this compound (*37,38*). If this absorption feature correlates with a $\sigma_{Sn\text{-}Sn} \rightarrow \sigma^{*}_{Sn\text{-}Sn}$ transition, then one might expect to observe a Sn–Sn bond length dependence on its strength and position. Indeed, the observation that it has shifted to 287 nm (ε_{max} 20,382) in the sterically encumbered

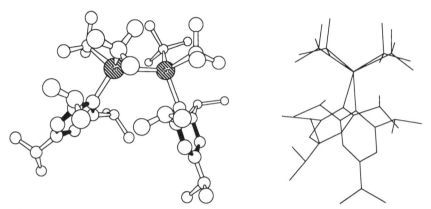

FIG. 3. Molecular structure of distannane **13**. Structure redrawn from coordinates provided in Weidenbruch *et al.* *(18)*.

distannane **16** (Sn–Sn; 3.052Å) seems to lend support to this assignment (Table II) *(39)*. For peralkyl distannane derivatives, this absorption maximum is blue-shifted (*cf.* for R = Me and cyclohexyl in Table II), possibly as a result of a smaller degree of polarizability of the alkyl vs aryl substituents. However, it is interesting to note that for hexabenzyldistannane (**21**) it is almost identical to that of compound **18** (Table II). Clearly, the origin and nature of substituent effects on this unique electronic transition in distannanes is a subject for more thorough theoretical and experimental investigation.

As mentioned previously, the calculated force constant for Sn–Sn bond stretching in distannane (**1**) is quite small, and the harmonic frequency associated with this bond length deformation is predicted to lie at 189 cm^{-1} *(29)*. This value is in quite excellent agreement with the value of 208 cm^{-1} that was obtained from the Raman spectrum of hexamethyldistannane (**22**) *(40)*. However, a dramatic perturbation occurs upon phenyl substitution, as revealed by the value of 138 cm^{-1} that is experimentally obtained for hexaphenyldistannane **18** (Table II). Not surprisingly, there also appears to be a strong correlation between nonbonded steric interactions and bond stretching frequency in distannanes. This is most readily observable in making the comparison between compounds **18** and **20** (*cf.* ν(Sn–Sn) 138 cm^{-1} vs 123 cm^{-1}, respectively) (Table II). Indeed, Neumann and co-workers *(41 a,b)* correlated bond-stretch frequency with Sn–Sn bond dissociation energies for the peraryl distannanes **23–25**, which they showed dissociated reversibly to the corresponding triaryltin radicals at decreasing temperatures along the series, due to an increasing amount of nonbonded interactions between the organic substituents (Scheme 4).

TABLE II

ELECTRONIC AND VIBRATIONAL SPECTRA FOR SELECTED DISTANNANES

Compound: $R_3Sn-SnR_3$	UV		Raman (cm^{-1}) $\nu(Sn-Sn)$	Sn-Sn (Å)
	λ_{max} (nm)	ε_{max}		
R 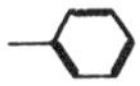 = (18)	248	40,400	138	2.770
R  = Me (19)	248	40,400	131	2.777
R 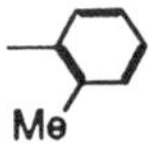 = (20)	—	—	123	2.883
R 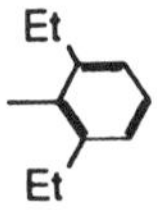 = (16)	287	20,382	—	3.052
R = Me (22)	<215	—	208	—
R = cyclohexyl	<220	>34,500	133	—
R = t-butyl (5)	222	59,000	115	2.894
R = benzyl (21)	247	40,600	—	2.823

$$R_3Sn-SnR_3 \quad \underset{}{\overset{T}{\rightleftharpoons}} \quad 2\ R_3Sn\cdot$$

$R_3Sn-SnR_3$	ΔH (kcal mol^{-1})	T (° C)	$\nu(Sn-Sn)$ (cm^{-1})
R 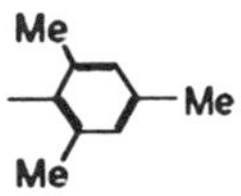 = Me (23)	49 ± 2	180	102
R 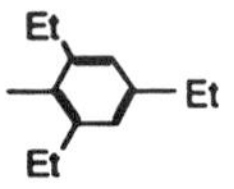 = Et (24)	26.6 ± 2	100	92
R 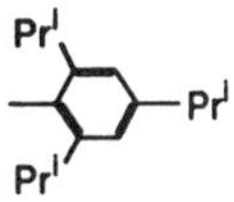 = Pr^i (25)	8.5 ± 1	20	--

SCHEME 4

Thus, compound **23**, with a $\nu(\text{Sn–Sn})$ of 102 cm^{-1}, appears to thermally dissociate beginning at 180°C, compound **24** at 100°C [$\nu(\text{Sn–Sn})$ 92 cm^{-1}], and finally, compound **25**, for which $\nu(\text{Sn–Sn})$ cannot be measured, at only 20°C. Unfortunately, the structural parameters and the electronic spectra for compounds **23–25** have not yet been obtained, so a correlation between these parameters and their Sn–Sn bond lengths cannot be made.

2. ^{119}Sn Nuclear Magnetic Resonance (NMR) Spectroscopy

Due to the occurrence of two isotopes of tin with $I = 1/2$, ^{119}Sn and ^{117}Sn, of relatively high natural abundance (8.58% and 7.61%, respectively) and high receptivity (natural abundance times sensitivity) (25.6 and 19.9, respectively) relative to ^{13}C (= 1), ^{119}Sn NMR spectroscopy is a powerful tool for the investigation of tin-containing compounds (*42 a,b*). This is especially true for polystannanes, where $^nJ(^{119}\text{Sn}–^{117}\text{Sn})$ coupling constants between magnetically equivalent tin atoms can still be obtained and the corresponding $^nJ(^{119}\text{Sn}–^{119}\text{Sn})$ values derived [via the relationship $^nJ(^{119}\text{Sn}–^{119}\text{Sn})/^nJ(^{119}\text{Sn}–^{117}\text{Sn}) = 1.046$].

As Table III indicates, the nature of the substituents in distannanes can influence both the chemical shift (δ; referenced relative to tetramethyltin) and the direct one-bond coupling constant, $^1J(^{119}\text{Sn}–^{119}\text{Sn})$. The influence of substituents on chemical shifts has been discussed in detail elsewhere (*42 a*); therefore, an in-depth analysis of the nature of this influence will not be repeated here. However, the influence of substituents on the $^1J(^{119}\text{Sn}–^{119}\text{Sn})$ values deserves some comment (*43–44 a–c*). In the case of peralkyldistannanes (Table III), it has been proposed that from an apparent linear correlation between $^1J(^{119}\text{Sn}–^{119}\text{Sn})$ and $\Sigma\sigma^*$ (the sum of the Taft σ^* constants of the six alkyl groups bound to the tin atoms), the main factor involved in determining the value of $^1J(^{119}\text{Sn}–^{119}\text{Sn})$ in these compounds is Z_{eff}, the effective nuclear charge at the tin nucleus (*43 a,b*). However, an alternative proposal to explain the trend observed in Table III is based on the equation derived by Pople and Santry (*45*) for the coupling mechanism in a two-body spin system [Eq. (7)]. Here, $\Psi_{\text{ns}}(0)$

$$^1J_{XY} = \frac{16\beta^2}{9\gamma_X\gamma_Y h} \cdot \psi^2_{\text{ns(X)}}(0) \cdot \psi^2_{\text{ns(Y)}}(0) \cdot \pi_{XY} \tag{7}$$

is the valence s electron density at the nucleus and π_{XY} is the mutual polarizability of the atoms X and Y. Since π_{XY} is related to the valence s–overlap integral β between X and Y, it has been argued that the increase in nonbonded interactions that occurs with increasing α-branching of the

TABLE III

^{119}SN NMR DATA FOR DISTANNANES

Compound:$R_3Sn-SnR_3$	δ (ppm)	$^1J(^{119}Sn-^{119}Sn)$ (Hz)
R  = ⬡ (18)	−144.7	4480
R  = ⬡−Me (19)	−141.9	4570
R 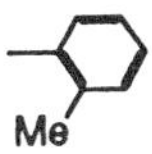 = ⬡ (20), Me	−136.8	4297
R_3 = Br, 2 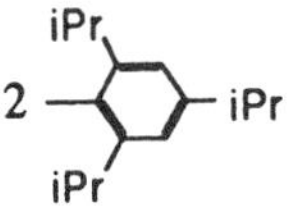 iPr−⬡−iPr (10), iPr	−118.3	5212
R = Me (22)	−108.7	4211
R = Bu	−83.2	2625
R = Pr^i	−29.1	1155
R_3 = Pr^i_2, Bu^t	−21.5	730
R = Bu^t (5)	−6.1	57
R = Benzyl (21)	−44.6	1780

alkyl substituents in distannanes leads to a diminishing of β, and in turn, to the magnitude of $^1J(^{119}Sn-^{119}Sn)$ (44 a–c). Interestingly, in the case of peraryldistannanes, $^1J(^{119}Sn-^{119}Sn)$ seems to be much less sensitive to an increase in Sn–Sn bond length brought about by an increase in steric interactions [cf. $^1J(^{119}Sn-^{119}Sn)$ = 4480 Hz for hexaphenyldistannane 18 vs a value of 4297 Hz for hexakis(2-methylphenyl)distannane 20 (Table III)].

D. *Reactivity*

The chemical reactivity of the Sn–Sn bond has also been amply reviewed, and some of the more common reactions involving this moiety are shown in Scheme 5 (13). Of these, the reductive cleavage of distannanes with alkaline metals has been used as a convenient source of the correspondinng triorganotin anions, R_3SnM, and these have found utility in the preparation of a number of organotin compounds that are useful for organic synthesis, especially for the case where R = Me or Bu (46). Concerning the strength of the Sn–Sn bond in distannanes, ΔH_{diss} has

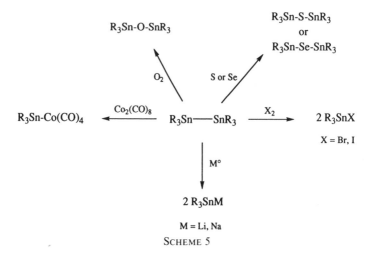

SCHEME 5

been estimated to be about 66 kcal mol^{-1} in hexamethyldistannane (**22**) (*47*), and simple hexaalkydistannanes can be distilled without decomposition. As previously discussed, only when severe nonbonded interactions between substituents are present, as in the compounds **23–25**, does one observe facile bond dissociation in solution. On the other hand, homolysis of the Sn–Sn bond occurs readily when distannanes are irradiated with ultraviolet light. It has also been observed that the halide-containing distannanes, R_3SnSnR_2Br (R = Me and Et) and 1,2-dichloro-1,1,2,2-tetramethyldistannane (**8**), thermally decompose at fairly low temperatures (*48*). With oxygen, distannanes react to provide the corresponding distannoxanes, $R_3Sn–O–SnR_3$. Sulphur and selenium can similarly be inserted into the Sn–Sn bond, and while peralkyl distannanes are typically more reactive than peraryl ones, bulky substituents can retard these reactions by raising energy barriers for bimolecular processes. The use of sterically demanding substituents is, in fact, the basis for a strategy known as *kinetic stabilization*, which has been successfully employed for the synthesis of polystannane frameworks that would normally have highly reactive Sn–Sn bonds. Finally, a few other reactions of distannanes deserve comment. First, halogens typically react rapidly and quantitatively (*33a*), and titration with iodine or bromine can be used for estimating the Sn–Sn bond content of polystannanes. In addition, a variety of transition-metal compounds react with distannanes to form triorganotin–transition-metal complexes (*42b*).

III

LINEAR POLYSTANNANES

A. Synthesis

1. Tri- and Tetrastannanes

The synthesis of tristannanes and linear tetrastannanes can be achieved in a straightforward manner by employing the hydrostannolysis reaction and diorganotin dihydrides (26) and 1,1,2,2-tetraorganodistannanes (9) according to Eq. (8) (49). As shown, this methodology can be used to

$$2\,R'_3Sn\text{-}NMe_2 \quad + \quad H\text{-}(R_2Sn)_n\text{-}H \quad \longrightarrow \quad R'_3Sn\text{-}(R_2Sn)_n\text{-}SnR'_3 \quad (8)$$

$$\begin{array}{ll} \mathbf{26}\,(n=1) & n=1\text{ and }2 \\ \mathbf{9}\,(n=2) & R=R';\,R\neq R' \end{array}$$

prepare a range of compounds with a mixture of substituents, although in the case of tetrastannanes, this range is limited, since only two tetraorganodistannanes are readily available (R = Bu and Bu^i in 9). Interestingly, the alternative reaction of diorganotin bisamides with triorganotin hydrides leads only to a mixture of oligomeric products, presumably as a result of a transamination process. A mixture of oligomeric perbutylated linear polystannanes is also obtained by the reaction of hexabutyldistannoxane (27) with formic acid at elevated temperature (50). From this mixture, Jousseaume and co-workers (50) separated octabutytristannane (28) and decabutyltetrastannane (29) in pure form through the use of reverse-phase high-pressure liquid chromatography (HPLC). These researchers also report that the coupling of diorganotin dihyrides with triorgano(N-phenylformamido)stannanes, a procedure originally introduced by Creemers and Noltes (51) for the synthesis of linear polystannanes, actually produces a mixture of products as determined by HPLC analysis.

For the stepwise construction of homologously pure linear polystannane oligomers via the hydrostannolysis reaction, a new strategy that makes use of a "masked" Sn–H functional group has recently been introduced (52). As shown in Scheme 6, the reagents 30, bearing a β-ethoxyethyl substituent, are prepared through the hydrostannation of ethylvinyl ether with the corresponding diorganotin hydridochloride 7, followed by reaction of the intermediate product with lithium dimethylamide. Coupling of 30a (R = Bu) with tributyltin hydride provides the distannane 31, which is treated with diisobutylaluminum hydride (DIBAL-H) to remove the

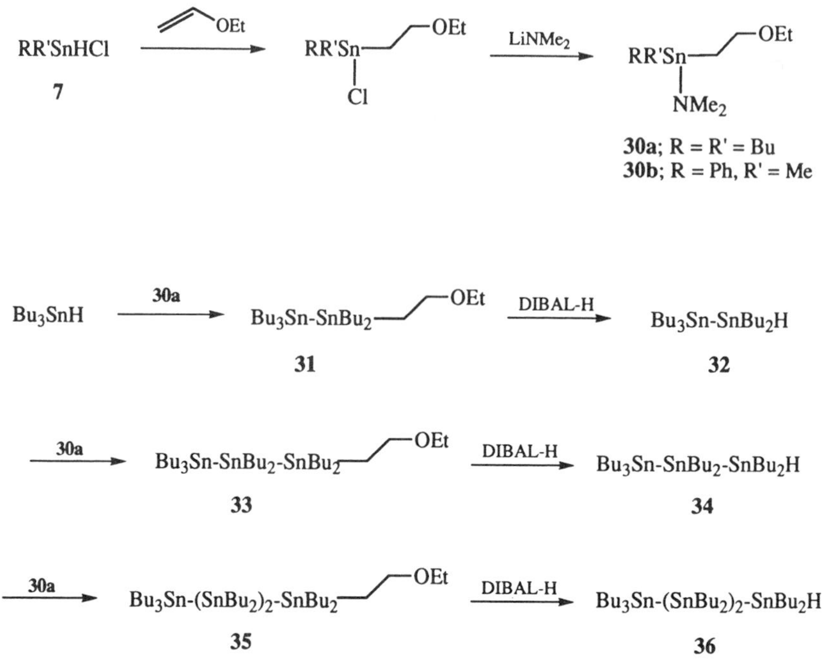

$$RR'SnHCl \xrightarrow{\quad \diagup OEt \quad} RR'Sn\diagup^{OEt}_{Cl} \xrightarrow{LiNMe_2} RR'Sn\diagup^{OEt}_{NMe_2}$$

7

30a; R = R' = Bu
30b; R = Ph, R' = Me

$$Bu_3SnH \xrightarrow{\textbf{30a}} Bu_3Sn\text{-}SnBu_2\diagup^{OEt} \xrightarrow{DIBAL\text{-}H} Bu_3Sn\text{-}SnBu_2H$$

31   **32**

$$\xrightarrow{\textbf{30a}} Bu_3Sn\text{-}SnBu_2\text{-}SnBu_2\diagup^{OEt} \xrightarrow{DIBAL\text{-}H} Bu_3Sn\text{-}SnBu_2\text{-}SnBu_2H$$

33   **34**

$$\xrightarrow{\textbf{30a}} Bu_3Sn\text{-}(SnBu_2)_2\text{-}SnBu_2\diagup^{OEt} \xrightarrow{DIBAL\text{-}H} Bu_3Sn\text{-}(SnBu_2)_2\text{-}SnBu_2H$$

35   **36**

SCHEME 6

β-ethoxyethyl group and generate the new distannane **32**. Repetition of this two-step sequence of reactions then provides the butylated tri- (**33** and **34**) and tetrastannanes (**35** and **36**). Most importantly, however, through the use of conventional column chromatography, each of the polystannanes can be purified to homogenity, as determined by ^1H and ^{119}Sn NMR and HPLC analysis, after each step in Scheme 6. Finally, by employing different coupling reagents (i.e., **30a** and **30b**), the programed synthesis of isomeric linear polystannanes can be carried out, as in the case of the two tristannanes **37** and **38** shown in Scheme 7 (52).

In addition to the hydrostannolysis reaction, another procedure that has been used for the synthesis of tristannanes, such as compound **39**, is through the reaction of the triorganotin anions, R_3SnLi, with diorganotin dihalides according to Eq. (9) in Scheme 8. This method has also been

$$Bu_3Sn\text{-}SnPhMe\text{-}SnBu_2\diagup^{OEt} \qquad\qquad Bu_3Sn\text{-}SnBu_2\text{-}SnPhMe\diagup^{OEt}$$

37   **38**

SCHEME 7

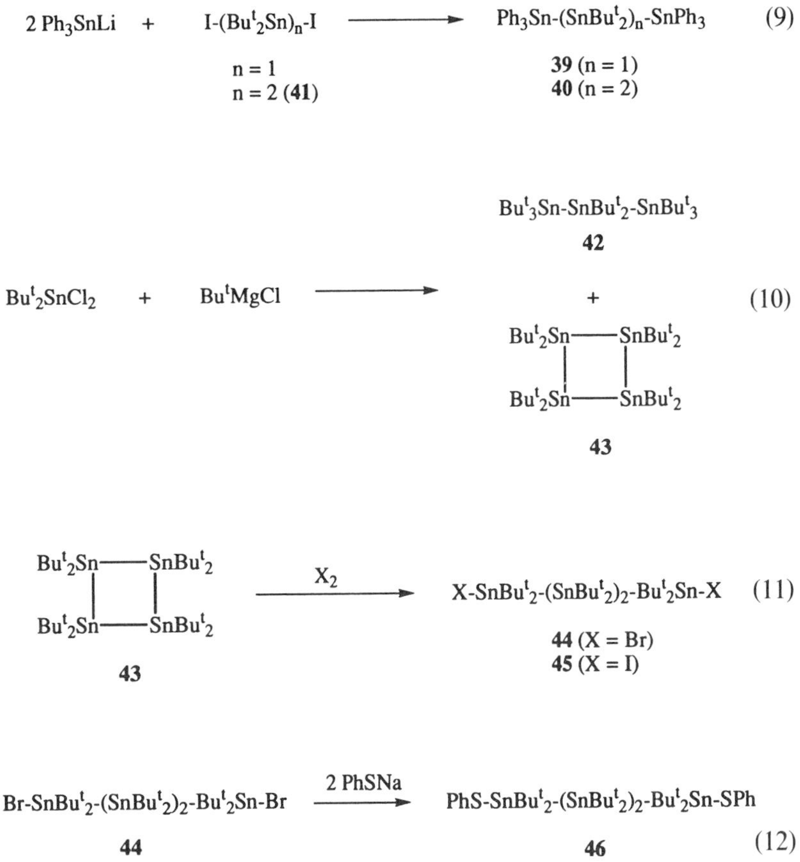

$$2\,Ph_3SnLi \quad + \quad I\text{-}(Bu^t_2Sn)_n\text{-}I \quad \longrightarrow \quad Ph_3Sn\text{-}(SnBu^t_2)_n\text{-}SnPh_3 \quad (9)$$

$$n = 1 \qquad \mathbf{39}\,(n = 1)$$
$$n = 2\,(\mathbf{41}) \qquad \mathbf{40}\,(n = 2)$$

$$Bu^t_3Sn\text{-}SnBu^t_2\text{-}SnBu^t_3$$
$$\mathbf{42}$$

$$Bu^t_2SnCl_2 \quad + \quad Bu^tMgCl \quad \longrightarrow \qquad + \qquad (10)$$

$$\begin{array}{ccc} Bu^t_2Sn & \!\!\!\!-\!\!\!\! & SnBu^t_2 \\ | & & | \\ Bu^t_2Sn & \!\!\!\!-\!\!\!\! & SnBu^t_2 \end{array}$$

$$\mathbf{43}$$

$$\begin{array}{ccc} Bu^t_2Sn & \!\!\!\!-\!\!\!\! & SnBu^t_2 \\ | & & | \\ Bu^t_2Sn & \!\!\!\!-\!\!\!\! & SnBu^t_2 \end{array} \quad \xrightarrow{\;X_2\;} \quad X\text{-}SnBu^t_2\text{-}(SnBu^t_2)_2\text{-}Bu^t_2Sn\text{-}X \quad (11)$$

$$\mathbf{43} \qquad\qquad\qquad \mathbf{44}\,(X = Br)$$
$$\mathbf{45}\,(X = I)$$

$$Br\text{-}SnBu^t_2\text{-}(SnBu^t_2)_2\text{-}Bu^t_2Sn\text{-}Br \quad \xrightarrow{\;2\,PhSNa\;} \quad PhS\text{-}SnBu^t_2\text{-}(SnBu^t_2)_2\text{-}Bu^t_2Sn\text{-}SPh$$
$$\mathbf{44} \qquad\qquad\qquad\qquad \mathbf{46} \qquad (12)$$

SCHEME 8

extended to the synthesis of the tetrastannane **40** through the addition of Ph$_3$SnLi to 1,2-diiodo-1,1,2,2-tetra(t-butyl)distannane **41** (*53*), as shown in Eq. (9) (*54*). Finally, as in the case of distannanes, a few tri- and tetrastannanes have been prepared through some novel routes. These include octa(t-butyl)tristannane **42**, which is obtained as a by-product in the reaction of di(t-butyl)tin dichloride with t-BuMgCl [Eq. (10), Scheme 8] (*24*). The main product of this latter reaction is actually the cyclotetrastannane **43** (*54*), which can be selectively ring-cleaved with one equivalent of either bromine or iodine to provide the 1,2-dihalo-1,1,2,2,3,3,4,4-octa(t-butyl)tetrastannanes **44** and **45** (X = Br and I, respectively) [Eq. (11), Scheme 8] (*24,55,56*). Finally, compound **44** has been used for the synthesis of the tetrastannane **46** [Eq. (12)] (*24*).

$Bu_3Sn\text{-}(SnBu_2)_n\text{-}SnBu_2\diagup^{OEt}$ $Ph_3Sn\text{-}(SnBu^t_2)_n\text{-}SnPh_3$

48 (n = 3)   **50** (n = 3)
49 (n = 4)   **51** (n = 4)

SCHEME 9

2. Higher-Order Polystannane Oligomers

Very few studies of fully characterized linear polystannanes with chain lengths of greater than four tin atoms have appeared. As mentioned previously, each of the compounds in the perethylated series, $Et_3Sn(Et_2Sn)_n$-$SnEt_3$ (n = 0–4) originally reported by Ceemers and Noltes (51) is likely to be a mixture of oligomeric products as revealed by modern methods of analysis. In fact, through the use of HPLC, Jousseaume and co-workers (50) were also able to separate the perbutylated pentastannane, Bu_3-$Sn(Bu_2Sn)_2SnBu_3$ (47), in pure form from a similar type of mixture. Extension of the synthetic methodolgy presented in Scheme 6 has also provided the homologously pure pentastannane and hexastannane, **48** and **49** (Scheme 9), respectively, and the potential exists for the synthesis of longer linear polystannanes by this route. Finally, using methodology presented in Scheme 8, Adams and Dräger (54) have extended their series of crystalline polystannanes to include the penta- and hexastannanes, **50** and **51**, respectively (Scheme 9).

B. Structural Studies

Relatively few linear polystannanes have been the subject of crystallographic analysis; Table IV and Fig. 4 provide some selected structural

TABLE IV

SELECTED BOND LENGTHS AND BOND ANGLES FOR LINEAR POLYSTANNANES

	Sn–Sn (Å)^a			
Compound	Sn_b–Sn_t	Sn_b–Sn_b	Sn–Sn–Sn (°)	Ref.
$Bu^t_3Sn\text{-}SnBu^t_2\text{-}SnBu^t_3$ (**42**)	2.966	—	122.1	(24)
$X\text{-}Bu^t_2Sn\text{-}(SnBu^t_2)_2\text{-}SnBu^t_2\text{-}X$				
X = Br (**44**)	2.887	2.928	117.3	(24)
X = I (**45**)	2.895	2.924	117.0	(56)
X = SPh (**46**)	2.910	2.928	118.7	(24)

^{a} b = bridging, t = terminal.

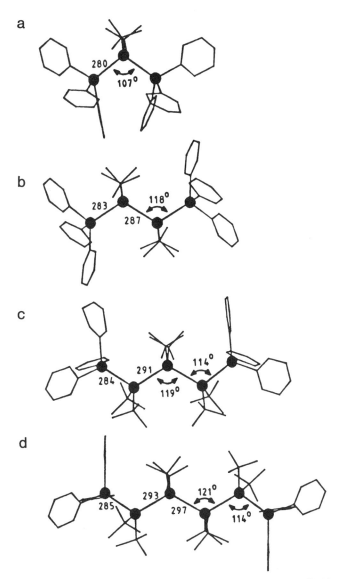

FIG. 4. Molecular structures for: (a) tristannane **39**, (b) tetrastannane **40**, (c) pentastannane **50**, and (d) hexastannane **51**. Reprinted with permission from Adams and Dräger (54).

parameters of those that have been reported to date. From this tabulation, many of the trends that were observed for substituted distannanes are preserved for the longer linear polystannanes. For instance, severe nonbonded interactions can greatly perturb both Sn–Sn bond lengths and

Sn–Sn–Sn bond angles as revealed by the solid-state structure of the tristannane **42**. In this compound, the Sn–Sn bond length has been elongated to 2.966 Å and the Sn–Sn–Sn bond angle has been opened up to an incredible 122.1° (Table IV). These values can be contrasted with the structural parameters for the tristannane **39**, which possesses more-normal Sn–Sn bond length and Sn–Sn–Sn bond angle values of 2.80 Å and 107°, respectively (Fig. 4). Compound **39** is the first in a homologous series of linear polystannanes, i.e., **39, 40, 50,** and **51,** that have been subjected to crystallographic analysis. As shown in Fig. 4, an interesting chain-length dependence can be observed in both the Sn–Sn bond length and the Sn–Sn–Sn bond angle values; namely, both of these structural parameters increase as the chain length gets longer. For example, in going from the tristannane **39** to the tetrastannane **40,** the Sn_b–Sn_b bond length value increases to 2.87 Å, while the Sn–Sn–Sn bond angle increases to 117.3°. Undoubtedly, these increased distortions are dictated by an increase in nonbonded interactions, and these are probably magnified further for the central tin atom as one goes to the pentastannane **50** (Sn_b–Sn_c, 2.91 Å; Sn–Sn–Sn, 119°). Quite curious, however, is the observation that the trend continues for the hexastannane **51** (Sn_b–Sn_c, 2.97 Å; Sn–Sn–Sn, 121°), even though it could be argued that the nonbonded interactions experienced for the central tin atom should have reached a maximun in **50.** Unfortunately, longer crystalline polystannanes have not yet been prepared that could be used to determine where these parameters reach their maximum values. Concerning other polystannanes, the series of tetrastannanes represented by compounds **44–46** have been studied by crystallographic analysis, and their structural parameters all fall within the range that is now expected (Table IV).

C. Properties

1. Electronic Spectroscopy

It has long been known that linear polysilane oligomers of the general formula $R–(R_2Si)_n–R$ exhibit absorption maxima that shift to longer wavelengths with an increase in chain length (57), and there has been much interest in developing theoretical models that can be used to explain the origin of this property (11,58). Polygermanes and polystannanes have also been shown to share this phenomenon. For the latter class of compounds, early work by Drenth, Noltes, and co-workers (59) showed that for the perethylated series of linear polystannanes. $Et_3Sn(Et_2Sn)_nSnEt_3$ ($n = 0–4$), the lowest energy-absorption maxima red-shifted from λ_{max} 232 nm for the distannane ($n = 0$) to 325 nm for the hexastannane ($n = 6$). These

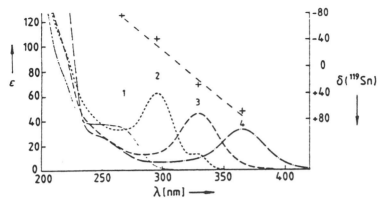

FIG. 5.   Electronic spectra for: (1) tristannane **39**, (2) tetrastannane **40**, (3) pentastannane **50**, and (4) hexastannane **51**. Also shown is a plot of the ^{119}Sn NMR chemical shift (ppm) of the central tin atom versus the corresponding excitation wavelength of the linear polystannane. Reprinted with permission from Adams and Dräger (*54*).

values are also red-shifted relative to polysilanes and polygermanes. Two other series of fully characterized linear polystannane oligomers have been spectroscopically investigated. One of these is the series of Adams and Dräger, who reported the collection of electronic spectra shown in Fig. 5 for the compounds **39, 40, 50,** and **51.** Once more, one observes a red-shifting of the lowest energy absorption with increasing chain length and a broadening that results in an apparent decrease in intensity for this transition. These investigators interpret the origin of this electronic property as arising from the "onset of formation of a metal-like electronic band and thus propose the term 'molecular metals' for these compounds." More recently, Takeda and Shiraishi (*60*) have investigated this concept through a theoretical approach, and they find that the band gap in the linear polystannane, $H-(H_2Sn)_n-H$, should decrease significantly as a function of increasing Sn–Sn bond length and increasing Sn–Sn–Sn (α) and H–Sn–H (β) bond angles. Furthermore, at fairly large values of α (e.g., 160°), they suggest that a metallic state for the polystannane may exist and that this state might be stabilized by an increase in pressure.

The series of linear polystannanes represented by compounds **31, 33, 35, 48,** and **49** have also been the subject of spectroscopic analysis. In this case, a similar trend is observed, although the magnitude of the red-shift per tin atom is less than that observed for the previous series (Fig. 6). This discrepency between the two series of linear polystannanes is likely to be a result of the strong nonbonded interactions in the former (i.e., **39, 40, 50,** and **51**), which severely restrict conformational mobility

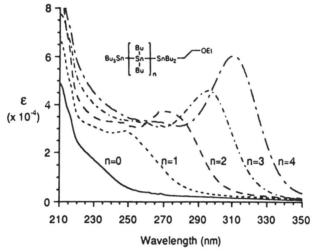

FIG. 6.   Electronic spectra for the linear polystannanes represented by compound **31** ($n = 0$), compound **33** ($n = 1$), compound **35** ($n = 2$), compound **48** ($n = 3$), and compound **49** ($n = 4$). Reprinted with permission from Sita (52).

and produce an increase in the mean Sn–Sn bond length with increasing chain length. For polysilanes, it has been observed that a red-shifting occurs when the Si–Si σ-bonded backbone adopts an all-*trans* conformation (10,11). Finally, concerning the electronic spectra of other linear polystannanes, unfortunately, only that of the tetrastannane **45** with a λ_{max} of 340 nm has been reported. This absorption is shifted by nearly 50 nm relative to the tetrastannane **40**, and it will be interesting to see whether this effect of halogen substitution will be a general one. In summary, it is clear from the preceding discussion that there is still much left to be done concerning the investigation of the electronic properties of linear polystannane oligomers. In this regard, the synthesis of well-defined linear oligomers with chain lengths of greater than six tin atoms and with a variety of substituents is of paramount importance.

2. ^{119}Sn NMR Spectroscopy

The ^{119}Sn NMR parameters [δ and $^nJ(^{119}Sn-^{119}Sn)$ values] for several selected linear tri- and tetrastannanes and for one pentastannane are provided in Table V. As with distannanes, some general trends can be observed from this compilation. To begin with, an increase in the α-branching of alkyl substituents has a pronounced effect on the chemical shifts of tristannanes for both the terminal and the bridging tin atoms, as seen in

TABLE V

^{119}Sn NMR Data for Selected Linear Polystannanes

Compound	δ (ppm)^a		$^nJ(^{119}\text{Sn}_b-^{119}\text{Sn}_b)$ (Hz)^a				Ref.
	Sn_t	Sn_b	$^1J(^{119}\text{Sn}_t-^{119}\text{Sn}_b)$	$^1J(^{119}\text{Sn}_b-^{119}\text{Sn}_b)$	$^2J(^{119}\text{Sn}_t-^{119}\text{Sn}_b)$	$^3J(^{119}\text{Sn}_t-^{119}\text{Sn}_t)$	
$R_3Sn-SnR_2-SnR_3$							
R = Me	−99.5	−261.7	2873	—	763	—	(43b)
R = Bu (28)	−75.6	−226.6	1596	—	417	—	(50)
R = Pr^i	−35.0	−132.9	403	—	188	—	(43b)
R = Bu^t (42)	20.4	17.0	774^b	—	(?)^c	—	(24)
$X-R_2Sn-(SnR_2)_2-SnR_2-X$							
R = X = Bu (29)	−74.3	−212.6	1596	601	295	233	(48)
R = Bu^t, X = Br (44)	102.8	25.2	1991^b	1475^b	25	271	(24)
R = Bu^t, X = I (45)	67.7	17.4	2199^b	1575^b	20	307	(56)
R = Bu^t, X = SPh (46)	33.8	40.9	1536/1655^b	1428^b	—	250	(24)
$Bu_3Sn-(SnBu_2)_3-SnBu_3$ (47)	−75.6	−199 (Sn$_b$) −210 (Sn$_c$)	1368	(?)^c	261	220	(48)

^{a} b = bridging, t = terminal, c = central.

^{b} Might be negative in sign.

^{c} Not reported or not observable.

going from $R = Me$ to $R = Bu^t$. In this series, the chemical shift for the bridging tin atom moves over a remarkably large range by shifting from a value of -261.7 ppm to a value of 17.0 ppm. For the first three members of this series (i.e., $R = Me$, Bu, and Pr^i), a steady decrease in the $^nJ(^{119}Sn-^{119}Sn)$ ($n = 1$ and 2) values is also observed. This latter trend is surprisingly broken, however, by a jump in magnitude for $^1J(^{119}Sn-^{119}Sn)$ that is observed for the case where $R = Bu^t$ (Table V). In this regard, it is important to mention that what are reported in Table 5 are actually the absolute magnitudes for the coupling constants, $|^nJ(^{119}Sn-^{119}Sn)|$. Since it has been previously shown that $^1J(^{119}Sn-^{119}Sn)$ values can be negative in sign and that a change in sign from positive to negative occurs around a Sn–Sn bond length of 2.85 Å, the value reported for $R = Bu^t$ (Sn–Sn, 2.966 Å) might, in fact, be a negative value (44a,54). If this correction is made, then the original trend observed for the series is preserved.

In comparing the values of $^nJ(^{119}Sn-^{119}Sn)$ reported for the perbutylated polystannane series, $Bu_3Sn-(Bu_2Sn)_n-SnBu_3$ ($n = 1-3$) (compounds **28, 29,** and **47**), it is interesting to note the chain-length dependence of the $^1J(^{119}Sn_t-^{119}Sn_b)$ values, which decrease from a value of 2873 Hz to a value of 1368 Hz as n goes from 1 to 3 (Table V). It is difficult to imagine that for $R = Bu$ there is a significant change in the Sn_t-Sn_b bond length for the series of compounds. Thus, it seems probable that the $^1J(^{119}Sn_t-^{119}Sn_b)$ values for these compounds are reflecting a subtle change in the electronic structure of the polystannane backbone as chain length increases. Another interesting observation can be made by comparing the $^nJ(^{119}Sn-^{119}Sn)$ values obtained for the series of tetrastannanes listed in Table V. In this comparison, it is interesting to note that the $^1J(^{119}Sn-^{119}Sn)$ values appear to be significantly enhanced for the compounds **44–46** relative to those of the perbutylated compound **47**. However, all of the Sn–Sn bonds for **44–46** are over 2.85 Å; therefore, it is possible that the values of $^1J(^{119}Sn-^{119}Sn)$ that are reported for these compounds should be negative as well. As mentioned previously, a detailed understanding of all of these phenomenon awaits the synthesis of additional linear polystannanes derivatives.

D. *Reactivity*

Much less has been documented concerning the chemical reactivity of linear polystannane oligomers, but, at the outset, it can be expected to closely parallel that reported for distannanes. It is known that these compounds are air-sensitive when the substituents posses small steric bulk. But so far, no attempts have been conveyed concerning the controlled oxidation or insertion of sulfur and selenium into the Sn–Sn bonds of these compounds. In the absence of disproportionation, this latter reaction might lead to oligomers of the general formula $R_3Sn-(R_2SnE)_n-SnR_3$ (E = S and Se), which might possess interesting properties (61). Linear polystannanes are known to undergo photolytic Sn–Sn bond dissociation; however, a detailed investigation of their thermal stability has yet to be conducted (62). Finally, the 1,4-diiodotetra-stannane 45 has found utility as a precursor to the cyclic compounds 52–54 via the route shown in Eq. (13) (63).

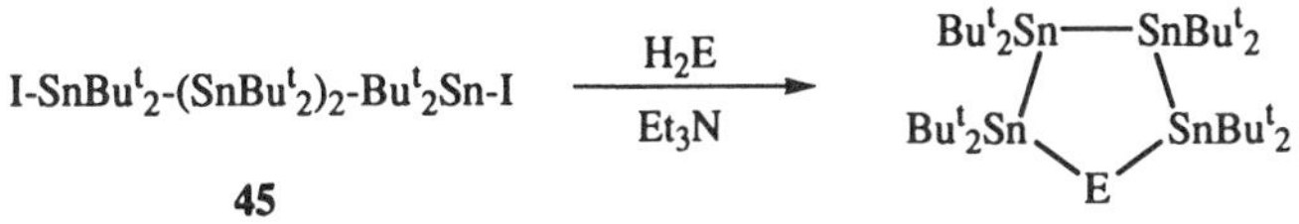

$$I\text{-}SnBu^t_2\text{-}(SnBu^t_2)_2\text{-}Bu^t_2Sn\text{-}I \xrightarrow[\text{Et}_3\text{N}]{\text{H}_2\text{E}} \begin{array}{c} Bu^t_2Sn\text{---}SnBu^t_2 \\ Bu^t_2Sn \diagdown \diagup SnBu^t_2 \\ E \end{array} \qquad (13)$$

45

52 (E = S)
53 (E = Se)
54 (E = Te)

IV

BRANCHED POLYSTANNANES

A. *Synthesis*

The first synthesis of a branched polystannane was achieved by Willemsens and van der Kerk (64) through the reaction of Ph_3SnLi with tin tetrachloride to produce compound 55, as shown in Scheme 10 [Eq. (14)]. A more general route, however, is represented by the hydrostannolysis reaction between organotin trihydrides 56 and (dimethylamino)trimethylstannane (57) according to Eq. (15) (65). In fact, extension of this methodology was also successfully used to prepare the permethylpentastannane 58 [Eq. (16)] (65). Finally, it has been shown that a solution of Me_3SnLi disproportionates to produce $(Me_3Sn)_3SnLi$ (59) in solution [Eq. (17)] (44c,66), and this compound has in turn been used to prepare branched polystannane transition-metal complexes (67).

B. *Structural Studies, Properties, and Reactivity*

Unfortunately, very little has been done to date regarding the investigation of branched polystannanes. No structural studies have been performed, and only the electronic spectrum of compound 55 (λ_{max} 277 nm;

$$Ph_3SnLi \ + \ SnCl_4 \ \longrightarrow \ (Ph_3Sn)_3Sn \qquad (14)$$

$$\mathbf{55}$$

$$RSnH_3 \ + \ 3 \ Me_2N\text{-}SnMe_3 \ \longrightarrow \ (Me_3Sn)_3SnR \qquad (15)$$

$$\mathbf{56} \qquad\qquad \mathbf{57}$$

$$R = Me, Et, Bu, Bu^i,$$
$$\text{n-Pentyl, Ph}$$

$$H_4Sn \ + \ 4 \ Me_2N\text{-}SnMe_3 \ \longrightarrow \ (Me_3Sn)_4Sn \qquad (16)$$

$$\mathbf{58}$$

$$Me_3SnLi \ \xrightarrow{\quad O \quad} \ (Me_3Sn)_3SnLi \qquad (17)$$

$$\mathbf{59}$$

SCHEME 10

ε_{max} 79,000) has been reported (37). Most of the compounds shown in Scheme 10 have, however, been the subject of investigation by ^{119}Sn NMR spectroscopy; the data collected from these studies are listed in Table VI. As can be seen in this compilation, an increase in the number of tin atoms bonded to the tin center dramatically alters its chemical shift to higher field [cf. δ −108.1 ppm for $Me_3SnSnMe_3$, −261.7 ppm for $(Me_3Sn)_2SnMe_2$, −489.7 ppm for $(Me_3Sn)_3SnMe$, and −739 ppm for $(Me_3Sn)_4Sn$]. Another feature is that as more tin atoms are bonded to the central tin, the $^1J(^{119}Sn-^{119}Sn)$ values progressively decrease in magnitude [cf. 4404 Hz for $Me_3SnSnMe_3$, 2873 ppm for $(Me_3Sn)_2SnMe_2$, 1733 Hz for $(Me_3Sn)_3$-SnMe, and 881 Hz for $(Me_3Sn)_4Sn$]. A similar relationship is also observed for the $^2J(^{119}Sn-^{119}Sn)$ values (Table VI). Since structural parameters for these compounds do not yet exist, it is not clear whether some of the trends can be attributed to an increase in Sn–Sn bond lengths as the degree of branching at the central tin atom increases. It is likely, however, that the effective nuclear charge at the central tin atom, Z_{eff}, does vary with branching, and this then may be involved, at least to some degree, in defining the ^{119}Sn spectral parameters. Finally, while the chemical reactivity of branched polystannanes has not been investigated, Mitchell and El-Behairy (65) do report that heating a solution of $(Me_3Sn)_3SnMe$ in benzene at 80°C resulted in decomposition to produce hexamethyldistannane and some as-yet-unidentified products.

TABLE VI

^{119}Sn NMR Data for Branched Polystannanes

Compound	δ (ppm)^a		$^nJ(^{119}Sn_b-^{119}Sn_b)$ (Hz)^a		Ref.
	Sn_t	Sn_b	$^1J(^{119}Sn_t-^{119}Sn_b)$	$^2J(^{119}Sn_t-^{119}Sn_t)$	
RSn(SnMe₃)₃					
R = Me	−89.5	−489.7	1733	287	(65)
R = Et	−89.3	−440.9	1538	259	(65)
R = Bu	−90.3	−459.9	1548	265	(65)
R = Bu^i	−90.3	−480.4	1546	262	(65)
R = n-Pentyl	−90.1	−460.5	1535	262	(65)
R = Ph	−83.2	−434.2	1670	271	(65)
R = Li (59)	−101.7	−1044	−5737	908	(44a−c)
Sn(SnMe₃)₄ (58)	−80.0	−739.0	881	20	(44a−c)

^{a} b = bridging, t = terminal.

V

CYCLOPOLYSTANNANES

A. *Synthesis*

Of all the different structural classes for polystannanes, cyclic structures have received by far the most attention. Much of this interest emanates from early reports claiming the successful synthesis of monomeric divalent diorganotin species (stannylenes) of the general formula R₂Sn for the specific cases of R = cyclohexyl and phenyl (68). Subsequent investigations have proven, however, that only through the use of organic substituents that are extremely sterically encumbered can one stabilize the monomeric form. This concept was demonstrated by the classical work of Lappert and co-workers (25,69), who synthesized compound 15 and found that, while dimeric in the solid state in the form of the distannene 60, it was present in solution to a large extent as the monomer (Scheme 11). This solution equilibrium between stannylene and distannene was later confirmed by variable-temperature NMR and ultraviolet (UV) spectroscopic studies (23,70). By fine-tuning the steric envelope surrounding the tin atom, Sakurai and co-workers (71) recently succeeded in synthesizing the first divalent diorganotin compound, compound 61, which is monomeric both in solution and in the solid state, as revealed by crystallographic analysis. The molecular structure of 61 as provided by this latter study is shown in Fig. 7. Finally, Weidenbruch and co-workers (72) were also

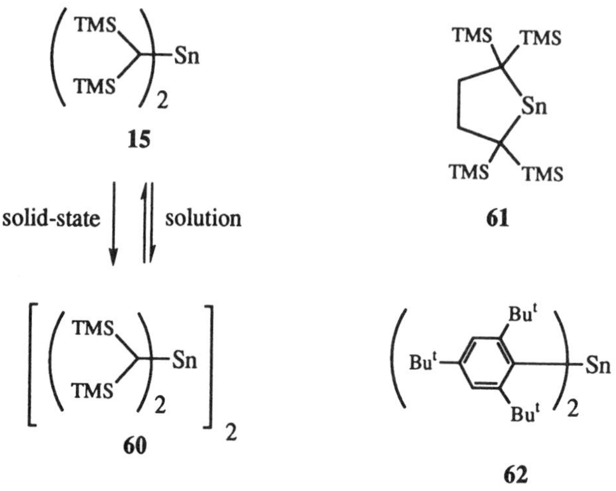

SCHEME 11

subsequently able to show that the diarylstannylene **62** is monomeric in solution and the solid state (Scheme 11).

Regarding cyclopolystannanes, it was first proposed by Neumann and Konig (73) that the "diphenyltin," which was previously prepared from either reduction of diphenyltin dichloride with alkaline metals or reaction of tin dichloride with either phenyllithium or phenylmagnesium bromide, was, in fact, dodecaphenylcyclohexastannane **63** (Scheme 12). This was later confirmed by the partial structural analysis of **63**, which revealed the tin atoms to be arranged in a six-membered ring that adopts a chair conformation (74). Neumann and Konig also determined that **63** could be produced through the pyridine-catalysed dehydrogenative coupling of diphenyltin dihydride according to Scheme 12. Through extension of this latter methodology, Neumann and co-workers (75) reported the synthesis of a number of other peraryl cyclohexastannanes, as well as the synthesis of a number of peralkyl cyclopolystannanes, $(R_2Sn)_n$, of varying ring sizes. For example, where R = ethyl, cyclopolystannanes for $n = 6$ and 9 are supposedly formed. Additional cyclic compounds reported by this group include a cyclotetrastannane for R = benzyl and a cyclopentastannane for R = cyclohexyl. Furthermore, by varying the nature of the amine used for the dehydrogenative coupling reaction, it was claimed that decaphenylcyclopentastannane could be formed. Subsequent reinvestigations have now revealed by crystallographic analysis that dodecabenzylcyclohexastannane **64** is formed as the major product under the conditions reported for the synthesis of the cyclotetrastannane (Scheme 12) (76), and

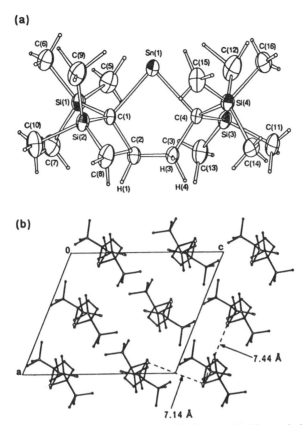

FIG. 7. Molecular structure of the stannylene **61**. Reprinted with permission from Kira *et al.* (*71*).

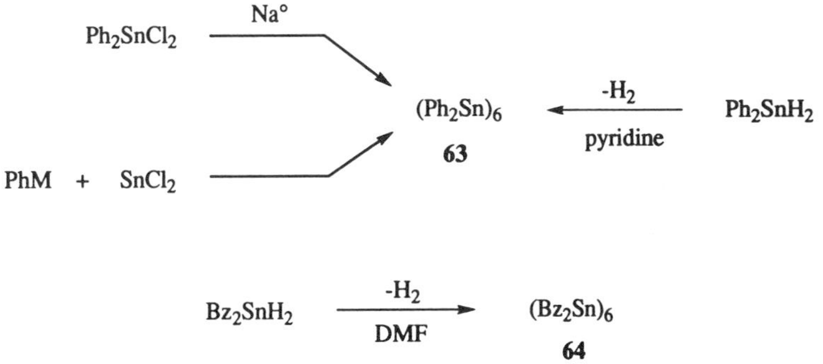

$$Ph_2SnCl_2 \xrightarrow{Na^\circ}$$

$$PhM + SnCl_2 \longrightarrow$$

$$(Ph_2Sn)_6 \quad \xleftarrow[\text{pyridine}]{-H_2} \quad Ph_2SnH_2$$
$$\mathbf{63}$$

$$Bz_2SnH_2 \xrightarrow[\text{DMF}]{-H_2} (Bz_2Sn)_6$$
$$\mathbf{64}$$

M = Li or MgBr; Bz = benzyl; DMF = dimethylformamide

SCHEME 12

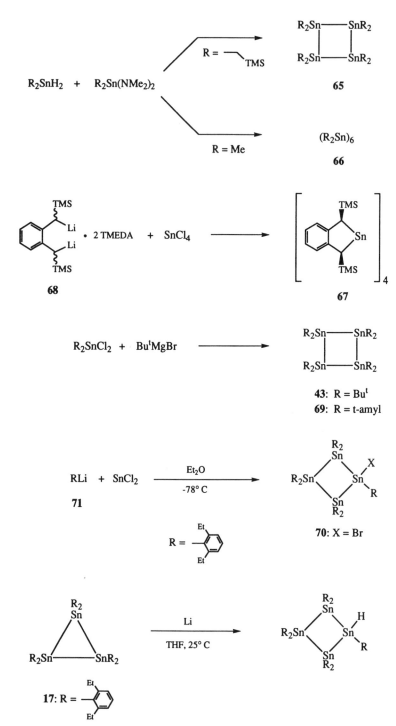

SCHEME 13

the cyclohexastannane **63** is still the major product when the original synthesis of decaphenylcyclopentastannane is carried out (*77*). To date, unequivocal determination of the structure reported for decacyclohexyl-cyclopentastannane by either crystallographic analysis or ^{119}Sn NMR spectroscopy has not been reported. However, Jousseaume and co-workers (*78*) have recently used this latter technique to verify the structures of decabutylcyclopentastannane, $(Bu_2Sn)_5$, and dodecabutylcyclohexastannane, $(Bu_2Sn)_6$, both of which were isolated in pure form from a mixture of cyclopolystannanes prepared from the palladium-catalyzed dehydrogenative cyclization of dibutylstannane.

In addition to the methods shown in Scheme 12 for the synthesis of cyclopolystannanes, this class of compound has been prepared by a hydrostannolysis procedure that utilizes diorganotin dihydrides and diorganotin bisamides. In this way, the cyclotetrastannane **65** was prepared in high yield according to Scheme 13 (*79*). This strategy has also been used to prepare dodecamethylcyclohexastannane **66** as well as its perdeuterated analog (*80*). For the synthesis of other cyclotetrastannanes, Lappert and co-workers (*81*) isolated the four-membered ring compound **67** through reaction of tin tetrachloride with the organolithium reagent **68** at $-78°C$ (Scheme 13). In addition, Puff and co-workers (*53*) utilized the same procedure as for the synthesis of octa-*t*-butylcyclotetrastannane **43** to prepare the cyclotetrastannane **69** that possesses *t*-amyl (i.e., 1,1-dimethylpropyl) substituents. The cyclotetrastannane **70** bearing one bromine substituent was produced unexpectedly from the reaction of 1-lithio-2,6-diethylbenzene **71** with tin dichloride at $-78°C$ in diethyl ether (*82*). Finally, it has been reported that the reaction of a solution of the cyclotristannane **17** in THF with lithium metal, followed by quenching with ammonium

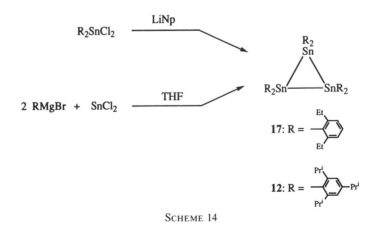

SCHEME 14

chloride, results in a very good yield of the substitutionally unsaturated cyclotetrastannane (see Scheme 13) (39).

Concerning the synthesis of the smallest cyclopolystannane ring system, the first synthesis of a cyclotristannane was achieved in 1983 through the reduction of bis(2,6-diethylphenyl)tin dichloride with lithium naphthalenide at low temperature according to Scheme 14 (27). The cyclotristannane 12 was prepared in a similar fashion a few years later (21). Since then, several groups have reported that these cyclotristannanes can be prepared in a straightforward fashion through addition of the corresponding arylmagnesium bromide reagents to a solution of tin dichloride in tetrahydrofuran (83–85). Paralleling these observations, it has also been reported that the synthesis of the cyclopolystannanes, $(\text{Phen}_2\text{Sn})_n$ (n = 3, 4, and 6) (Phen = 9-phenanthryl) can be achieved via similar procedures (86).

B. Structural Studies

Dodecaphenylcyclohexastannane 63 has been the subject of several crystallographic analyses (74,76,77). Surprisingly, as Table VII reveals, the structural parameters for this compound, such as Sn–Sn bond lengths and Sn–Sn–Sn bond angles, are remarkably dependent on its crystalline form. Thus, without inclusion of solvent molecules in the unit cell, 63 adopts a conformation with Sn–Sn–Sn bond angles ranging from 108.3° to a very wide value of 121.6°, whereas when two molecules of toluene are included, these bond angles are in the more expected range of 109.5° to 114.0°. In the latter structure, the Sn–Sn bonds are also slightly reduced to 2.775 Å from the 2.781 Å value observed for the former crystalline form. Altogether, this structural information supports the view that due to small-force constants for Sn–Sn bond length and Sn–Sn–Sn bond angle

TABLE VII

SELECTED BOND LENGTHS AND BOND ANGLES
FOR $[\text{PH}_2\text{SN}]_6$ (63) (76,77)

Parameter	Crystal Form		
	A	B	2 toluene
Sn–Sn bond lengths (Å)	2.781	2.787	2.775
Sn–Sn–Sn bond angles (°)			
Sn1–Sn2–Sn3	108.3	109.0	109.5
Sn2–Sn3–Sn(1')	121.6	107.4	114.0
Sn3–Sn(1')–Sn(2')	111.7	116.0	113.9

distortions, the equilibrium geometry of polystannane frameworks are easily perturbed by both internal forces (e.g., nonbonded interactions) and external forces (e.g., crystal packing). The crystal structure of dodecabenzylcyclohexastannane **64** reveals very few surprises, with a mean Sn–Sn bond length of 2.802 Å and Sn–Sn–Sn bond angles in the range 105°–111.7° (76).

Forced to adopt formal bond angles of 60° and 90° degrees, respectively, the skeletal frameworks of cyclotristannanes and cyclotetrastannanes are of interest with regard to the information they could provide concerning the origins and manifestations of strain in polystannane structures. The

TABLE VIII

SELECTED BOND LENGTHS AND BOND ANGLES FOR CYCLOTETRASTANNANES AND CYCLOTRISTANNANES

Compound	Sn–Sn (Å)	Sn–Sn–Sn (°)	Ring	Ref.
$[R_2Sn]_4$				
R = Bu^t (**43**)	2.887	90	Planar	(53)
R = t-amyl (**69**)	2.919	89.1	Puckered	(53)
$R_2 =$ (**67**)	2.852	88.2	Planar	(81)
R = trimethylsilylmethyl (**65**)	2.831	90	Planar	(79)
BrSnR[SnR$_2$]$_3$ (**70**)				
R =	2.899 2.910 2.843 2.831		Puckered	(82)
$[R_2Sn]_3$				
R = (**17**)	2.854 2.856 2.870	60 (avg)	Isoceles	(27)
R = (**12**)	2.916 2.947 2.963	60.16 60.73 59.13	Scalene	(84)

(a) (b)

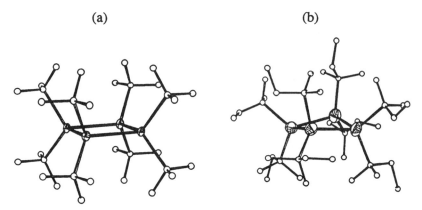

FIG. 8. Molecular structures of: (a) the cyclotetrastannane **43**, and (b) the cyclotetrastannane **69**. Reprinted with permission from Puff *et al.* (*53*).

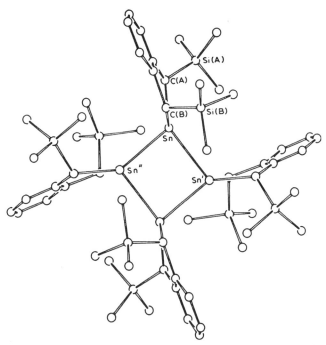

FIG. 9. Molecular structure of the cyclotetrastannane **67**. Reprinted with permission from Lappert *et al.* (*81*).

first cyclotetrastannane to be structurally characterized was the t-butylated derivative **43**, and this compound was subsequently subjected to full crystallographic analysis along with the cyclotetrastannane **69** (*53*). As Table VIII and Fig. 8 indicate, the two cyclotetrastannanes adopt different ring conformations, with **43** being planar and **69** being puckered. The mean Sn–Sn bond length of the latter, 2.919 Å, is slightly longer than the 2.887 Å found for **43**.

A very interesting cyclotetrastannane structure is that of compound **67** reported by Lappert and co-workers (*81*). During the synthesis of this cyclotetrastannane, several different diastereomers could have been formed; however, only the diastereomer shown in Fig. 9 of C_{4h} symmetry was produced. The four-membered ring in this structure is planar, with a Sn–Sn bond length of 2.852 Å and a mean Sn–Sn–Sn bond angle of 88.2°. Another planar cyclotetrastannane is compound **65**, which has a slightly shorter mean Sn–Sn bond length, 2.831 Å (Table VIII) (*79*). Finally, Cardin and co-workers (*82*) have reported the structure of the novel bromocyclotetrastannane **70**, which adopts a substantially puckered conformation in the solid state. In addition, most of the Sn–Sn–Sn bond angles in this structure deviate significantly from 90°, and the Sn–Sn bond

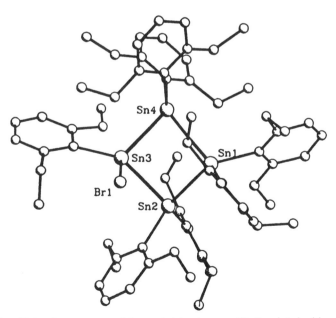

FIG. 10. Molecular structure of the cyclotetrastannane **70**. Reprinted with permission from Cardin *et al.* (*82*).

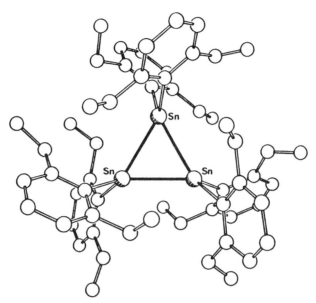

FIG. 11. Molecular structure of the cyclotristannane **17**. Reprinted with permission from Masamune *et al.* (*27*).

lengths fall into two categories, long and short, as Table VIII and Fig. 10 reveal.

Concerning cyclotristannanes, both the derivatives **17** and **12** have been the subjects of crystallographic analysis. As shown in Fig. 11, compound **17** forms an isoceles triangle in the solid state, with Sn–Sn bond lengths of 2.854, 2.856, and 2.870 Å, respectively (Table 8) (*27*). However, this distortion from the expected equilateral structure produces very minor deviations in the observed Sn–Sn–Sn bond angles. Several years after its synthesis was first reported, the crystal structure of the more sterically encumbered cyclotristannane **12** was also determined (*84*). Although the refinement for this preliminary structure is less than optimal, owing to the large size and complexity of the unit cell, several important features can still be discerned. First, the Sn–Sn bond lengths for **12** form a scalene triangle, with distinct values of 2.916, 2.947, and 2.963 Å, respectively. Most importantly, due to severe nonbonded steric interactions present in this compound, the planes defined by the Sn_3 ring and the C–Sn–C bond angles deviate from orthogonality by between 8.1° and 11.5°. It is likely that this pronounced structural distortion is responsible for the low barrier to Sn–Sn bond dissociation present in the molecule (see later).

C. Properties

The electronic spectrum of dodecaphenylcyclohexastannane **63** exhibits an intense absorption maximum at λ_{max} 280 nm (ε_{max} ~45,000). In the two cyclotristannanes, this characteristic feature is shifted to longer wavelengths and now appears at λ_{max} 295 nm (ε_{max} 45,000) for **17** and at λ_{max} 300 nm (ε_{max} 55,000) for **17** (*21,27*). Whether this shift is a result of longer Sn–Sn bond lengths in the three-membered ring system or is an inherent property of the cyclotristannane framework itself remains a question until further studies with these compounds are performed. In this regard, it is unfortunate that the electronic spectra of other cyclopolystannanes have not been reported.

As Table IX reveals, a number of cyclopolystannanes have been subjected to analysis by ^{119}Sn NMR spectroscopy. For these compounds, the values for the various $^{n}J(^{119}$Sn–^{117}Sn) coupling constants can be unequivocally established via the relationship of the relative intensities of the satel-

TABLE IX
^{119}Sn NMR Data for Selected Cyclopolystannanes

Compound	δ (ppm)	$^{1}J(^{119}$Sn–^{117}Sn)	$^{2}J(^{119}$Sn–^{117}Sn)	$^{3}J(^{119}$Sn–^{117}Sn)	Ref.
		$^{n}J(^{119}$Sn–^{117}Sn) (Hz)			
[R$_2$Sn]$_6$					
R = Ph (**63**)	-208	1117	813	71	(*77*)
R = Bz (**64**)	-140	769	362	(?)^a	(*76*)
R = Bu	-202	462	386	81	(*78*)
[R$_2$Sn]$_5$					
R = Bu	-201	476	461	—	(*78*)
[R$_2$Sn]$_4$					
R = Bu^t (**43**)	99	1195	1658	—	(*53*)
R = t-amyl (**69**)	101	1339	1573	—	(*53*)
R = trimethylsilylmethyl (**65**)	-77	624	3850	—	(*21*)
[R$_2$Sn]$_3$					
R = (Et, Et ring) (**17**)	-416	2285	—	—	(*26*)
R = (Pr^i, Pr^i, Pr^i ring) (**12**)	-379	3017	—	—	(*20*)

^{a} Not observed or not reported.

lites to the main resonance. In fact, as long as one is observing all the coupling constants that are possible, ^{119}Sn NMR spectroscopy is a powerful tool for establishing the ring size of cyclopolystannanes. For cyclohexastannanes, the absolute magnitudes for the $^nJ(^{119}Sn-^{117}Sn)$ constants fall off with increasing n. However, for cyclotetrastannanes, it is curious to note that the two-bond coupling constant, $^2J(^{119}Sn-^{117}Sn)$, is routinely larger than the corresponding one-bond coupling constant, $^1J(^{119}Sn-^{117}Sn)$. One possible explanation for this trend is that for cyclotetrastannanes, the two two-bond pathways both make positive additive contributions to the value of $^2J(^{119}Sn-^{117}Sn)$. Finally, it is interesting to note that for both of the cyclotristannanes, 17 and 12, the values for $^1J(^{119}Sn-^{117}Sn)$ fall within the range expected for simple compounds such as distannanes.

Perhaps the most important question concerning the properties of cyclopolystannanes is in regard to their thermodynamic stability. In the case of cyclohexastannanes, it can be anticipated that the six-membered rings in these compounds are relatively ring-strain free, and therefore they should be thermodynamically stable. For compounds 63 and 64, this assumption appears to be valid. However, for dodecamethylcyclohexastannane 66, it has been reported that this compound undergoes facile disproportionation in solution at room temperature to produce an equilibrium mixture of 66 and the corresponding five-, seven-, and eight-membered cyclopolystannanes, as suggested by ^{119}Sn NMR spectroscopy (80). In this particular case, it is possible that the methyl substituents cannot provide the minimum steric barrier required to prevent the bimolecular interactions that may be responsible for initiating disproportionation.

Regarding the three- and four-membered rings of cyclotristannanes and cyclotetrastannanes, it is interesting to note that the cyclotetrastannane 43 is apparently thermally robust in solution to temperatures of at least 100°C while there is evidence to suggest that the cyclotristannane 17 begins to disproportionate in solution above 80°C. This difference in stability is in keeping with recent theoretical calculations that predict that, in contrast to cyclopropane and cyclobutane, which possess similar strain-energy (SE) values, the parent cyclotristannane, $(H_2Sn)_3$, is considerably more strained (SE = 36.6 kcal mol^{-1}) than the parent cyclotetrastannane, $(H_2Sn)_4$ (SE = 12.2 kcal mol^{-1}) as a result of a greater reluctance of the heavier Group 14 elements to engage in the sp hybridization that is required for the formation of three-membered rings (87,88). With the cyclotristannane 12, the barriers to disproportionation have now been lowered to the extent that this process occurs readily at room temperature (21). In this system, it has been unequivocally established that 12 is in rapid thermal equilibrium with the distannene 72, according to Scheme 15 as supported by both variable-temperature UV spectroscopy (Fig. 12) and ^{119}Sn NMR spectroscopy. Through this latter method, it was also established that

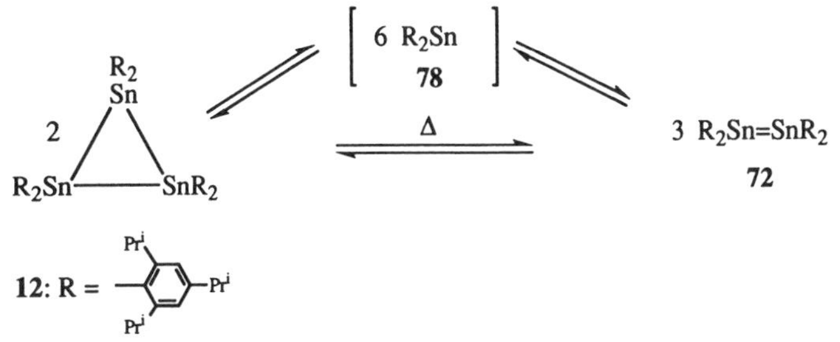

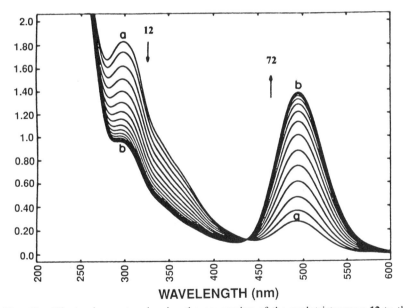

$$2 \begin{array}{c} R_2 \\ Sn \\ \triangle \\ R_2Sn \overline{\qquad} SnR_2 \end{array} \rightleftarrows \left[\begin{array}{c} 6\ R_2Sn \\ \mathbf{78} \end{array} \right] \rightleftarrows 3\ R_2Sn{=}SnR_2$$

$$\mathbf{12}: R = \underset{\underset{Pr^i}{}}{\overset{Pr^i}{\text{—}}}\!\!\!\!\!\!\!\!\text{—}Pr^i$$

SCHEME 15

compound **12** can be converted quantitatively to the distannene **72** through photolysis at $-78°C$.

D. Reactivity

Depending on the steric bulk of the substituents, cyclopolystannanes are stable towards air oxidation to varying degrees. Thus, whereas dodeca-methylcyclohexastannane **66** oxidizes almost spontaneously, the dodeca-

FIG. 12. Electronic spectra showing the conversion of the cyclotristannane **12** to the distannene **72**. Trace (a) is the spectrum taken after near thermal equilibrium of the cell at room temperature, and trace (b) is the spectrum taken after near thermal equilibrium of the cell at 90°C. Spectra taken every 2 seconds. Reprinted with permission from Masamune and Sita (*21*).

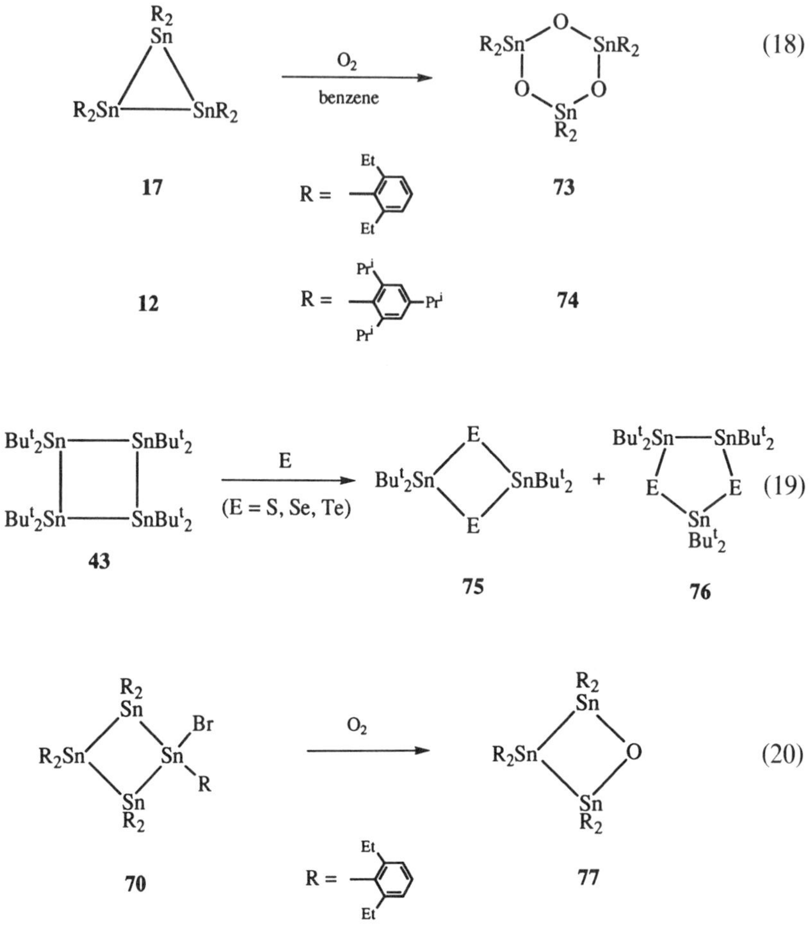

(18)

17

$R =$

73

12

$R =$

74

(19)

43

75

76

(20)

70

$R =$

77

SCHEME 16

phenyl derivative **63** appears to be indefinitely stable in both solution and the solid state. Except for compound **70**, all of the cyclotetrastannanes reported to date also appear to be stable towards air oxidation as well as the cyclotristannanes, **17** and **12**, in solid form. However, in solution, these latter two compounds slowly oxidize to the corresponding cyclotristannoxanes **73** and **74**, respectively, which have been structurally characterized [Eq. (18), Scheme 16] (21,27). With other chalcogens (S, Se, and Te), Puff and co-workers (89) were able to isolate the new compounds **75** and **76** starting from the cyclotetrastannane **43** according to Eq. (19). Interestingly, the product of a chalcogen atom inserting into a single Sn–Sn

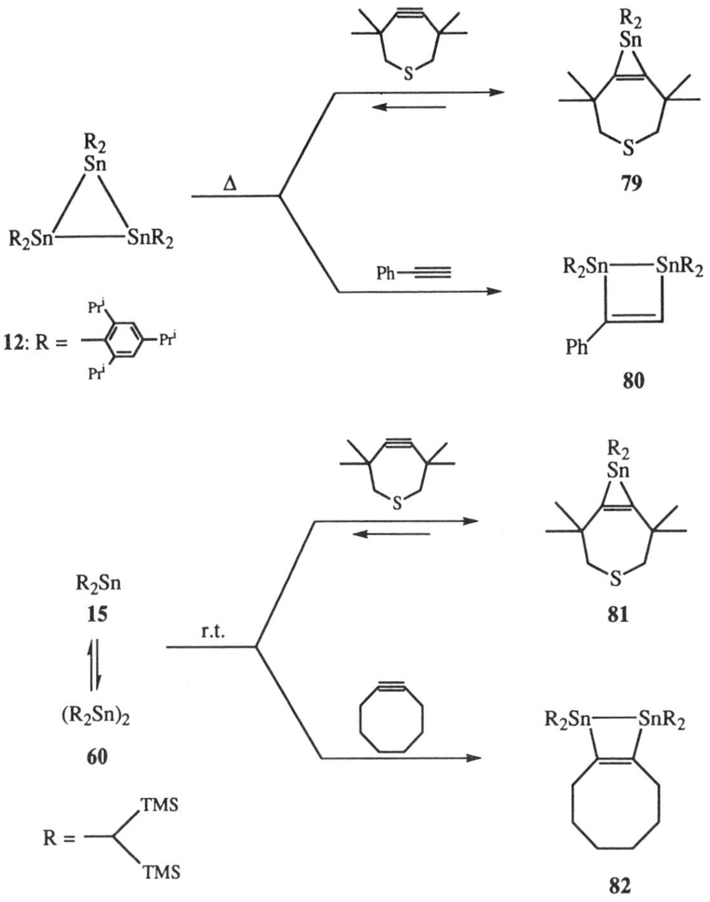

SCHEME 17

bond of **43** [i.e., compounds **52–54**, Eq. (13)] could not be observed. Finally, it has been observed that the cyclotetrastannane **70** oxidizes in solution to produce the new four-membered ring compound **77**, which has been structurally characterized [Eq. (20)] (*90*).

As noted earlier, the cyclotristannane **12** is a good thermal source for the distannene **72**, and presumably also for the diarylstannylene **78**, R_2Sn (R = 2,4,6-triisopropylphenyl), which is not spectroscopically observed but which must be involved in the equilibrium between **12** and **72**. Given this, several groups have employed **12** as a precursor to **72** and **78** in order to study their chemical reactivity. The results of these studies are compiled in Scheme 17, which shows that both of these reactive species can be

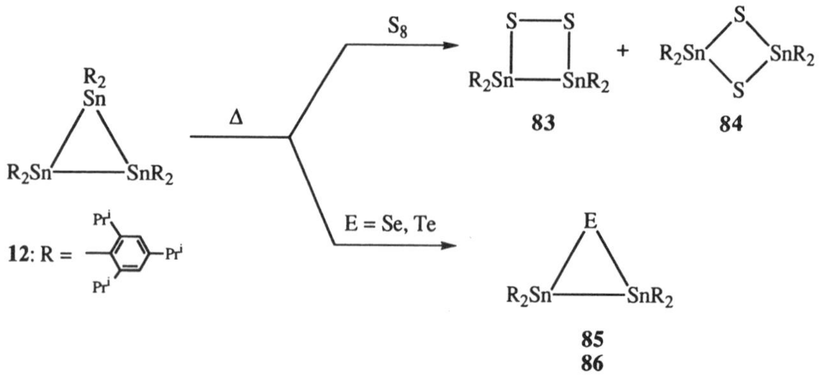

SCHEME 18

trapped by alkynes to produce the stannacyclopropene **79** (*91*) and the 1,2-distannacyclobut-3-ene **80** (*92*). It has been proposed that the formation of this latter compound proceeds via 2 + 2 addition of the distannene **72** to phenylacetylene. However, the sequential addition of the stannylene monomer **78** to the alkyne, followed by insertion of another stannylene equivalent into the Sn–C bond of the intermediate stannacyclopropene, cannot be ruled out. Indeed, this latter mechanism was previously proposed to account for the formation of both the stannacyclopropene **81** (*93*) and the 1,2-distannacyclobut-3-ene **82** (*94*) from the stannylene **15** according to Scheme 17. Perhaps more compelling evidence for the direct addition to the distannene **72** is provided by Weidenbruch and co-workers (*95,96*), who utilized the cyclotristannane **12** once more to generate **72** in situ, which was then reacted with various chalcogens, as shown in Scheme 18. With sulfur, the 1,2-dithiadistannetane **83** was produced, along with its isomer **84** (*95*), whereas with selenium and tellurium, only the corresponding three-membered ring compounds **85** and **86**, respectively, were isolated (*96*).

VI

Polycyclic Polystannanes

A. *Synthesis*

As late as 1988, there were no known examples of polycyclic polystannanes reported in the literature. However, through the careful fractionation of complex mixtures of products obtained from various oligomerization reaction processes, a number of colored compounds were isolated, characterized, and found to be derivatives of a structurally diverse array of polycyclic polystannanes. These studies have been the subject of earlier reviews (97), and thus only the salient features will be presented here.

As shown in Scheme 19, the first example of a polycyclic polystannane, the bicyclo[2.2.0]hexastannane **87**, was obtained from fractionation of the

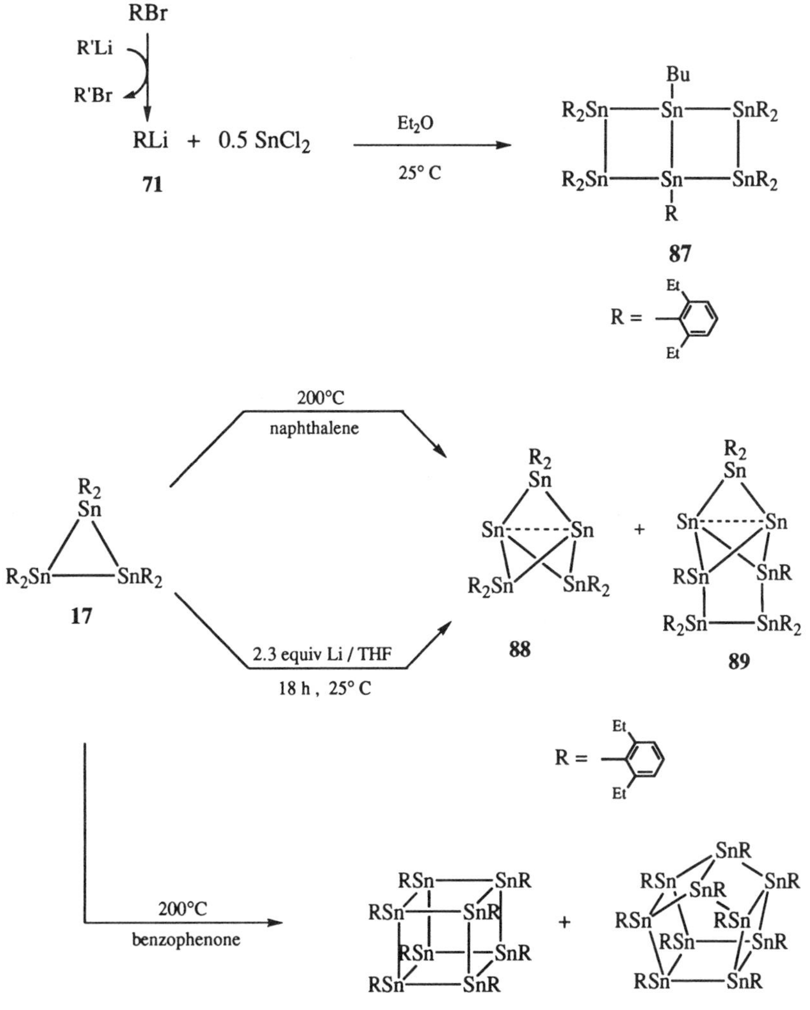

SCHEME 19

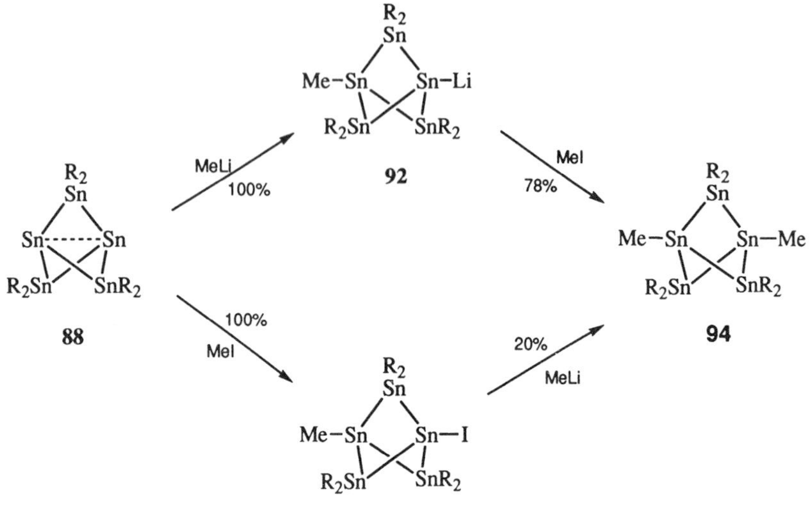

SCHEME 20

mixture of compounds obtained from the reaction of the aryllithium reagent **71** with tin dichloride in diethyl ether at 0°C (*83*). Interestingly, this compound was found to be stable towards air oxidation in both solution and in the solid state. A more surprising result, however, is the discovery that simple thermolysis of the cyclotristannane **17** at 200°C in various solvents, such as xylenes, naphthalene, and benzophenone, provides a mixture of compounds from which the perstanna[1.1.1]propellanes **88** and **89** and the perstanna-[*n*]-prismanes **90** (*n* = 4) and **91** (*n* = 5) can be isolated (*28,39,98,99*). Later it was discovered that the [1.1.1]propellanes **88** and **89** could also be obtained through reaction of the cyclotristannane **17** with lithium metal in tetrahydrofuran, as shown in Scheme 19 (*39*). Finally, it has been demonstrated that the pentastanna[1.1.1]propellane is a good precursor for the synthesis of the bicyclo[1.1.1]pentastannanes **92–94** via the reactions shown in Scheme 20 (*100,101*).

B. *Structural Studies*

Given the unique frameworks associated with the polycyclic polystannanes shown in Scheme 19, it is fortunate that each of these compounds could be obtained in crystalline form and subjected to crystallographic analysis. It is also fortunate that, in each case, the parent compounds for these frameworks have been the subject of theoretical studies, and this

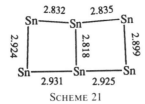

SCHEME 21

allows then for much-needed correlations between theory and experiment. Regarding the bicyclo[2.2.0]hexastannane **87**, crystallographic analysis provides a mean Sn–Sn bond length of 2.881 Å, which is within the expected range (*83*). However, inspection of the individual Sn–Sn bond lengths in this molecule reveal how relative differences in nonbonded interactions between the organic substituents can severely distort the σ-bonded framework away from its anticipated C_{2v} symmetric geometry, as shown in Scheme 21. Surprisingly, the bridgehead–bridgehead Sn–Sn bond in this compound is the shortest, at 2.818 Å, whereas calculations predict that it should be the longest, at 2.883 Å (*102*). In this case, it is likely that the nonbonded interactions in **87** enforce a "steric compression" of the central Sn–Sn bond rather than an elongation, which is their usual mode of action. Each of the four-membered rings in **87** are puckered rather than planar, and the mean of the Sn–Sn–Sn angles that define these rings is 89.9°. Of more important note is the average bond angle between the mean planes formed by these rings, which is 132.1° as opposed to the corresponding value of 113.6° that is calculated for the parent system. This structural discrepancy is once again most likely the result of the collaboration between strong nonbonded interactions and small-force constants for Sn–Sn–Sn bond angle deformations.

[1.1.1]Propellanes are a class of nonclassical structures, since they possess bridgehead atoms having formally inverted tetrahedral geometries that force the four bonds from each of the bridgehead atoms to lie within the same hemisphere. Given this configuration, theoretical investigations have long sought to determine the stability and reactivity of these molecular frameworks prior to the availability of representative synthetic derivatives. The isolation and structural characterization of the two perstanna[1.1.1]propellanes, **88** and **89**, thus represent a unique opportunity to test the accuracy of these predictions. The molecular structure of **88** is shown in Fig. 13; in the solid state, this compound is chiral, with D_3 symmetry. Due to hindered rotation about the Sn–C bonds, the ^{1}H NMR spectrum of **88** taken at 20°C show multiple resonances due to inequivalencies of the ethyl and phenyl protons, although at 100°C, partial rotations

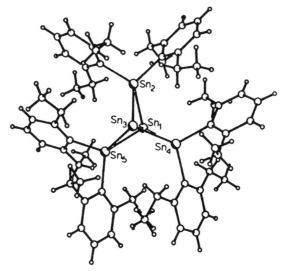

FIG. 13. Molecular structure of the pentastanna[1.1.1]propellane **88**. Reprinted with permission from Sita and Bickerstaff (*28*).

of the phenyl substitutents serve to remove some but not all of these inequivalencies. (i.e., the D_3 symmetry is preserved). The mean $Sn_{br}-Sn_{bh}$ (br = bridging, bh = bridgehead) bond length in **88** is 2.858 Å, whereas from the crystal structure of **89**, the corresponding value is determined to be slightly contracted, at 2.845 Å. Most importantly, the $Sn_{bh}-Sn_{bh}$ bond length in **88** is 3.367 Å, while once again this same value is slightly smaller for the heptastanna[1.1.1]propellane **89**, at 3.348 Å. Both of these values are much longer than even that of the distannane **16** (*cf.* 3.052 Å, Table I), and it is therefore natural to ask whether these former distances allow for true Sn–Sn bonding. In this regard, it is important to point out that the crystal structure reported for the bicyclo[1.1.1]pentastannane **93** reveals that the structural parameters for the tin-bonded framework of this compound are virtually identical to those of **88**, with a $Sn_{bh}-Sn_{bh}$ nonbonded distance of 3.361 Å (*100*). Overall, then, the data tend to suggest that the bridgehead–bridgehead interaction in compounds **88** and **89** is either nonexistent or very weak at best. This view is supported by recent theoretical studies of the parent pentastanna[1.1.1] propellane compound, which provide a calculated $Sn_{bh}-Sn_{bh}$ distance of between 3.463 and 3.495 Å and conclude that no "formal" bond connects these two atoms (*87,103*).

The crystal structures of the perstanna-[*n*]-prismanes **90** and **91** are

(a) (b)

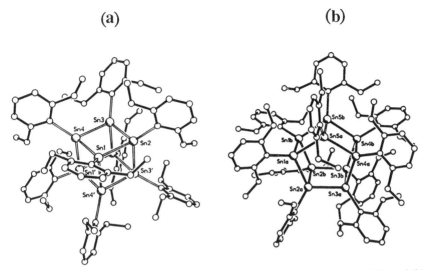

FIG. 14. Molecular structures of the perstanna-[n]-prismanes: (a) compound **90**, and (b) compound **91**. Reprinted with permission from Sita and Kinoshita (98,99).

shown in Fig. 14. In compound **90** the mean Sn–Sn bond length is 2.854 Å, whereas for compound **91** the mean Sn–Sn bond length within, and between, the five-membered rings is 2.856 Å. All of the Sn–Sn–Sn bond angles in these two compounds are very close to those expected on the basis of geometric considerations, and this lends support to the view that steric interactions between the 2,6-diethylphenyl substituents are minimal and do not contribute to significant distortions of the polystannane frameworks. Interestingly, the mean Sn–Sn bond length determined for **90** is slightly smaller than that predicted for the parent system (104).

C. Properties and Reactivity

All of the polycyclic polystannanes reported to date are thermally robust in solution to temperatures of at least 200°C. For the bicyclo[2.2.0]hexastannane derivative **87**, this is not too surprising, since the calculated strain-energy value for the parent compound is still ca. 12 kcal mol^{-1} less than the parent cyclotristannane system (102). However, the parent perstanna-[n]-prismanes for $n = 4$ and 5 are expected to be more ring-strained than cyclotristannane (cf. 73.2 and 61.0 kcal mol^{-1}, respectively) (99,104); therefore, the thermal stability of the two compounds, **90** and **91**, suggest that these compounds are kinetically stable by virtue of significantly high barriers toward Sn–Sn bond dissociation. Although thermally robust, the bicyclo[1.1.1]pentastannane **94** has been shown to regenerate the pentastanna[1.1.1]propellane **88** upon photolysis, and the driving force for this process has been interpreted as arising from the release of strong steric interactions at the bridgehead positions (*101*).

All of the polycyclic polystannanes reported to date are visibly colored, and their electronic spectra accordingly exhibit low energy-absorption maxima. Perhaps the most striking examples are the perstanna[1.1.1]propellanes, **88** and **89**, solutions of which are blue-violet and burgundy in color, and which show corresponding absorption maxima at λ_{max} 558 (ε_{max} 1540) and 528 nm (4541), respectively. A correlation between these absorption maxima and Sn_{bh}–Sn_{bh} bond length has been used to suggest that a significant bridgehead–bridgehead bonding interaction in perstanna[1.1.1]-propellanes does in fact exist. For the two perstanna-[n]-prismanes, **90** and **91**, the electronic spectra are quite different, with unique absorption maxima for **90** occurring at λ_{max} 275 nm (ε_{max} 112,000) and 450 nm (ε_{max} 2,000), while for **91** no distinct absorptions can be discerned [λ 270 nm (ε 83,000), 350 nm (ε 6,100)]. The bicyclo[2.2.0]hexastannane **87** is also highly colored, with absorptions occurring at λ_{max} 307 nm (ε_{max} 37,900) and 360 nm (13,690). Finally, the bicyclo[1.1.1]pentastannanes **93** and **94** display significantly different electronic spectra, with the former being red-colored and having an unusual absorption maxima at λ_{max} 500 nm (ε_{max} 1700) that is absent in the latter.

Given the unique structures of the polycyclic polystannanes that have been isolated to date, these compounds provide an opportunity to explore the relationships between structure and ^{119}Sn NMR parameters. For the [1.1.1]propellane **88**, the chemical shifts of Sn_{br} and Sn_{bh} occur at 356 and -1751 ppm, respectively. The values for the coupling constants, $^1J(^{119}Sn_{br}-^{117}Sn_{bh})$ and $^2J(^{119}Sn_{br}-^{117}Sn_{br})$ in this compound are 3975 and 262 Hz, respectively. Unfortunately $^1J(^{119}Sn_{bn}-^{117}Sn_{bh})$ could not be observed for this compound. The values obtained above for **88** can be contrasted to those determined for the bicyclo[1.1.1]pentastannane **94** by ^{119}Sn NMR spectroscopy. In this compound the chemical shifts for Sn_{br} and Sn_{bh} are now at -334 and -247, respectively (*105*), and the values for the coupling constants, $^1J(^{119}Sn_{br}-^{117}Sn_{bh})$, $^2J(^{119}Sn_{br}-^{117}Sn_{br})$, and $^2J(^{119}Sn_{bh}-^{117}Sn_{bh})$, are 125, 1387, and 7390 Hz, respectively. These latter values are fairly unusual, and the large size for $^2J(^{119}Sn_{bh}-^{117}Sn_{bh})$ suggests that additive contributions from multiple pathways within a polycyclic framework play an important role in determining the magnitude of coupling constants in polycyclic polystannanes. The value for the one-bond coupling constant, $^1J(^{119}Sn_{br}-^{117}Sn_{bh})$, is small but still within the range expected on the basis of a bond-length dependence for this parameter. What is surprising, however, is that this coupling constant is dramatically temperature dependent, starting at a value of 200 Hz at 214 K and decreasing to a value of 38 Hz at 373 K. This behavior is quite unusual for a stereochemically rigid polyatomic molecule. It has been interpreted as arising from the increase in nonbonded steric interactions at elevated temperatures, which serve to facilitate a breathing-mode expansion of the bicyclo[1.1.1]pentastannane framework and thereby to shift the Sn–Sn bond lengths and Sn–Sn–Sn bond angles away from their equilibrium values. As a final note concerning the ^{119}Sn NMR spectra of polycyclic polystannanes, the spectra for the perstanna-[n]-prismanes, **90** and **91**, reveal chemical shifts for these compounds of 44.3 and -21.3 ppm, respectively. For compound **90**, the observed coupling constants, $^1J(^{119}Sn-^{117}Sn)$ and $^2J(^{119}Sn-^{117}Sn)$, have values of 1576 and 1345 Hz, respectively. In compound **91**, the coupling constants between tin atoms in different five-membered rings are $^1J(^{119}Sn-^{117}Sn) = 693$ Hz and $^2J(^{119}Sn-^{117}Sn) = 1224$ Hz, whereas those between tin atoms in the same five-membered rings are $^1J(^{119}Sn-^{117}Sn) = 3312$ Hz and $^2J(^{119}Sn-^{117}Sn) = 90$ Hz. It is interesting to note that in both **90** and **91**, the expected $^3J(^{119}Sn-^{117}Sn)$ coupling constants are not observed, presumably due to a Karplus correlation between dihedral angle and vicinal coupling constants that predicts that the magnitude of these particular coupling constants should be near zero due to the 90° dihedral angle in both molecular frameworks.

Regarding the chemical reactivity of polycyclic polystannanes, to date this subject has not been thoroughly investigated. However, it can be anticipated that this will be a rich field for study. In particular, these compounds should be interesting starting materials for the synthesis of new classes of compounds. Some demonstration of this was shown in Scheme 20 by the synthesis of bicyclo[1.1.1]pentastannanes from the pentastanna[1.1.1]propellane **88**. It has also been shown that both **88** and **89** can be chemically reduced to the corresponding dianions without cleaving any of the Sn–Sn bonds of their polycyclic frameworks. These dianions might then serve to generate other compounds through their reaction with various electrophiles. Finally, another class of reaction that should prove interesting is the insertion of chalcogen atoms into the tin-bonded frameworks of the perstanna-[n]-prismanes, **90** and **91**.

VII

CONCLUSION

From the review of our knowledge regarding polystannanes presented here, several points become obvious. First, a wide structural diversity for polystannanes does exist. Second, by virtue of the unique properties associated with their Sn–Sn σ-bonded frameworks, polystannanes are an important class of compound to study. Finally, there remains several large gaps in our knowledge concerning the structure/property relationships of polystannanes. However, it remains the hope that as new synthetic methodologies are developed, most of these holes will be filled through the acquisition and investigation of new derivatives of several different structural classes.

REFERENCES

(1) Rundle, R. E. J. Phys. Chem. **1957,** *61* 45.

(2) Miller, J. R. J. Chem. Soc. **1961,** 4452; **1965,** 713.

(3) Little, W. A. Phys. Rev. **1964,** *134,* 1416.

(4) Ferrell, R. A. Phys. Rev. Lett. **1964,** *13,* 330.

(5) Pitt, C. G.; Monteith, L. K.; Ballard, L. F.; Collman, J. P.; Morrow, J. C.; Roper, W. R.; Ulku, D. J. Am. Chem. Soc. **1966,** *88,* 4286; Collman, J. P.; Hoyano, J. K.; Murphy, D. W. ibid. **1973,** *95,* 3424; Collman, J. P.; Murphy, D. W.; Fleischer, E. B.; Swift, D. Inorg. Chem. **1974,** *13,* 1.

(6) Jetz, W.; Graham, W. A. G. Inorg. Chem. **1971,** *10,* 1647.

(7) A novel solid-phase approach to the synthesis of catenated metal oligomers has been presented, however, to date, demonstration of its utility is confined to a single three-metal example, see Burlitch, J. M.; Winterton, R. C. J. Am. Chem. Soc. **1975,** *97,* 5605.

(8) Creemer, H. M. J. C.; Verbeek, F.; Noltes, J. G. J. Organomet. Chem. **1968,** *15,* 125.

(9) For some recent reviews, see (a) Tsumuraya, T.; Batcheller, S. A.; Masamune, S. Agnew. Chem., Int. Ed. Engl. **1991,** *30,* 902; (b) Sekiguchi, A.; Sakurai, H. In "The Chemistry of Inorganic Ring Systems"; Steudel, R., Ed., Elsevier: New York, 1992, pp. 101–125.

(10) Miller, R. D.; Michl, J. J. Chem. Rev. **1989,** *89,* 1359; Zeigler, J. M.; Fearson, R. W. G. Adv. Chem. Seri. **1990,** *224.*

(11) Bigelow, R. W. Chem. Phys. Lett. **1986,** *126,* 63; Balaji, V.; Michl, J. Polyhedron **1991,** *10,* 1265; Ortiz, J. V. ibid., 1285.

(12) For a previous review, see: Adams, S.; Dräger, M. Main Group Metal Chemistry **1988,** *XI,* 151.

(13) Jolly, W. L. Angew. Chem. **1960,** *72,* 268; J. Chem. Soc. **1961,** *83,* 335.

(14) Neumann, W. P. "The Organic Chemistry of Tin"; Wiley: New York, 1970; Davies, A. G.; Smith, P. J. In "Comprehensive Organometallic Chemistry"; Wilkinson, G.; Stone, F. G. A.; Abel, E. W., Eds.; Pergamon: New York, 1982, pp. 519–627 in Vol. 2.

(15) Neumann, W. P.; Schneider, B.; Sommer, R. Justus Liebigs Ann. Chem. **1966,** *692,* 1–11.

(16) Gielen, M.; Tondeur, Y. J. Chem. Soc., Chem. Commun. **1978,** 81.

(17) Kandil, S. A.; Allred, A. L. Chem. Soc. A **1970,** 2987.

(18) Weidenbruch, M.; Schlaefke, J.; Peters, K.; von Schnering, H. G. J. Organomet. Chem. **1991,** *414,* 319.

(19) Neumann, W. P.; Pedain, J. Tetrahedron Lett. **1964,** 2461.

(20) Mathiasch, B. Inorg. Nucl. Chem. Lett **1977,** *13,* 13; Mathiasch, B. Synth. React. Inorg. Met-Org. Chem. **1977,** *7,* 227.

(21) Masamune, S.; Sita, L. R. J. Am. Chem. Soc. **1985,** *107,* 6390.

(22) Brown, P.; Mahon, M. F.; Molloy, K. C. J. Organomet. Chem. **1992,** *435,* 265.

(23) Sita, L. R. Ph. D. Dissertation, MIT, Cambridge, MA, 1985.

(24) Puff, H.; Breuer, B.; Gehrke-Brinkmann, G.; Kind, P.; Reuter, H.; Schuh, W.; Wald, W.; Weidenbrüch, G. J. Organomet. Chem. 1989, 363, 265.

(25) Davidson, P. J.; Harris, D. H.; Lappert, M. F. J. Chem. Soc., Dalton Trans. 1976, 2268.

(26) Al-Allaf, T. A. K.; Eaborn, C.; Hitchcock, P. B.; Lappert, M. F.; Pidcock, A. J. Chem. Soc., Chem. Commun. 1985, 548.

(27) Masamune, S.; Sita, L. R.; Williams, D. J. J. Am. Chem. Soc. 1983, 105, 630.

(28) Sita, L. R.; Bickerstaff, R. D. J. Am. Chem. Soc. 1989, 111, 6454.

(29) Sanz, J. F.; Marquez, A. J. Phys. Chem. 1989, 93, 7328.

(30) Preut, H.; Haupt, H.-J.; Huber, F. Z. Anorg. Allg. Chem. 1973, 396, 81.

(31) Hounshll, W. D.; Dougherty, D. A.; Hummel, J. P.; Mislow, K. J. Am. Chem. Soc. 1977, 99, 1916.

(32) Schleyer, P. v. R.; Kaupp, M.; Hampel, F.; Bremer, M.; Mislow, K. J. Am. Chem. Soc. 1992, 114, 6791.

(33a) Tagliavani, G.; Faleschini, S.; Pilloni, G.; Plazzogna, G. J. Organomet. Chem. 1966, 5, 136 (synthesis).

(33b) Schneider, C.; Dräger, M. J. Organomet. Chem. 1991, 415, 349–362 (X-ray).

(34) Morris, H.; Byerly, W.; Selwood, P. W. J. Am. Chem. Soc. 1942, 64, 1727 (synthesis); Gilman, H.; Rosenberg, S. D. J. Org. Chem. 1953, 18, 1554 (synthesis); Schneider-Koglin, C.; Behrends, K.; Dräger, M. J. Organomet. Chem. 1993, 448, 29 (X-ray).

(35) Ref. 28; see supplementary material.

(36) Adams, S.; Dräger, M.; Mathiasch, B. Z. Anorg. Allg. Chem. 1986, 532, 81 (X-ray).

(37) Drenth, W.; Janssen, M. J.; Van Der Kerk, G. J. M.; Vliegenthart, J. A. J. Organomet. Chem. 1964, 2, 265.

(38) Hauge, D. N.; Prince, R. H. J. Chem. Soc. 1965, 4690.

(39) Sita, L. R.; Kinoshita, I. J. Am. Chem. Soc. 1992, 114, 7024.

(40) Gager, H. M.; Lewis, J.; Ware, M. J. J. Chem. Soc., Chem. Commun. 1966, 616.

(41a) Buschhaus, H.-U.; Neumann, W. P.; Apoussidis, T. Liebigs Ann. Chem. 1981, 1190.

(41b) El-Farargy, A. F.; Lehnig, M.; Neumann, W. P. Chem. Ber. 1982, 115, 2783.

(42a) Wrackmeyer, B. Annu. Rep. NMR Spectrosc. 1985, 16, 73.

(42b) Holt, M. S.; Wilson, W. L.; Nelson, J. H. Chem. Rev. 1989, 89, 11.

(43a) Mitchell, T. N. J. Organomet. Chem. 1974, 70, C1.

(43b) Mitchell, T. N.; Walter, G. J. Chem. Soc., Perkin Trans. 2, 1977, 1842.

(44a) Kennedy, J. D.; McFarlane, W. J. Chem. Soc., Chem. Commun. 1974, 983.

(44b) Kennedy, J. D.; McFarlane, W.; Pyne, G. S.; Wrackmeyer, B. J. Chem. Soc., Dalton Trans. 1975, 386.

(44c) Kennedy, J. D.; McFarlane, W. J. Chem. Soc., Dalton Trans. 1976, 1219–1223.

(44d) Biffar, W.; Ebeling-Gasparis, T.; Noth, H.; Storch, W.; Wrackmeyer, B. J. Magn. Reson. 1981, 44, 54.

(45) Pople, J. A.; Santry, D. P. Mol. Phys. 1964, 8, 1.

(46) Pereyre, M. "Tin in Organic Synthesis"; Butterworth: Boston, 1987.

(47) See Table 1 in Ref. 41a.

(48) Grugel, C.; Neumann, W. P.; Seifert, P. Tetrahedron Lett. 1977, 2205.

(49) Sommer, R.; Schneider, B.; Neumann, W. P. Justus Liebigs Ann. Chem. 1966, 692, 12.

(50) Jousseaume, B.; Chanson, E.; Bevilacqua, M.; Saux, A.; Pereyre, M.; Barbe, B.; Petraud, M. J. Organomet. Chem. 1985, 294, C41.

(51) Creemers, H. M. J. C.; Noltes, J. G. Recl. Trav. Chim. Pays-Bas 1965, 84, 382; Creemers, H. M. J. C.; Noltes, J. G. J. Organomet. Chem. 1967, 7, 237.

(52) Sita, L. R. Organometallics 1992, 11, 1442.

(53) Puff, H.; Bach, C.; Schuh, W.; Zimmer, R. J. Organomet. Chem. 1986, 312, 313.

(54) Adams, S.; Dräger, M. Agnew. Chem., Int. Ed. Engl. 1987, 26, 1255.

(55) Farrar, W. V.; Skinner, H. A. *J. Organomet. Chem.* **1964**, *1*, 434.

(56) Adams, S.; Dräger, M. *J. Organomet. Chem.* **1985**, *288*, 295.

(57) Gilman, H.; Artwell, W. H.; Schwebke, G. L. *J. Organomet. Chem.* **1964**, *2*, 369; Sakurai, H.; Kumada, M. *Bull. Chem. Soc. Jpn.* **1964**, *37*, 1894.

(58) Pitt, C. G.; Jones, L. L.; Ramsey, B. G. *J. Am. Chem. Soc.* **1967**, *89*, 5471.

(59) Drenth, W.; Noltes, J. G.; Bulten, E. J.; Creemers, H. M. J. C. *J. Organomet. Chem.* **1969**, *17*, 173.

(60) Takeda, K.; Shiraishi, K. *Chem. Phys. Lett.* **1992**, *195*, 121.

(61) Puff, H.; Bongartz, A.; Sievers, R.; Zimmer, R. *Agnew. Chem., Int. Ed. Engl.* **1978**, *17*, 939.

(62) Grugel, C.; Lehnig, M.; Neumann, W. P.; Sauer, J. *Tetrahedron Lett.* **1980**, 273.

(63) Puff, H.; Bongartz, A.; Schuh, W.; Zimmer, R. *J. Organomet. Chem.* **1983**, *248*, 61.

(64) Willemsens, L. C.; van der Kerk, G. J. M. *J. Organomet. Chem.* **1964**, *2*, 260.

(65) Mitchell, T. N.; El-Behairy M. *J. Organomet. Chem.* **1977**, *141*, 43.

(66) Kobayashi, K.; Kawanisi, M.; Hitomi, T.; Kozima, S. *J. Organomet. Chem.* **1982**, *233*, 299–311.

(67) Klaui, W.; Werner, H. *J. Organomet. Chem.* **1973**, *54*, 331.

(68) For early work related to the synthesis of monomeric stannylene species, see (a) Krause, E.; Becker, R. *Ber. Dtsch. Chem. Ges.* **1920**, *53*, 173; (b) Krause, E.; Pohland, R.; *Ber. Dtsch. Chem. Ges.* **1924**, *57*, 532; (c) Baker, G.; Gelius, R. *Chem. Ber.* **1958**, *91*, 829; (d) Jensen, K. A.; Clauson-Kaas, N. *Z. Anorg. Allg. Chem.* **1943**, *250*, 277; (e) Kuivila, H. G.; Beumel, O. F., Jr. *J. Am. Chem. Soc.* **1958**, *80*, 3250.

(69) Goldberg, D. E.; Hitchcock, P. B.; Lappert, M. F.; Thomas, K. M.; Thorne, A. J.; Fjeldberg, T.; Haaland, A.; Schilling, B. E. R. *J. Chem. Soc., Dalton Trans.* **1986**, 2387, and references cited therein.

(70) Zilm, K.; Lawless, G. A.; Merrill, R. M.; Millar, J. M.; Webb, G. G. *J. Am. Chem. Soc.* **1987**, *109*, 7236.

(71) Kira, M.; Yauchibara, R.; Hirano, R.; Kabuto, C.; Sakurai, H. *J. Am. Chem. Soc.* **1991**, *113*, 7785.

(72) Weidenbruch, M.; Schlaefke, J.; Schäfer, A.; Peters, K.; von Schnering, H. G.; Marsman, H. *Angew. Chem. Int. Ed. Engl.* **1994**, *33*, 1846.

(73) Neumann, W. P.; Konig, K. *Angew. Chem., Int. Ed. Engl.* **1962**, *1*, 212; Neumann, W. P.; Konig, K. *Justus Liebigs Ann. Chem.* **1964**, *677*, 1.

(74) Olson, D. H.; Rundle, R. E. *Inorg. Chem.* **1963**, *2*, 1310.

(75) Neumann, W. P.; Konig, K. *Justus Liebigs Ann. Chem.* **1964**, *677*, 12; Neumann, W. P.; Pedain, J. *ibid. 672*, 34; Neumann, W. P.; Pedain, J.; Sommer, R. *ibid.* **1966**, *694*, 9; Neumann, W. P.; Konig, K. *Angew. Chem., Int. Ed. Engl.* **1964**, *3*, 751.

(76) Puff, H.; Bach, C.; Reuter, H.; Schuh, W. *J. Organomet. Chem.* **1984**, *277*, 17.

(77) Dräger, M.; Mathiasch, B.; Ross, L.; Ross, M. *Z. Anorg. Allg. Chem.* **1983**, *506*, 99.

(78) Jousseaume, B.; Noriet, N.; Pereyre, M.; Saux, A.; Francès, J.-M. *Organometallics* **1994**, *13*, 1034.

(79) Belsky, V. K.: Zemlyansky, N. N.; Kolosova, N. D.; Borisova, I. V. *J. Organomet. Chem.* **1981**, *215*, 41.

(80) Watta, B.; Neumann, W. P.; Sauer, J. *Organometallics* **1985**, *4*, 1954.

(81) Lappert, M. F.; Leung, W. P.; Raston, C. L.; Thorne, A. J.; Skelton, B. W.; White, A. H. *J. Organomet. Chem.* **1982**, *233*, C28.

(82) Cardin, C. J.; Cardin, D. J.; Convery, M. A.; Devereaux, M. M.; Kelly, N. B. *J. Organomet. Chem.* **1991**, *414*, C9.

(83) Sita, L. R.; Bickerstaff, R. D. *J. Am. Chem. Soc.* **1989**, *111*, 3769.

(84) Brady, F. J.; Cardin, C. J.; Cardin, D. J.; Convery, M. A.; Devereux, M. M.; Lawless, G. A. *J. Organomet. Chem.* **1991**, *241*, 199.

(85) Schäfer, A.; Weidenbruch, M.; Saak, W.; Pohl, S.; Marsmann, H. *Angew. Chem., Int. Ed. Engl.* **1991,** *30,* 834.

(86) Neumann, W. P.; Fu, J. *J. Organomet. Chem.* **1984,** *273,* 295–302; Fu, J.; Neumann, W. P. *ibid.,* *272,* C5–C9.

(87) Nagase, S. *Polyhedron* **1991,** *10,* 1299.

(88) For another theoretical study of cyclotristannane and cyclotetrastannane, see Rubio, J.; Illas, F. *J. Mol. Struct. (Theochem.)* **1984,** *110,* 131.

(89) Puff, H.; Gattermayer, R.; Hundt, R.; Zimmer, R. *Angew. Chem., Int. Ed. Engl.* **1977,** *16,* 547.

(90) Cardin, C. J.; Cardin, D. J.; Convery, M. A.; Devereux, M. M. *J. Organomet. Chem.* **1991,** *411,* C3.

(91) Sita, L. R.; Bickerstaff, R. D. *Phosphorus, Sulfur Silicon Relat. Elem.* **1989,** *41,* 31.

(92) Weidenbruch, M.; Schäfer, A.; Kilian, H.; Pohl, S.; Saak, W.; Marsmann, H. *Chem. Ber.* **1992,** *125,* 563.

(93) Sita, L. R.; Bickerstaff, R. D. *J. Am. Chem. Soc.* **1988,** *110,* 5208.

(94) Sita, L. R.; Kinoshita, I.; Lee, S. P. *Organometallics* **1990,** *9,* 1644.

(95) Schäfer, A.: Weidenbruch, M.; Saak, W.; Pohl, S.; Marsmann, H. *Angew. Chem., Int. Ed. Engl.* **1991,** *30,* 962.

(96) Schäfer, A.; Weidenbruch, M.; Saak, W.; Pohl, S.; Marsmann, H. *Angew. Chem., Int. Ed. Engl.* **1991,** *30,* 834.

(97) Sita, L. R.; Kinoshita, I. In "Inorganic and Organometallic Oligomers and Polymers"; Harrod, J. F.; Laine, R. M., Eds.; Kluwer Academic Publishers: Dordrecht, The Netherlands, 1991; Sita, L. R. *Acc. Chem. Res.* **1994,** *27,* 191.

(98) Sita, L. R.; Kinoshita, I. *Organometallics* **1990,** *9,* 2865.

(99) Sita, L. R.; Kinoshita, I. *J. Am. Chem. Soc.* **1991,** *113,* 1856.

(100) Sita, L. R.; Kinoshita, I. *J. Am. Chem. Soc.* **1990,** *112,* 8839.

(101) Sita, L. R.; Kinoshita, I. *J. Am. Chem. Soc.* **1991,** *113,* 5070.

(102) Nagase, S.; Kudo, T. *J. Chem. Soc., Chem. Commun.* **1990,** 630.

(103) Gordon, M. S.; Nguyen, K. A.; Carroll, M. T. *Polyhedron* **1991,** *10,* 1247.

(104) Nagase, S. *Angew. Chem., Int. Ed. Engl.* **1989,** *28,* 329.

(105) Ref. *101,* see supplementary material.

Index

Cumulative List of Contributors for Volumes 1–36

Abel, E. W., **5,** 1; **8,** 117
Aguiló, A., **5,** 321
Akkerman, O. S., **32,** 147
Albano, V. G., **14,** 285
Alper, H., **19,** 183
Anderson, G. K., **20,** 39; **35,** 1
Angelici, R. J., **27,** 51
Aradi, A. A., **30,** 189
Armitage, D. A., **5,** 1
Armor, J. N., **19,** 1
Ash, C. E., **27,** 1
Ashe III, A. J., **30,** 77
Atwell, W. H., **4,** 1
Baines, K. M., **25,** 1
Barone, R., **26,** 165
Bassner, S. L., **28,** 1
Behrens, H., **18,** 1
Bennett, M. A., **4,** 353
Bickelhaupt, F., **32,** 147
Birmingham, J., **2,** 365
Blinka, T. A., **23,** 193
Bockman, T. M., **33,** 51
Bogdanović, B., **17,** 105
Bottomley, F., **28,** 339
Bradley, J. S., **22,** 1
Brew, S. A., **35,** 135
Brinckman, F. E., **20,** 313
Brook, A. G., **7,** 95; **25,** 1
Bowser, J. R., **36,** 57
Brown, H. C., **11,** 1
Brown, T. L., **3,** 365
Bruce, M. I., **6,** 273, **10,** 273; **11,** 447; **12,** 379; **22,** 59
Brunner, H., **18,** 151
Buhro, W. E., **27,** 311
Byers, P. K., **34,** 1
Cais, M., **8,** 211
Calderon, N., **17,** 449
Callahan, K. P., **14,** 145
Canty, A. J., **34,** 1
Cartledge, F. K., **4,** 1
Chalk, A. J., **6,** 119
Chanon, M., **26,** 165

Chatt, J., **12,** 1
Chini, P., **14,** 285
Chisholm, M. H., **26,** 97; **27,** 311
Chiusoli, G. P., **17,** 195
Chojinowski, J., **30,** 243
Churchill, M. R., **5,** 93
Coates, G. E., **9,** 195
Collman, J. P., **7,** 53
Compton, N. A., **31,** 91
Connelly, N. G., **23,** 1; **24,** 87
Connolly, J. W., **19,** 123
Corey, J. Y., **13,** 139
Corriu, R. J. P., **20,** 265
Courtney, A., **16,** 241
Coutts, R. S. P., **9,** 135
Coville, N. J., **36,** 95
Coyle, T. D., **10,** 237
Crabtree, R. H., **28,** 299
Craig, P. J., **11,** 331
Csuk, R., **28,** 85
Cullen, W. R., **4,** 145
Cundy, C. S., **11,** 253
Curtis, M. D., **19,** 213
Darensbourg, D. J., **21,** 113; **22,** 129
Darensbourg, M. Y., **27,** 1
Davies, S. G., **30,** 1
Deacon, G. B., **25,** 237
de Boer, E., **2,** 115
Deeming, A. J., **26,** 1
Dessy, R. E., **4,** 267
Dickson, R. S., **12,** 323
Dixneuf, P. H., **29,** 163
Eisch, J. J., **16,** 67
Ellis, J. E., **31,** 1
Emerson, G. F., **1,** 1
Epstein, P. S., **19,** 213
Erker, G., **24,** 1
Ernst, C. R., **10,** 79
Errington, R. J., **31,** 91
Evans, J., **16,** 319
Evans, W. J., **24,** 131
Faller, J. W., **16,** 211
Farrugia, L. J., **31,** 301

Faulks, S. J., **25**, 237
Fehlner, T. P., **21**, 57; **30**, 189
Fessenden, J. S., **18**, 275
Fessenden, R. J., **18**, 275
Fischer, E. O., **14**, 1
Ford, P. C., **28**, 139
Forniés, J., **28**, 219
Forster, D., **17**, 255
Fraser, P. J., **12**, 323
Friedrich, H., **36**, 229
Friedrich, H. B., **33**, 235
Fritz, H. P., **1**, 239
Fürstner, A., **28**, 85
Furukawa, J., **12**, 83
Fuson, R. C., **1**, 221
Gallop, M. A., **25**, 121
Garrou, P. E., **23**, 95
Geiger, W. E., **23**, 1; **24**, 87
Geoffroy, G. L., **18**, 207; **24**, 249; **28**, 1
Gilman, H., **1**, 89; **4**, 1; **7**, 1
Gladfelter, W. L., **18**, 207; **24**, 41
Gladysz, J. A., **20**, 1
Glänzer, B. I., **28**, 85
Green, M. L. H., **2**, 325
Grey, R. S., **33**, 125
Griffith, W. P., **7**, 211
Grovenstein, Jr., E., **16**, 167
Gubin, S. P., **10**, 347
Guerin, C., **20**, 265
Gysling, H., **9**, 361
Haiduc, I., **15**, 113
Halasa, A. F., **18**, 55
Hamilton, D. G., **28**, 299
Handwerker, H., **36**, 229
Harrod, J. F., **6**, 119
Hart, W. P., **21**, 1
Hartley, F. H., **15**, 189
Hawthorne, M. F., **14**, 145
Heck, R. F., **4**, 243
Heimbach, P., **8**, 29
Helmer, B. J., **23**, 193
Henry, P. M., **13**, 363
Heppert, J. A., **26**, 97
Herberich, G. E., **25**, 199
Herrmann, W. A., **20**, 159
Hieber, W., **8**, 1
Hill, A. F., **36**, 131
Hill, E. A., **16**, 131
Hoff, C., **19**, 123
Hoffmeister, H., **32**, 227

Holzmeier, P., **34**, 67
Honeyman, R. T., **34**, 1
Horwitz, C. P., **23**, 219
Hosmane, N. S., **30**, 99
Housecroft, C. E., **21**, 57; **33**, 1
Huang, Y. Z., **20**, 115
Hughes, R. P., **31**, 183
Ibers, J. A., **14**, 33
Ishikawa, M., **19**, 51
Ittel, S. D., **14**, 33
Jain, L., **27**, 113
Jain, V. K., **27**, 113
James, B. R., **17**, 319
Janiak, C., **33**, 291
Jastrzebski, J. T. B. H., **35**, 241
Jenck, J., **32**, 121
Jolly, P. W., **8**, 29; **19**, 257
Jonas, K., **19**, 97
Jones, M. D., **27**, 279
Jones, P. R., **15**, 273
Jordan, R. F., **32**, 325
Jukes, A. E., **12**, 215
Jutzi, P., **26**, 217
Kaesz, H. D., **3**, 1
Kalck, P., **32**, 121; **34**, 219
Kaminsky, W., **18**, 99
Katz, T. J., **16**, 283
Kawabata, N., **12**, 83
Kemmitt, R. D. W., **27**, 279
Kettle, S. F. A., **10**, 199
Kilner, M., **10**, 115
Kim, H. P., **27**, 51
King, R. B., **2**, 157
Kingston, B. M., **11**, 253
Kisch, H., **34**, 67
Kitching, W., **4**, 267
Kochi, J. K., **33**, 51
Köster, R., **2**, 257
Kreiter, C. G., **26**, 297
Krüger, G., **24**, 1
Kudaroski, R. A., **22**, 129
Kühlein, K., **7**, 241
Kuivila, H. G., **1**, 47
Kumada, M., **6**, 19; **19**, 51
Lappert, M. F., **5**, 225; **9**, 397; **11**, 253; **14**, 345
Lawrence, J. P., **17**, 449
Le Bozec, H., **29**, 163
Lednor, P. W., **14**, 345
Linford, L., **32**, 1

Stone, F. G. A., **1**, 143; **31**, 53; **35**, 135
Su, A. C. L., **17**, 269
Suslick, K. M., **25**, 73
Süss-Fink, G., **35**, 41
Sutin, L., **28**, 339
Swincer, A. G., **22**, 59
Tamao, K., **6**, 19
Tate, D. P., **18**, 55
Taylor, E. C., **11**, 147
Templeton, J. L., **29**, 1
Thayer, J. S., **5**, 169; **13**, 1; **20**, 313
Theodosiou, I., **26**, 165
Timms, P. L., **15**, 53
Todd, L. J., **8**, 87
Touchard, D., **29**, 163
Traven, V. F., **34**, 149
Treichel, P. M., **1**, 143; **11**, 21
Tsuji, J., **17**, 141
Tsutsui, M., **9**, 361; **16**, 241
Turney, T. W., **15**, 53
Tyfield, S. P., **8**, 117
Usón, R., **28**, 219
Vahrenkamp, H., **22**, 169
van der Kerk, G. J. M., **3**, 397

van Koten, G., **21**, 151; **35**, 241
Veith, M., **31**, 269
Vezey, P. N., **15**, 189
von Ragué Schleyer, P., **24**, 353; **27**, 169
Vrieze, K., **21**, 151
Wada, M., **5**, 137
Walton, D. R. M., **13**, 453
Wailes, P. C., **9**, 135
Webster, D. E., **15**, 147
Weitz, E., **25**, 277
West, T., **5**, 169; **16**, 1; **23**, 193
Werner, H., **19**, 155
White, D., **36**, 95
Wiberg, N., **23**, 131; **24**, 179
Wiles, D. R., **11**, 207
Wilke, G., **8**, 29
Williams, R. E., **36**, 1
Winter, M. J., **29**, 101
Wojcicki, A., **11**, 87; **12**, 31
Yamamoto, A., **34**, 111
Yashina, N. S., **14**, 63
Ziegler, K., **6**, 1
Zuckerman, J. J., **9**, 21
Zybill, C., **36**, 229

Cumulative Index
Volumes 37 and 38

ISBN 0-12-031138-0